Chapter 1
Systems of Measurement

Conceptual Problems

1 • **[SSM]** Which of the following *is not* one of the base quantities in the SI system? (*a*) mass, (*b*) length, (*c*) energy, (*d*) time, (*e*) All of the above are base quantities.

Determine the Concept The base quantities in the SI system include mass, length, and time. Force is not a base quantity. $\boxed{(c)}$ is correct.

5 • **[SSM]** Show that there are 30.48 cm per foot. How many centimeters are there in one mile?

Picture the Problem We can use the facts that there are 2.540 centimeters in 1 inch and 12 inches in 1 foot to show that there are 30.48 cm per ft. We can then use the fact that there are 5280 feet in 1 mile to find the number of centimeters in one mile.

Multiply 2.540 cm/in by 12 in/ft to find the number of cm per ft:

$$\left(2.540\frac{\text{cm}}{\text{in}}\right)\left(12\frac{\text{in}}{\text{ft}}\right)=\boxed{30.48\,\text{cm/ft}}$$

Multiply 30.48 cm/ft by 5280 ft/mi to find the number of centimeters in one mile:

$$\left(30.48\frac{\text{cm}}{\text{ft}}\right)\left(5280\frac{\text{ft}}{\text{mi}}\right)=\boxed{1.609\times10^{5}\,\text{cm/mi}}$$

Remarks: Because there are exactly 2.54 cm in 1 in and exactly 12 inches in 1 ft, we are justified in reporting four significant figures in these results.

11 • **[SSM]** A vector $\vec{A}$ points in the $+x$ direction. Show graphically at least three choices for a vector $\vec{B}$ such that $\vec{B}+\vec{A}$ points in the $+y$ direction.

Determine the Concept The following figure shows a vector $\vec{A}$ pointing in the positive x direction and three unlabeled possibilities for vector $\vec{B}$. Note that the choices for $\vec{B}$ start at the end of vector $\vec{A}$ rather than at its initial point. Note further that this configuration could be in any quadrant of the reference system shown.

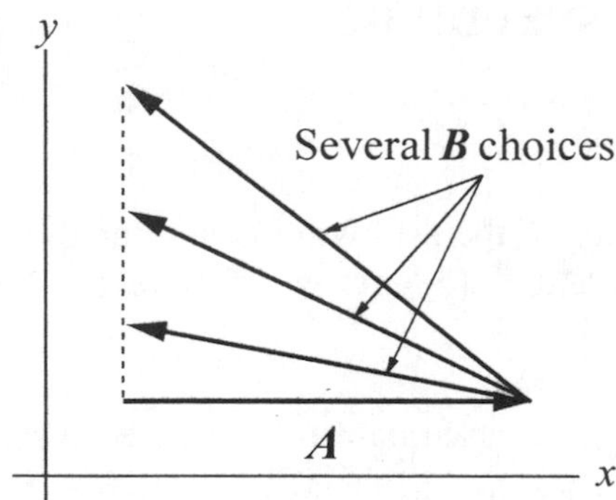

13 • **[SSM]** Is it possible for three equal magnitude vectors to add to zero? If so, sketch a graphical answer. If not, explain why not.

Determine the Concept In order for the three equal magnitude vectors to add to zero, the sum of the three vectors must form a triangle. The equilateral triangle shown to the right satisfies this condition for the vectors $\vec{A}$, $\vec{B}$, and $\vec{C}$ for which it is true that $A = B = C$, whereas $\vec{A}+\vec{B}+\vec{C}=0$.

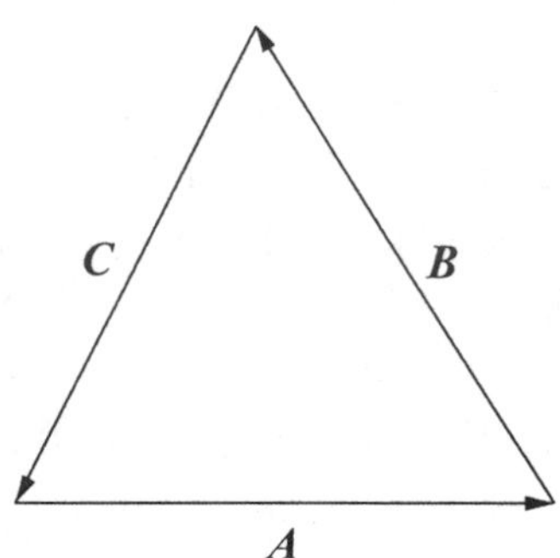

Estimation and Approximation

15 • **[SSM]** Some good estimates about the human body can be made if it is assumed that we are made mostly of water. The mass of a water molecule is 29.9×10^{-27} kg. If the mass of a person is 60 kg, estimate the number of water molecules in that person.

Picture the Problem We can estimate the number of water molecules in a person whose mass is 60 kg by dividing this mass by the mass of a single water molecule.

Letting N represent the number of water molecules in a person of mass $m_{\text{human body}}$, express N in terms of $m_{\text{human body}}$ and the mass of a water molecule $m_{\text{water molecule}}$:

$$N=\frac{m_{\text{human body}}}{m_{\text{water molecule}}}$$

Substitute numerical values and evaluate N:

$$N = \frac{60\ \text{kg}}{29.9\times10^{-27}\ \dfrac{\text{kg}}{\text{molecule}}} = \boxed{2.0\times10^{27}\ \text{molecules}}$$

19 •• **[SSM]** A megabyte (MB) is a unit of computer memory storage. A CD has a storage capacity of 700 MB and can store approximately 70 min of high-quality music. (*a*) If a typical song is 5 min long, how many megabytes are required for each song? (*b*) If a page of printed text takes approximately 5 kilobytes, estimate the number of novels that could be saved on a CD.

Picture the Problem We can set up a proportion to relate the storage capacity of a CD to its playing time, the length of a typical song, and the storage capacity required for each song. In (*b*) we can relate the number of novels that can be stored on a CD to the number of megabytes required per novel and the storage capacity of the CD.

(*a*) Set up a proportion relating the ratio of the number of megabytes on a CD to its playing time to the ratio of the number of megabytes N required for each song:

$$\frac{700\ \text{MB}}{70\ \text{min}} = \frac{N}{5\ \text{min}}$$

Solve this proportion for N to obtain:

$$N = \left(\frac{700\ \text{MB}}{70\ \text{min}}\right)(5\ \text{min}) = \boxed{50\ \text{MB}}$$

(*b*) Letting n represent the number of megabytes per novel, express the number of novels N_{novels} that can be stored on a CD in terms of the storage capacity of the CD:

$$N_{\text{novels}} = \frac{700\ \text{MB}}{n}$$

Assuming that a typical page in a novel requires 5 kB of memory, express n in terms of the number of pages p in a typical novel:

$$n = \left(5\ \frac{\text{kB}}{\text{page}}\right)p$$

Substitute for n in the expression for N_{novels} to obtain:

$$N_{\text{novels}} = \frac{700\ \text{MB}}{\left(5\ \dfrac{\text{kB}}{\text{page}}\right)p}$$

Assuming that a typical novel has 200 pages:

$$N_{\text{novels}} = \frac{700\ \text{MB} \times \dfrac{10^3\ \text{kB}}{\text{MB}}}{\left(5\dfrac{\text{kB}}{\text{page}}\right)\left(200\dfrac{\text{pages}}{\text{novel}}\right)} = \boxed{7\times 10^2\ \text{novels}}$$

Units

23 •• **[SSM]** In the following equations, the distance x is in meters, the time t is in seconds, and the velocity v is in meters per second. What are the SI units of the constants C_1 and C_2? (*a*) $x = C_1 + C_2t$, (*b*) $x = \frac{1}{2}C_1t^2$, (*c*) $v^2 = 2C_1x$, (*d*) $x = C_1 \cos C_2t$, (*e*) $v^2 = 2C_1v - (C_2x)^2$.

Picture the Problem We can determine the SI units of each term on the right-hand side of the equations from the units of the physical quantity on the left-hand side.

(*a*) Because x is in meters, C_1 and C_2t must be in meters:

$$\boxed{C_1 \text{ is in m}; C_2 \text{ is in m/s}}$$

(*b*) Because x is in meters, $\frac{1}{2}C_1t^2$ must be in meters:

$$\boxed{C_1 \text{ is in m/s}^2}$$

(*c*) Because v^2 is in m^2/s^2, $2C_1x$ must be in m^2/s^2:

$$\boxed{C_1 \text{ is in m/s}^2}$$

(*d*) The argument of a trigonometric function must be dimensionless; i.e. without units. Therefore, because x is in meters:

$$\boxed{C_1 \text{ is in m}; C_2 \text{ is in s}^{-1}}$$

(*e*) All of the terms in the expression must have the same units. Therefore, because v is in m/s:

$$\boxed{C_1 \text{ is in m/s}; C_2 \text{ is in s}^{-1}}$$

Conversion of Units

33 •• **[SSM]** You are a delivery person for the Fresh Aqua Spring Water Company. Your truck carries 4 pallets. Each pallet carries 60 cases of water. Each case of water has 24 one-liter bottles. You are to deliver 10 cases of water to each convenience store along your route. The dolly you use to carry the water into the stores has a weight limit of 250 lb. (*a*) If a milliliter of water has a mass of 1 g, and a kilogram has a weight of 2.2 lb, what is the weight, in pounds, of all the water in your truck? (*b*) How many full cases of water can you carry on the cart?

Picture the Problem The weight of the water in the truck is the product of the volume of the water and its weight density of 2.2 lb/L.

(*a*) Relate the weight w of the water on the truck to its volume V and weight density (weight per unit volume) D:

$$w = DV$$

Find the volume V of the water:

$$V = (4\ \text{pallets})\left(60\,\frac{\text{cases}}{\text{pallet}}\right)\left(24\,\frac{\text{L}}{\text{case}}\right)$$

$$= 5760\ \text{L}$$

Substitute numerical values for D and L and evaluate w:

$$w = \left(2.2\,\frac{\text{lb}}{\text{L}}\right)(5760\ \text{L}) = 1.267\times10^{4}\ \text{lb}$$

$$= \boxed{1.3\times10^{4}\ \text{lb}}$$

(*b*) Express the number of cases of water in terms of the weight limit of the cart and the weight of each case of water:

$$N = \frac{\text{weight limit of the cart}}{\text{weight of each case of water}}$$

Substitute numerical values and evaluate N:

$$N = \frac{250\ \text{lb}}{\left(2.2\,\frac{\text{lb}}{\text{L}}\right)\left(24\,\frac{\text{L}}{\text{case}}\right)} = 4.7\ \text{cases}$$

$$\boxed{\text{You can carry 4 cases.}}$$

33 •• **[SSM]** You are a delivery person for the Fresh Aqua Spring Water Company. Your truck carries 4 pallets. Each pallet carries 60 cases of water. Each case of water has 24 one-liter bottles. You are to deliver 10 cases of water to each convenience store along your route. The dolly you use to carry the water into the stores has a weight limit of 250 lb. (*a*) If a milliliter of water has a mass of 1 g, and a kilogram has a weight of 2.2 lb, what is the weight, in pounds, of all the water in your truck? (*b*) How many full cases of water can you carry on the cart?

Picture the Problem The weight of the water in the truck is the product of the volume of the water and its weight density of 2.2 lb/L.

(*a*) Relate the weight *w* of the water on the truck to its volume *V* and weight density (weight per unit volume) *D*:

$$w = DV$$

Find the volume *V* of the water:

$$V = (4\text{ pallets})(60\,\frac{\text{cases}}{\text{pallet}})(24\,\frac{\text{L}}{\text{case}})$$

$$= 5760\text{ L}$$

Substitute numerical values for *D* and *L* and evaluate *w*:

$$w = \left(2.2\,\frac{\text{lb}}{\text{L}}\right)(5760\text{ L}) = 1.267\times10^4\text{ lb}$$

$$= \boxed{1.3\times10^4\text{ lb}}$$

(*b*) Express the number of cases of water in terms of the weight limit of the cart and the weight of each case of water:

$$N = \frac{\text{weight limit of the cart}}{\text{weight of each case of water}}$$

Substitute numerical values and evaluate *N*:

$$N = \frac{250\text{ lb}}{\left(2.2\,\frac{\text{lb}}{\text{L}}\right)\left(24\,\frac{\text{L}}{\text{case}}\right)} = 4.7\text{ cases}$$

$$\boxed{\text{You can carry 4 cases.}}$$

35 •• **[SSM]** In the following, *x* is in meters, *t* is in seconds, *v* is in meters per second, and the acceleration *a* is in meters per second squared. Find the SI units of each combination: (*a*) v^2/x, (*b*) $\sqrt{x/a}$, (*c*) $\frac{1}{2}at^2$.

Picture the Problem We can treat the SI units as though they are algebraic quantities to simplify each of these combinations of physical quantities and constants.

(*a*) Express and simplify the units of v^2/x:

$$\frac{(\text{m/s})^2}{\text{m}} = \frac{\text{m}^2}{\text{m}\cdot\text{s}^2} = \boxed{\frac{\text{m}}{\text{s}^2}}$$

(*b*) Express and simplify the units of $\sqrt{x/a}$:

$$\sqrt{\frac{\text{m}}{\text{m/s}^2}} = \sqrt{\text{s}^2} = \boxed{\text{s}}$$

(*c*) Noting that the constant factor $\frac{1}{2}$ has no units, express and simplify the units of $\frac{1}{2}at^2$:

$$\left(\frac{\text{m}}{\text{s}^2}\right)(\text{s})^2 = \boxed{\text{m}}$$

Dimensions of Physical Quantities

41 •• **[SSM]** The momentum of an object is the product of its velocity and mass. Show that momentum has the dimensions of force multiplied by time.

Picture the Problem The dimensions of mass and velocity are M and L/T, respectively. We note from Table 1-2 that the dimensions of force are ML/T^2.

Express the dimensions of momentum:

$$[mv] = \text{M} \times \frac{\text{L}}{\text{T}} = \frac{\text{ML}}{\text{T}}$$

From Table 1-2:

$$[F] = \frac{\text{ML}}{\text{T}^2}$$

Express the dimensions of force multiplied by time:

$$[Ft] = \frac{\text{ML}}{\text{T}^2} \times \text{T} = \frac{\text{ML}}{\text{T}}$$

Comparing these results, we see that momentum has the dimensions of force multiplied by time.

43 •• **[SSM]** When an object falls through air, there is a drag force that depends on the product of the cross sectional area of the object and the square of its velocity, that is, $F_{\text{air}} = CAv^2$, where C is a constant. Determine the dimensions of C.

Picture the Problem We can find the dimensions of C by solving the drag force equation for C and substituting the dimensions of force, area, and velocity.

Solve the drag force equation for the constant C:

$$C = \frac{F_{\text{air}}}{Av^2}$$

Express this equation dimensionally:

$$[C] = \frac{[F_{\text{air}}]}{[A][v]^2}$$

Substitute the dimensions of force, area, and velocity and simplify to obtain:

$$[C]=\frac{\frac{\text{ML}}{\text{T}^2}}{\text{L}^2\left(\frac{\text{L}}{\text{T}}\right)^2}=\boxed{\frac{\text{M}}{\text{L}^3}}$$

Scientific Notation and Significant Figures

45 • **[SSM]** Express as a decimal number without using powers of 10 notation: (*a*) 3×10^4, (*b*) 6.2×10^{-3}, (*c*) 4×10^{-6}, (*d*) 2.17×10^5.

Picture the Problem We can use the rules governing scientific notation to express each of these numbers as a decimal number.

(*a*) $3\times10^4=\boxed{30{,}000}$

(*b*) $6.2\times10^{-3}=\boxed{0.0062}$

(*c*) $4\times10^{-6}=\boxed{0.000004}$

(*d*) $2.17\times10^5=\boxed{217{,}000}$

47 • **[SSM]** Calculate the following, round off to the correct number of significant figures, and express your result in scientific notation: (*a*) $(1.14)(9.99 \times 10^4)$, (*b*) $(2.78 \times 10^{-8}) - (5.31 \times 10^{-9})$, (*c*) $12\pi/(4.56 \times 10^{-3})$, (*d*) $27.6 + (5.99 \times 10^2)$.

Picture the Problem Apply the general rules concerning the multiplication, division, addition, and subtraction of measurements to evaluate each of the given expressions.

(*a*) The number of significant figures in each factor is three; therefore the result has three significant figures:

$$(1.14)(9.99\times10^4)=\boxed{1.14\times10^5}$$

(*b*) Express both terms with the same power of 10. Because the first measurement has only two digits after the decimal point, the result can have only two digits after the decimal point:

$$\begin{aligned}(2.78\times10^{-8})-(5.31\times10^{-9})\\ &=(2.78-0.531)\times10^{-8}\\ &=\boxed{2.25\times10^{-8}}\end{aligned}$$

(*c*) We'll assume that 12 is exact. Hence, the answer will have three significant figures:

$$\frac{12\pi}{4.56\times10^{-3}}=\boxed{8.27\times10^3}$$

(*d*) Proceed as in (*b*):

$$27.6+\left(5.99\times10^{2}\right)=27.6+599$$
$$=627$$
$$=\boxed{6.27\times10^{2}}$$

49 • **[SSM]** A cell membrane has a thickness of 7.0 nm. How many cell membranes would it take to make a stack 1.0 in high?

Picture the Problem Let N represent the required number of membranes and express N in terms of the thickness of each cell membrane.

Express N in terms of the thickness of a single membrane:

$$N=\frac{1.0\,\text{in}}{7.0\,\text{nm}}$$

Convert the units into SI units and simplify to obtain:

$$N=\frac{1.0\,\text{in}}{7.0\,\text{nm}}\times\frac{2.540\,\text{cm}}{\text{in}}\times\frac{1\,\text{m}}{100\,\text{cm}}\times\frac{1\,\text{nm}}{10^{-9}\,\text{m}}=3.63\times10^{6}=\boxed{3.6\times10^{6}}$$

51 •• **[SSM]** A square peg must be made to fit through a square hole. If you have a square peg that has an edge length of 4.29×10^{-2} m, and the square hole has an edge length of 4.32×10^{-2} m, (*a*) what is the area of the space available when the peg is in the hole? (*b*) If the peg is made rectangular by removing 1.0×10^{-4} m of material from one side, what is the area available now?

Picture the Problem Let s_h represent the side of the square hole and s_p the side of the square peg. We can find the area of the space available when the peg is in the hole by subtracting the area of the peg from the area of the hole.

(*a*) Express the difference between the two areas in terms of s_h and s_p:

$$\Delta A=s_h^2-s_p^2$$

Substitute numerical values and evaluate ΔA:

$$\Delta A=\left(4.32\times10^{-2}\,\text{m}\right)^2-\left(4.29\times10^{-2}\,\text{m}\right)^2$$
$$=2.58\times10^{-5}\,\text{m}^2\times\left(\frac{10^{3}\,\text{mm}}{\text{m}}\right)^2$$
$$\approx\boxed{26\,\text{mm}^2}$$

(*b*) Express the difference between the area of the square hole and the rectangular peg in terms of s_h, s_p and the new length of the peg ℓ_p:

$$\Delta A'=s_h^2-s_p\ell_p$$

Substitute numerical values and evaluate $\Delta A'$:

$$\Delta A' = \left(4.32\times10^{-2}\text{ m}\right)^2 - \left(4.29\times10^{-2}\text{ m}\right)\left(4.29\times10^{-2}\text{ m} - 1.0\times10^{-4}\text{ m}\right)$$

$$= 3.01\times10^{-5}\text{ m}^2 \times \left(\frac{10^3\text{ mm}}{\text{m}}\right)^2$$

$$\approx \boxed{30\text{ mm}^2}$$

Vectors and Their Properties

53 • **[SSM]** Rewrite the following vectors in terms of their magnitude and angle (counterclockwise with the $+x$ direction). (*a*) A displacement vector with an x component of +8.5 m and a y component of -5.5 m. (*b*) A velocity vector with an x component of -75 m/s and a y component of +35 m/s. (*c*) A force vector with a magnitude of 50 lb that is in the third quadrant with an x component whose magnitude is 40 lb.

Picture the Problem The x and y components of these vectors are their projections onto the x and y axes. Note that the components are calculated using the angle each vector makes with the $+x$ axis.

(*a*) Sketch the displacement vector (call it $\vec{A}$) and note that it makes an angle of 60° with the $+x$ axis:

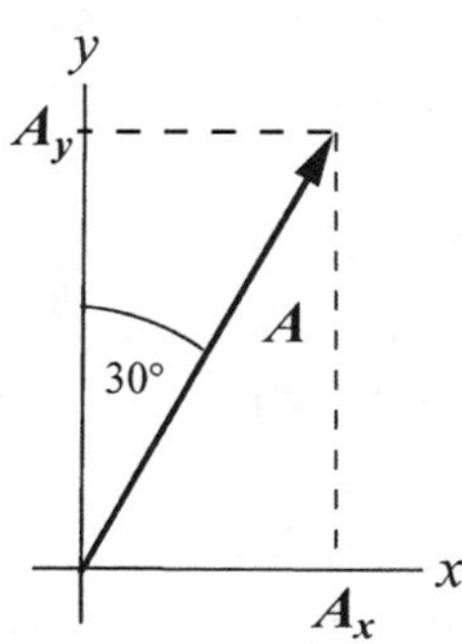

Find the x and y components of $\vec{A}$:

$$A_x = (10\text{ m})\cos 60° = \boxed{5.0\text{ m}}$$

and

$$A_y = (10\text{ m})\sin 60° = \boxed{8.7\text{ m}}$$

(*b*) Sketch the velocity vector (call it $\vec{v}$) and note that it makes an angle of 220° with the +*x* axis:

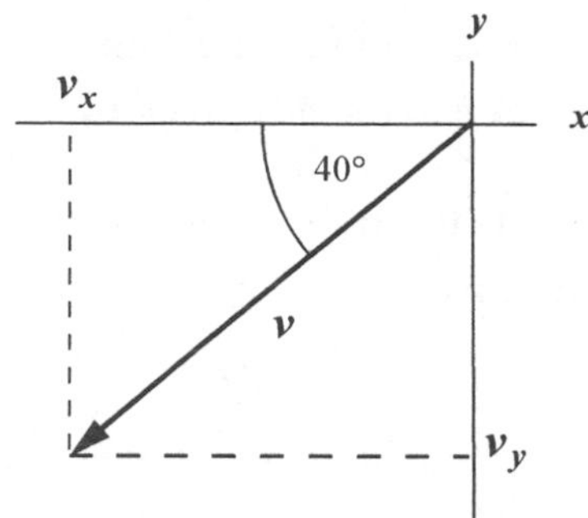

Find the *x* and *y* components of $\vec{v}$:

$$v_x = (25\text{ m/s})\cos 220° = \boxed{-19\text{ m/s}}$$

and

$$v_y = (25\text{ m/s})\sin 220° = \boxed{-16\text{ m/s}}$$

(*c*) Sketch the force vector (call it $\vec{F}$) and note that it makes an angle of 30° with the +*x* axis:

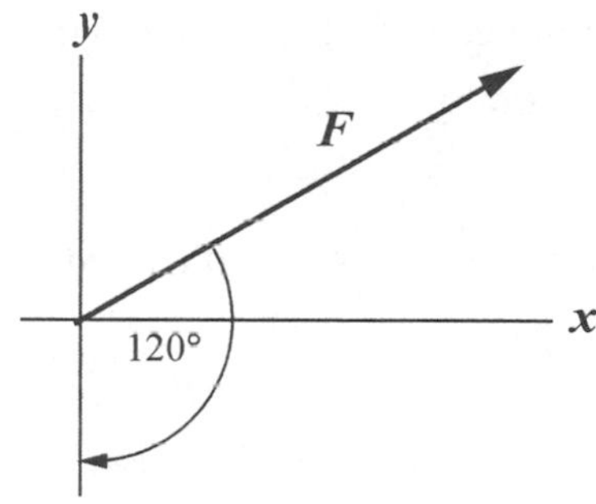

Find the *x* and *y* components of $\vec{F}$:

$$F_x = (40\text{ lb})\cos 30° = \boxed{35\text{ lb}}$$

and

$$F_y = (40\text{ lb})\sin 30° = \boxed{20\text{ lb}}$$

59 •• **[SSM]** Calculate the unit vector (in terms of $\hat{i}$ and $\hat{j}$) in the direction opposite to the direction of each of the vectors in Problem 57.

Picture the Problem The unit vector in the direction opposite to the direction of a vector is found by taking the negative of the unit vector in the direction of the vector. The unit vector in the direction of a given vector is found by dividing the given vector by its magnitude.

The unit vector in the direction of $\vec{A}$ is given by:

$$\hat{A} = \frac{\vec{A}}{A} = \frac{\vec{A}}{\sqrt{A_x^2 + A_y^2}}$$

Substitute for $\vec{A}$, A_x, and A_y and evaluate $\hat{A}$:

$$\hat{A} = \frac{3.4\hat{i} + 4.7\hat{j}}{\sqrt{(3.4)^2 + (4.7)^2}} = 0.59\hat{i} + 0.81\hat{j}$$

The unit vector in the direction opposite to that of $\vec{A}$ is given by:

$$-\hat{A} = \boxed{-0.59\hat{i} - 0.81\hat{j}}$$

The unit vector in the direction of $\vec{B}$ is given by:

$$\hat{B} = \frac{\vec{B}}{B} = \frac{\vec{B}}{\sqrt{B_x^2 + B_y^2}}$$

Substitute for $\vec{B}$, B_x, and B_y and evaluate $\hat{B}$:

$$\hat{B} = \frac{-7.7\hat{i} + 3.2\hat{j}}{\sqrt{(-7.7)^2 + (3.2)^2}} = -0.92\hat{i} + 0.38\hat{j}$$

The unit vector in the direction opposite to that of $\vec{B}$ is given by:

$$-\hat{B} = \boxed{0.92\hat{i} - 0.38\hat{j}}$$

The unit vector in the direction of $\vec{C}$ is given by:

$$\hat{C} = \frac{\vec{C}}{C} = \frac{\vec{C}}{\sqrt{C_x^2 + C_y^2}}$$

Substitute for $\vec{C}$, C_x, and C_y and evaluate $\hat{C}$:

$$\hat{C} = \frac{5.4\hat{i} - 9.1\hat{j}}{\sqrt{(5.4)^2 + (-9.1)^2}} = \boxed{0.51\hat{i} - 0.86\hat{j}}$$

The unit vector in the direction opposite that of $\vec{C}$ is given by:

$$-\hat{C} = \boxed{-0.51\hat{i} + 0.86\hat{j}}$$

General Problems

61 • **[SSM]** The Apollo trips to the moon in the 1960's and 1970's typically took 3 days to travel the Earth-moon distance once they left Earth orbit. Estimate the spacecraft's average speed in kilometers per hour, miles per hour, and meters per second.

Picture the Problem Average speed is defined to be the distance traveled divided by the elapsed time. The Earth-moon distance and the distance and time conversion factors can be found on the inside-front cover of the text. We'll assume that 3 days means exactly three days.

Express the average speed of Apollo as it travels to the moon:

$$v_{av} = \frac{\text{distance traveled}}{\text{elapsed time}}$$

Substitute numerical values to obtain:

$$v_{av} = \frac{2.39\times10^5\ \text{mi}}{3\ \text{d}}$$

Use the fact that there are 24 h in 1 d to convert 3 d into hours:

$$v_{av} = \frac{2.39\times10^5\ \text{mi}}{3\,\text{d}\times\dfrac{24\ \text{h}}{\text{d}}} = 3.319\times10^3\ \text{mi/h} = \boxed{3.32\times10^3\ \text{mi/h}}$$

Use the fact that 1 mi is equal to 1.609 km to convert the spacecraft's average speed to km/h:

$$v_{av} = 3.319\times10^3\,\frac{\text{mi}}{\text{h}}\times1.609\,\frac{\text{km}}{\text{mi}} = 5.340\times10^3\,\frac{\text{km}}{\text{h}} = \boxed{5.34\times10^3\ \text{km/h}}$$

Use the facts that there are 3600 s in 1 h and 1000 m in 1 km to convert the spacecraft's average speed to m/s:

$$v_{av} = 5.340\times10^6\,\frac{\text{km}}{\text{h}}\times\frac{10^3\ \text{m}}{\text{km}}\times\frac{1\ \text{h}}{3600\ \text{s}} = 1.485\times10^3\ \text{m/s} = \boxed{1.49\times10^3\ \text{m/s}}$$

Remarks: An alternative to multiplying by 10^3 m/km in the last step is to replace the metric prefix "k" in "km" by 10^3.

71 ••• **[SSM]** You are an astronaut doing physics experiments on the moon. You are interested in the experimental relationship between distance fallen, y, and time elapsed, t, of falling objects dropped from rest. You have taken some data for a falling penny, which is represented in the table below. You expect that a general relationship between distance y and time t is $y = Bt^C$, where B and C are constants to be determined experimentally. To accomplish this, create a *log-log* plot of the data: (*a*) graph log(y) vs. log(t), with log(y) the ordinate variable and log(t) the abscissa variable. (*b*) Show that if you take the log of each side of your equation, you get log(y) = log(B) + Clog(t). (*c*) By comparing this linear relationship to the graph of the data, estimate the values of B and C. (*d*) If you drop a penny, how long should it take to fall 1.0 m? (*e*) In the next chapter, we will show that the expected relationship between y and t is $y = \frac{1}{2}at^2$, where a is the acceleration of the object. What is the acceleration of objects dropped on the moon?

y (m)	10	20	30	40	50
t (s)	3.5	5.2	6.0	7.3	7.9

Picture the Problem We can plot log y versus log t and find the slope of the best-fit line to determine the exponent C. The value of B can be determined from the intercept of this graph. Once we know C and B, we can solve $y = Bt^C$ for t as a function of *y* and use this result to determine the time required for an object to fall a given distance on the surface of the moon.

(*a*) The following graph of log *y* versus log *t* was created using a spreadsheet program. The equation shown on the graph was obtained using Excel's "Add Trendline" function. (Excel's "Add Trendline" function uses regression analysis to generate the trendline.)

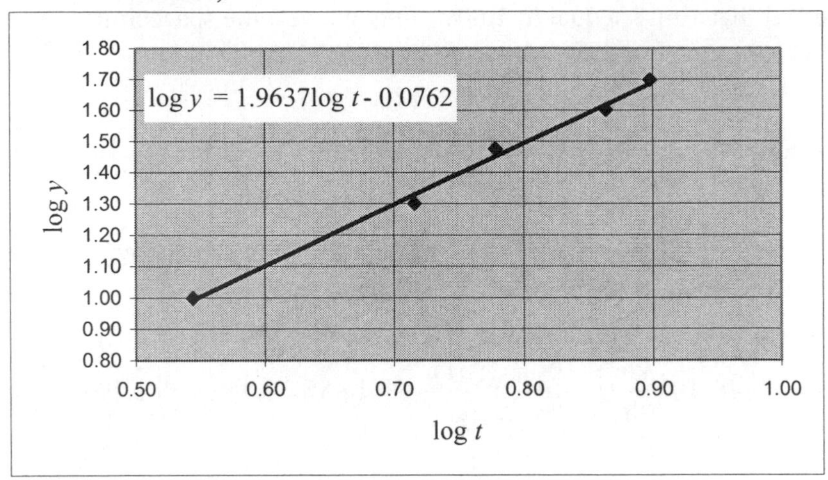

(*b*) Taking the logarithm of both sides of the equation $y = Bt^C$ yields:

$$\log y = \log\left(Bt^C\right) = \log B + \log t^C$$
$$= \boxed{\log B + C\log t}$$

(*c*) Note that this result is of the form:

$Y = b + mX$
where
$Y = \log y$, $b = \log B$, $m = C$, and
$X = \log t$

From the regression analysis (trendline) we have:

$\log B = -0.076$

Solving for *B* yields:

$B = 10^{-0.076} = \boxed{0.84\ \text{m/s}^2}$

where we have inferred the units from those given for $y = Bt^C$.

Also, from the regression analysis we have:

$C = 1.96 \approx \boxed{2.0}$

(*d*) Solve $y = Bt^C$ for t to obtain:

$$t = \left(\frac{y}{B}\right)^{\frac{1}{C}}$$

Substitute numerical values and evaluate t to determine how long it would take a penny to fall 1.0 m:

$$t = \left(\frac{1.0\text{ m}}{0.84\,\frac{\text{m}}{\text{s}^2}}\right)^{\frac{1}{2}} \approx \boxed{1.1\text{ s}}$$

(*e*) Substituting for B and C in $y = Bt^C$ yields:

$$y = \left(0.84\,\frac{\text{m}}{\text{s}^2}\right)t^2$$

Compare this equation to $y - \frac{1}{2}at^2$ to obtain:

$$\tfrac{1}{2}a = 0.84\,\frac{\text{m}}{\text{s}^2}$$

and

$$a = 2\left(0.84\,\frac{\text{m}}{\text{s}^2}\right) = \boxed{1.7\,\frac{\text{m}}{\text{s}^2}}$$

Remarks: One could use a graphing calculator to obtain the results in Parts (*a*) and (*c*).

73 ••• **[SSM]** The Super-Kamiokande neutrino detector in Japan is a large transparent cylinder filled with ultra pure water. The height of the cylinder is 41.4 m and the diameter is 39.3 m. Calculate the mass of the water in the cylinder. Does this match the claim posted on the official Super-K Web site that the detector uses 50000 tons of water?

Picture the Problem We can use the definition of density to relate the mass of the water in the cylinder to its volume and the formula for the volume of a cylinder to express the volume of water used in the detector's cylinder. To convert our answer in kg to lb, we can use the fact that 1 kg weighs about 2.205 lb.

Relate the mass of water contained in the cylinder to its density and volume:

$$m = \rho V$$

Express the volume of a cylinder in terms of its diameter d and height h:

$$V = A_{\text{base}}h = \frac{\pi}{4}d^2h$$

Substitute in the expression for m to obtain:

$$m = \rho\frac{\pi}{4}d^2h$$

Substitute numerical values and evaluate m:

$$m = \left(10^3\ \text{kg/m}^3\right)\left(\frac{\pi}{4}\right)(39.3\,\text{m})^2(41.4\,\text{m})$$
$$= 5.022 \times 10^7\ \text{kg}$$

Convert 5.02×10^7 kg to tons:

$$m = 5.022 \times 10^7\ \text{kg} \times \frac{2.205\,\text{lb}}{\text{kg}} \times \frac{1\,\text{ton}}{2000\,\text{lb}}$$
$$= 55.4 \times 10^3\ \text{ton}$$

The 50,000-ton claim is conservative. The actual weight is closer to 55,000 tons.

Chapter 2
Motion in One Dimension

Conceptual Problems

5 • **[SSM]** Stand in the center of a large room. Call the direction to your right "positive," and the direction to your left "negative." Walk across the room along a straight line, using a constant acceleration to quickly reach a steady speed along a straight line in the negative direction. After reaching this steady speed, keep your velocity negative but make your acceleration positive. (*a*) Describe how your speed varied as you walked. (*b*) Sketch a graph of x versus t for your motion. Assume you started at $x = 0$. (*c*) Directly under the graph of Part (*b*), sketch a graph of v_x versus t.

Determine the Concept The important concept is that when both the acceleration and the velocity are in the same direction, the speed increases. On the other hand, when the acceleration and the velocity are in opposite directions, the speed decreases.

(*a*) Your speed increased from zero, stayed constant for a while, and then decreased.

(*b*) A graph of your position as a function of time is shown to the right. Note that the slope starts out equal to zero, becomes more negative as the speed increases, remains constant while your speed is constant, and becomes less negative as your speed decreases.

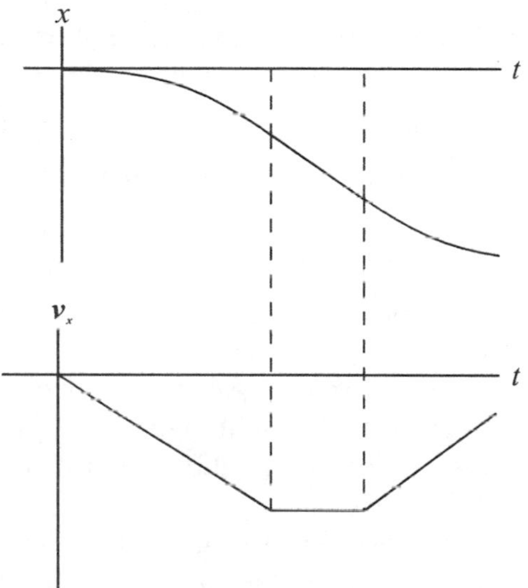

(*c*) The graph of $v(t)$ consists of a straight line with negative slope (your acceleration is constant and negative) starting at (0,0), then a flat line for a while (your acceleration is zero), and finally an approximately straight line with a positive slope heading to $v = 0$.

11 •• **[SSM]** Dr. Josiah S. Carberry stands at the top of the Sears Tower in Chicago. Wanting to emulate Galileo, and ignoring the safety of the pedestrians below, he drops a bowling ball from the top of the tower. One second later, he drops a second bowling ball. While the balls are in the air, does their separation (*a*) increase over time, (*b*) decrease, (*c*) stay the same? Ignore any effects due to air resistance.

Determine the Concept Neglecting air resistance, the balls are in free fall, each with the same free-fall acceleration, which is a constant.

At the time the second ball is released, the first ball is already moving. Thus, during any time interval their velocities will increase by exactly the same amount. What can be said about the speeds of the two balls? *The first ball will always be moving faster than the second ball.* This being the case, what happens to the separation of the two balls while they are both falling? *Their separation increases.* $\boxed{(a)}$ is correct.

13 •• [SSM] Which of the velocity-versus-time curves in figure 2-29 best describes the motion of an object (*a*) with constant positive acceleration, (*b*) with positive acceleration that is decreasing with time, (*c*) with positive acceleration that is increasing with time, and (*d*) with no acceleration? (There may be more than one correct answer for each part of the problem.)

Determine the Concept The slope of a $v(t)$ curve at any point in time represents the acceleration at that instant.

(*a*) The correct answer is $\boxed{(b)}$. The slope of curve (*b*) is constant and positive. Therefore the acceleration is constant and positive.

(*b*) The correct answer is $\boxed{(c)}$. The slope of curve (*c*) is positive and decreasing with time. Therefore the acceleration is positive and decreasing with time.

(*c*) The correct answer is $\boxed{(d)}$. The slope of curve (*d*) is positive and increasing with time. Therefore the acceleration is positive and increasing with time.

(*d*) The correct answer is $\boxed{(e)}$. The slope of curve (*e*) is zero. Therefore the velocity is constant and the acceleration is zero.

15 •• [SSM] An object moves along a straight line. Its position versus time graph is shown in Figure 2-30. At which time or times is its (*a*) speed at a minimum, (*b*) acceleration positive, and (*c*) velocity negative?

Determine the Concept Because this graph is of distance-versus-time we can use its instantaneous slope to describe the object's speed, velocity, and acceleration.

(*a*) The minimum speed is zero at $\boxed{\text{B, D, and E.}}$ In the one-dimensional motion shown in the figure, the velocity is a minimum when the slope of a position-versus-time plot goes to zero (i.e., the curve becomes horizontal). At these points, the slope of the position-versus-time curve is zero; therefore, the speed is zero.

(*b*) The acceleration is positive at points $\boxed{\text{A and D.}}$ Because the slope of the graph is increasing at these points, the velocity of the object is increasing and its acceleration is positive.

(*c*) The velocity is negative at point $\boxed{\text{C.}}$ Because the slope of the graph is negative at point C, the velocity of the object is negative.

19 •• **[SSM]** A ball is thrown straight up. Neglect any effects due to air resistance. (*a*) What is the velocity of the ball at the top of its flight? (*b*) What is its acceleration at that point? (*c*) What is different about the velocity and acceleration at the top of the flight if instead the ball impacts a horizontal ceiling very hard and then returns.

Determine the Concept In the absence of air resistance, the ball will experience a constant acceleration. In the graph that follows, a coordinate system was chosen in which the origin is at the point of release and the upward direction is positive. The graph shows the velocity of a ball that has been thrown straight upward with an initial speed of 30 m/s as a function of time.

(*a*) $\boldsymbol{v}_{\text{top of flight}} = \boxed{0}$

(*b*) Note that the acceleration of the ball is the same at every point of its trajectory, including the point at which $v = 0$ (at the top of its flight).
Hence $\boldsymbol{a}_{\text{top of flight}} = \boxed{-g}$

(c) If the ball impacts a horizontal ceiling very hard and then returns, its velocity at the top of its flight is still zero and its acceleration is still downward but greater than g in magnitude.

29 •• **[SSM]** The positions of two cars in parallel lanes of a straight stretch of highway are plotted as functions of time in the Figure 2-33.Take positive values of x as being to the right of the origin. Qualitatively answer the following: (*a*) Are the two cars ever side by side? If so, indicate that time (those times) on the axis. (*b*) Are they always traveling in the same direction, or are they moving in opposite directions for some of the time? If so, when? (*c*) Are they ever traveling at the same velocity? If so, when? (*d*) When are the two cars the farthest apart?

Determine the Concept Given the positions of the two cars as a function of time, we can use the intersections of the curves and their slopes to answer these questions.

(*a*) The positions of cars A and B are the same at two places where the graphs cross. These are at 1 s and 9 s.

(*b*) When the slopes of the curves have opposite signs, the velocities of the cars are oppositely directed. Thus, after approximately 7 s, car A is moving leftward while car B is moving rightward. Before $t = 7$ s, the two cars are traveling in the same direction.

(*c*) The two cars have the same velocity when their curves have the same slopes. This occurs at about 6 s.

(*d*) The time at which the two cars are farthest apart is roughly 6 s as indicated by the place at which, vertically, the two curves are farthest part.

Estimation and Approximation

37 • **[SSM]** Occasionally, people can survive falling large distances if the surface they land on is soft enough. During a traverse of the Eiger's infamous Nordvand, mountaineer Carlos Ragone's rock anchor gave way and he plummeted 500 feet to land in snow. Amazingly, he suffered only a few bruises and a wrenched shoulder. Assuming that his impact left a hole in the snow 4.0 ft deep, estimate his average acceleration as he slowed to a stop (that is, while he was impacting the snow).

Picture the Problem In the absence of air resistance, Carlos' acceleration is constant. Because all the motion is downward, let's use a coordinate system in which downward is the positive direction and the origin is at the point at which the fall began.

Using a constant-acceleration equation, relate Carlos' final velocity v_2 to his velocity v_1 just before his impact, his stopping acceleration a_s upon impact, and his stopping distance Δy:

$$v_2^2 = v_1^2 + 2a_s\Delta y \Rightarrow a_s = \frac{v_2^2 - v_1^2}{2\Delta y}$$

or, because $v_2 = 0$,

$$a_s = -\frac{v_1^2}{2\Delta y} \qquad (1)$$

Using a constant-acceleration equation, relate Carlos' speed just before impact to his acceleration during free-fall and the distance he fell h:

$$v_1^2 = v_0^2 + 2a_{\text{free-fall}}h$$

or, because $v_0 = 0$ and $a_{\text{free-fall}} = g$,

$$v_1^2 = 2gh$$

Substituting for v_1^2 in equation (1) yields:

$$a_s = -\frac{2gh}{2\Delta y}$$

Substitute numerical values and evaluate a_s:

$$a = -\frac{2(9.81\ \text{m/s}^2)(500\ \text{ft})}{2(4.0\ \text{ft})} = \boxed{-1.2\times10^3\,\text{m/s}^2}$$

Remarks: The magnitude of this acceleration is about 125*g*!

Speed, Displacement, and Velocity

43 • **[SSM]** A runner runs 2.5 km, in a straight line, in 9.0 min and then takes 30 min to walk back to the starting point. (*a*) What is the runner's average velocity for the first 9.0 min? (*b*) What is the average velocity for the time spent walking? (*c*) What is the average velocity for the whole trip? (*d*) What is the average speed for the whole trip?

Picture the Problem In this problem the runner is traveling in a straight line but not at constant speed - first she runs, then she walks. Let's choose a coordinate system in which her initial direction of motion is taken as the positive x direction.

(*a*) Using the definition of average velocity, calculate the average velocity for the first 9 min:

$$v_{av} = \frac{\Delta x}{\Delta t} = \frac{2.5\text{ km}}{9.0\text{ min}} = \boxed{0.28\text{ km/min}}$$

(*b*) Using the definition of average velocity, calculate her average speed for the 30 min spent walking:

$$v_{av} = \frac{\Delta x}{\Delta t} = \frac{-2.5\text{ km}}{30\text{ min}} = \boxed{-83\text{ m/min}}$$

(*c*) Express her average velocity for the whole trip:

$$v_{av} = \frac{\Delta x_{\text{round trip}}}{\Delta t} = \frac{0}{\Delta t} = \boxed{0}$$

(*d*) Finally, express her average speed for the whole trip:

$$\text{speed}_{av} = \frac{\text{distance traveled}}{\text{elapsed time}} = \frac{2(2.5\text{ km})}{30\text{ min} + 9.0\text{ min}} = \boxed{0.13\text{ km/min}}$$

47 • **[SSM]** Proxima Centauri, the closest star to us besides our own sun, is 4.1×10^{13} km from Earth. From Zorg, a planet orbiting this star, a Zorgian places an order at Tony's Pizza in Hoboken, New Jersey, communicating via light signals. Tony's fastest delivery craft travels at $1.00 \times 10^{-4}c$ (see Problem 46). (*a*) How long does it take Gregor's order to reach Tony's Pizza? (*b*) How long does Gregor wait between sending the signal and receiving the pizza? If Tony's has a "1000-years-or-it's-free" delivery policy, does Gregor have to pay for the pizza?

Picture the Problem In free space, light travels in a straight line at constant speed, c. We can use the definition of average speed to find the elapsed times called for in this problem.

(*a*) Using the definition of average speed (equal here to the assumed constant speed of light), solve for the time Δt required to travel the distance to Proxima Centauri:

$$\Delta t = \frac{\text{distance traveled}}{\text{speed of light}}$$

Substitute numerical values and evaluate Δt:

$$\Delta t = \frac{4.1\times10^{16}\text{ m}}{2.998\times10^{8}\text{ m/s}} = 1.37\times10^{8}\text{ s}$$
$$= \boxed{4.3\text{ y}}$$

(*b*) Traveling at $10^{-4}c$, the delivery time (t_{total}) will be the sum of the time for the order to reach Hoboken and the time for the pizza to be delivered to Proxima Centauri:

$$\Delta t_{\text{total}} = \Delta t_{\substack{\text{order to be}\\ \text{sent to Hoboken}}} + \Delta t_{\substack{\text{order to}\\ \text{be delivered}}} = 4.33\text{ y} + \frac{4.1\times10^{13}\text{ km}}{\left(1.00\times10^{-4}\right)\left(2.998\times10^{8}\text{ m/s}\right)}$$
$$= 4.33\text{ y} + 4.33\times10^{6}\text{ y} \approx 4.3\times10^{6}\text{ y}$$

Because 4.3×10^{6} y >> 1000 y, Gregor does not have to pay.

53 •• **[SSM]** The cheetah can run as fast as 113 km/h, the falcon can fly as fast as 161 km/h, and the sailfish can swim as fast as 105 km/h. The three of them run a relay with each covering a distance *L* at maximum speed. What is the average speed of this relay team for the entire relay? Compare this average speed with the numerical average of the three individual speeds. Explain carefully why the average speed of the relay team is *not* equal to the numerical average of the three individual speeds.

Picture the Problem We can find the average speed of the relay team from the definition of average speed.

Using its definition, relate the average speed to the total distance traveled and the elapsed time:

$$v_{\text{av}} = \frac{\text{distance traveled}}{\text{elapsed time}}$$

Express the time required for each animal to travel a distance *L*:

$$t_{\text{cheetah}} = \frac{L}{v_{\text{cheetah}}},\ t_{\text{falcon}} = \frac{L}{v_{\text{falcon}}}$$

and

$$t_{\text{sailfish}} = \frac{L}{v_{\text{sailfish}}}$$

Express the total time Δt:

$$\Delta t = L\left(\frac{1}{v_{\text{cheetah}}} + \frac{1}{v_{\text{falcon}}} + \frac{1}{v_{\text{sailfish}}}\right)$$

Use the total distance traveled by the relay team and the elapsed time to calculate the average speed:

$$v_{\text{av}} = \frac{3L}{L\left(\dfrac{1}{113\,\text{km/h}} + \dfrac{1}{161\,\text{km/h}} + \dfrac{1}{105\,\text{km/h}}\right)} = 122.03\ \text{km/h} = \boxed{122\,\text{km/h}}$$

Calculating the average of the three speeds yields:

$$\text{Average}_{\text{three speeds}} = \frac{113\,\text{km/h} + 161\,\text{km/h} + 105\,\text{km/h}}{3} = 126.33\ \text{km/h} = 126\ \text{km/h}$$

$$= \boxed{1.04 v_{\text{av}}}$$

Because we've ignored the time intervals during which members of this relay team get up to their running speeds, their accelerations are zero and the average speed of the relay team is *not* equal to (it is less than) the numerical average of the three individual speeds.

55 •• **[SSM]** A car traveling at a constant speed of 20 m/s passes an intersection at time $t = 0$. A second car traveling at a constant speed of 30 m/s in the same direction passes the same intersection 5.0 s later. (*a*) Sketch the position functions $x_1(t)$ and $x_2(t)$ for the two cars for the interval $0 \le t \le 20$ s. (*b*) Determine when the second car will overtake the first. (*c*) How far from the intersection will the two cars be when they pull even? (*d*) Where is the first car when the second car passes the intersection?

Picture the Problem One way to solve this problem is by using a graphing calculator to plot the positions of each car as a function of time. Plotting these positions as functions of time allows us to visualize the motion of the two cars relative to the (fixed) ground. More importantly, it allows us to see the motion of the two cars relative to each other. We can, for example, tell how far apart the cars are at any given time by determining the length of a vertical line segment from one curve to the other.

(*a*) Letting the origin of our coordinate system be at the intersection, the position of the slower car, $x_1(t)$, is given by:

$x_1(t) = 20t$
where x_1 is in meters if t is in seconds.

Because the faster car is also moving at a constant speed, we know that the position of this car is given by a function of the form:

$x_2(t) = 30t + b$

We know that when $t = 5.0$ s, this second car is at the intersection (that is, $x_2(5.0 \text{ s}) = 0$). Using this information, you can convince yourself that:

$$b = -150 \text{ m}$$

Thus, the position of the faster car is given by:

$$x_2(t) = 30t - 150$$

One can use a graphing calculator, graphing paper, or a spreadsheet to obtain the following graphs of $x_1(t)$ (the solid line) and $x_2(t)$ (the dashed line):

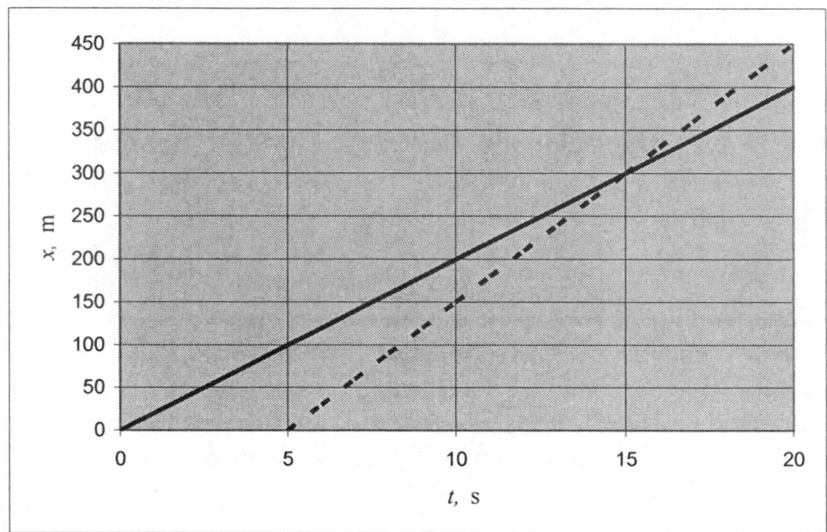

(*b*) Use the time coordinate of the intersection of the two lines to determine the time at which the second car overtakes the first:

From the intersection of the two lines, one can see that the second car will "overtake" (catch up to) the first car at $\boxed{t = 15 \text{ s.}}$

(*c*) Use the position coordinate of the intersection of the two lines to determine the distance from the intersection at which the second car catches up to the first car:

From the intersection of the two lines, one can see that the distance from the intersection is $\boxed{300 \text{ m.}}$

(*d*) Draw a vertical line from $t = 5$ s to the solid line and then read the position coordinate of the intersection of the vertical line and the solid line to determine the position of the first car when the second car went through the intersection. From the graph, when the second car passes the intersection, the first car was $\boxed{100 \text{ m ahead.}}$

Acceleration

59 • **[SSM]** An object is moving along the x axis. At $t = 5.0$ s, the object is at $x = +3.0$ m and has a velocity of +5.0 m/s. At $t = 8.0$ s, it is at $x = +9.0$ m and its velocity is –1.0 m/s. Find its average acceleration during the time interval $5.0\text{ s} \le t \le 8.0\text{ s}$.

Picture the Problem We can find the change in velocity and the elapsed time from the given information and then use the definition of average acceleration.

The average acceleration is defined as the change in velocity divided by the change in time:

$$a_{\text{av}} = \frac{\Delta v}{\Delta t}$$

Substitute numerical values and evaluate a_{av}:

$$a_{\text{av}} = \frac{(-1.0\text{ m/s}) - (5.0\text{ m/s})}{(8.0\text{ s}) - (5.0\text{ s})} = \boxed{-2.0\text{ m/s}^2}$$

61 •• **[SSM]** The position of a certain particle depends on time according to the equation $x(t) = t^2 - 5.0t + 1.0$, where x is in meters if t is in seconds. (*a*) Find the displacement and average velocity for the interval $3.0\text{ s} \le t \le 4.0\text{ s}$. (*b*) Find the general formula for the displacement for the time interval from t to $t + \Delta t$. (*c*) Use the limiting process to obtain the instantaneous velocity for any time t.

Picture the Problem We can closely approximate the instantaneous velocity by the average velocity in the limit as the time interval of the average becomes small. This is important because all we can ever obtain from any measurement is the average velocity, v_{av}, which we use to approximate the instantaneous velocity v.

(*a*) The displacement of the particle during the interval $3.0\text{ s} \le t \le 4.0\text{ s}$ is given by:

$$\Delta x = x(4.0\text{ s}) - x(3.0\text{ s}) \qquad (1)$$

The average velocity is given by:

$$v_{\text{av}} = \frac{\Delta x}{\Delta t} \qquad (2)$$

Find $x(4.0\text{ s})$ and $x(3.0\text{ s})$:

$x(4.0\text{ s}) = (4.0)^2 - 5(4.0) + 1 = -3.0\text{ m}$
and
$x(3.0\text{ s}) = (3.0)^2 - 5(3.0) + 1 = -5.0\text{ m}$

Substitute numerical values in equation (1) and evaluate Δx:

$$\Delta x = (-3.0\text{ m}) - (-5.0\text{ m}) = \boxed{2.0\text{ m}}$$

Substitute numerical values in equation (2) and evaluate v_{av}:

$$v_{av} = \frac{2.0\text{ m}}{1.0\text{ s}} = \boxed{2.0\text{ m/s}}$$

(*b*) Find $x(t + \Delta t)$:

$$x(t+\Delta t) = (t+\Delta t)^2 - 5(t+\Delta t) + 1 = (t^2 + 2t\Delta t + (\Delta t)^2) - 5(t+\Delta t) + 1$$

Express $x(t + \Delta t) - x(t) = \Delta x$:

$$\Delta x = \boxed{(2t-5)\Delta t + (\Delta t)^2}$$

where Δx is in meters if t is in seconds.

(*c*) From (*b*) find $\Delta x/\Delta t$ as $\Delta t \to 0$:

$$\frac{\Delta x}{\Delta t} = \frac{(2t-5)\Delta t + (\Delta t)^2}{\Delta t} = 2t - 5 + \Delta t$$

and

$$v = \lim_{\Delta t \to 0}(\Delta x/\Delta t) = \boxed{2t-5}$$

where v is in m/s if t is in seconds.

Alternatively, we can take the derivative of $x(t)$ with respect to time to obtain the instantaneous velocity.

$$v(t) = \frac{dx(t)}{dt} = \frac{d}{dt}(at^2 + bt + 1) = 2at + b = \boxed{2t - 5}$$

Constant Acceleration and Free-Fall

67 • **[SSM]** An object traveling along the x axis at constant acceleration has a velocity of +10 m/s when it is at $x = 6.0$ m and of +15 m/s when it is at $x = 10$ m. What is its acceleration?

Picture the Problem Because the acceleration of the object is constant we can use constant-acceleration equations to describe its motion.

Using a constant-acceleration equation, relate the velocity to the acceleration and the displacement:

$$v^2 = v_0^2 + 2a\Delta x \Rightarrow a = \frac{v^2 - v_0^2}{2\Delta x}$$

Substitute numerical values and evaluate a:

$$a = \frac{(15^2 - 10^2)\text{m}^2/\text{s}^2}{2(10\text{ m} - 6.0\text{ m})} = \boxed{16\text{ m/s}^2}$$

71 •• **[SSM]** A load of bricks is lifted by a crane at a steady velocity of 5.0 m/s when one brick falls off 6.0 m above the ground. (*a*) Sketch the position of the brick $y(t)$ versus time from the moment it leaves the pallet until it hits the ground. (*b*) What is the greatest height the brick reaches above the ground?

(*c*) How long does it take to reach the ground? (*d*) What is its speed just before it hits the ground?

Picture the Problem In the absence of air resistance, the brick experiences constant acceleration and we can use constant-acceleration equations to describe its motion. Constant acceleration implies a parabolic position-versus-time curve.

(*a*) Using a constant-acceleration equation, relate the position of the brick to its initial position, initial velocity, acceleration, and time into its fall:

$$y = y_0 + v_0 t + \tfrac{1}{2}(-g)t^2$$

Substitute numerical values to obtain:

$$y = 6.0\,\text{m} + (5.0\,\text{m/s})t - (4.91\,\text{m/s}^2)t^2 \qquad (1)$$

The following graph of $y = 6.0\,\text{m} + (5.0\,\text{m/s})t - (4.91\,\text{m/s}^2)t^2$ was plotted using a spreadsheet program:

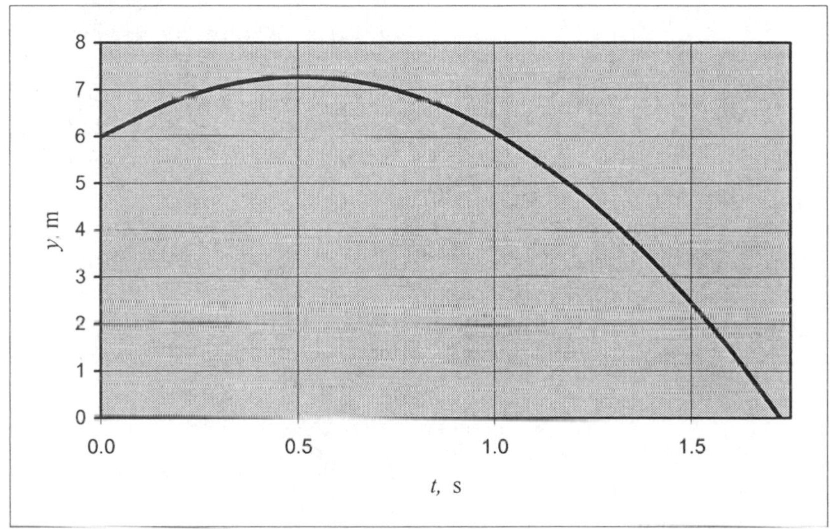

(*b*) Relate the greatest height reached by the brick to its height when it falls off the load and the additional height it rises Δy_{max}:

$$h = y_0 + \Delta y_{\text{max}} \qquad (2)$$

Using a constant-acceleration equation, relate the height reached by the brick to its acceleration and initial velocity:

$$v_{\text{top}}^2 = v_0^2 + 2(-g)\Delta y_{\text{max}}$$

or, because $v_{\text{top}} = 0$,

$$0 = v_0^2 + 2(-g)\Delta y_{\text{max}} \Rightarrow \Delta y_{\text{max}} = \frac{v_0^2}{2g}$$

Substitute numerical values and evaluate Δy_{max}:

$$\Delta y_{max} = \frac{(5.0\,\text{m/s})^2}{2(9.81\,\text{m/s}^2)} = 1.3\,\text{m}$$

Substitute numerical values in equation (2) and evaluate h:

$$h = 6.0\,\text{m} + 1.3\,\text{m} = \boxed{7.3\,\text{m}}$$

Note that the graph shown above confirms this result.

(*c*) Setting $y = 0$ in equation (1) yields:

$$0 = 6.0\,\text{m} + (5.0\,\text{m/s})t - (4.91\,\text{m/s}^2)t^2$$

Use the quadratic equation or your graphing calculator to obtain:

$t = \boxed{1.7\,\text{s}}$ and $t = -0.71$ s. Note that the second solution is nonphysical.

(*d*) Using a constant-acceleration equation, relate the speed of the brick on impact to its acceleration and displacement:

$$v^2 = v_0^2 + 2gh$$

or, because $v_0 = 0$,

$$v^2 = 2gh \Rightarrow v = \sqrt{2gh}$$

Substitute numerical values and evaluate v:

$$v = \sqrt{2(9.81\,\text{m/s}^2)(7.3\,\text{m})} = \boxed{12\,\text{m/s}}$$

75 •• **[SSM]** A stone is thrown vertically downward from the top of a 200-m cliff. During the last half second of its flight, the stone travels a distance of 45 m. Find the initial speed of the stone.

Picture the Problem In the absence of air resistance, the acceleration of the stone is constant. Choose a coordinate system with the origin at the bottom of the trajectory and the upward direction positive. Let $v_{\text{f-1/2}}$ be the speed one-half second before impact and v_f the speed at impact.

Using a constant-acceleration equation, express the final speed of the stone in terms of its initial speed, acceleration, and displacement:

$$v_f^2 = v_0^2 + 2a\Delta y \Rightarrow v_0 = \sqrt{v_f^2 + 2g\Delta y} \quad (1)$$

Find the average speed in the last half second:

$$v_{av} = \frac{v_{\text{f-1/2}} + v_f}{2} = \frac{\Delta x_{\text{last half second}}}{\Delta t} = \frac{45\,\text{m}}{0.5\,\text{s}} = 90\,\text{m/s}$$

and

$$v_{\text{f-1/2}} + v_f = 2(90\,\text{m/s}) = 180\,\text{m/s}$$

Using a constant-acceleration equation, express the change in speed of the stone in the last half second in terms of the acceleration and the elapsed time and solve for the change in its speed:

$$\Delta v = v_f - v_{f\text{-}1/2} = g\Delta t = (9.81\,\text{m/s}^2)(0.5\,\text{s}) = 4.91\,\text{m/s}$$

Add the equations that express the sum and difference of $v_{f-½}$ and v_f and solve for v_f:

$$v_f = \frac{180\,\text{m/s} + 4.91\,\text{m/s}}{2} = 92.5\,\text{m/s}$$

Substitute numerical values in equation (1) and evaluate v_0:

$$v_0 = \sqrt{(92.5\,\text{m/s})^2 + 2(9.81\,\text{m/s}^2)(-200\text{m})} = \boxed{68\,\text{m/s}}$$

Remarks: The stone may be thrown either up or down from the cliff and the results after it passes the cliff on the way down are the same.

81 •• **[SSM]** In a classroom demonstration, a glider moves along an inclined air track with constant acceleration. It is projected from the low end of the track with an initial velocity. After 8.00 s have elapsed, it is 100 cm from the low end and is moving along the track at a velocity of –15 cm/s. Find the initial velocity and the acceleration.

Picture the Problem The acceleration of the glider on the air track is constant. Its average acceleration is equal to the instantaneous (constant) acceleration. Choose a coordinate system in which the initial direction of the glider's motion is the positive direction.

Using the definition of acceleration, express the average acceleration of the glider in terms of the glider's velocity change and the elapsed time:

$$a = a_{av} = \frac{\Delta v}{\Delta t}$$

Using a constant-acceleration equation, express the average velocity of the glider in terms of the displacement of the glider and the elapsed time:

$$v_{av} = \frac{\Delta x}{\Delta t} = \frac{v_0 + v}{2} \Rightarrow v_0 = \frac{2\Delta x}{\Delta t} - v$$

Substitute numerical values and evaluate v_0:

$$v_0 = \frac{2(100\,\text{cm})}{8.00\,\text{s}} - (-15\,\text{cm/s}) = \boxed{40\,\text{cm/s}}$$

The average acceleration of the glider is:

$$a = \frac{-15\text{ cm/s} - (40\text{ cm/s})}{8.00\text{ s}} = \boxed{-6.9\text{ cm/s}^2}$$

83 •• **[SSM]** A typical automobile under hard braking loses speed at a rate of about 7.0 m/s^2; the typical reaction time to engage the brakes is 0.50 s. A local school board sets the speed limit in a school zone such that all cars should be able to stop in 4.0 m. (*a*) What maximum speed does this imply for an automobile in this zone? (*b*) What fraction of the 4.0 m is due to the reaction time?

Picture the Problem Assume that the acceleration of the car is constant. The total distance the car travels while stopping is the sum of the distances it travels during the driver's reaction time and the time it travels while braking. Choose a coordinate system in which the positive direction is the direction of motion of the automobile and apply a constant-acceleration equation to obtain a quadratic equation in the car's initial speed v_0.

(*a*) Using a constant-acceleration equation, relate the velocity of the car to its initial velocity, acceleration, and displacement during braking:

$$v^2 = v_0^2 + 2a\Delta x_{\text{brk}}$$

or, because the final velocity is zero,

$$0 = v_0^2 + 2a\Delta x_{\text{brk}} \Rightarrow \Delta x_{\text{brk}} = -\frac{v_0^2}{2a}$$

Express the total distance traveled by the car as the sum of the distance traveled during the reaction time and the distance traveled while slowing down:

$$\Delta x_{\text{tot}} = \Delta x_{\text{react}} + \Delta x_{\text{brk}} = v_0\Delta t_{\text{react}} - \frac{v_0^2}{2a}$$

Rearrange this quadratic equation to obtain:

$$v_0^2 - 2a\Delta t_{\text{react}} v_0 + 2a\Delta x_{\text{tot}} = 0$$

Substitute numerical values and simplify to obtain:

$$v_0^2 + (7.0\text{ m/s})v_0 - 56\text{ m}^2/\text{s}^2 = 0$$

Use your graphing calculator or the quadratic formula to solve the quadratic equation for its positive root:

$$v_0 = 4.76\text{ m/s}$$

Convert this speed to mi/h:

$$v_0 = (4.76\text{ m/s})\left(\frac{1\text{ mi/h}}{0.4470\text{ m/s}}\right) = \boxed{11\text{ mi/h}}$$

(*b*) Find the reaction-time distance:

$$\Delta x_{\text{react}} = v_0\Delta t_{\text{react}} = (4.76\text{ m/s})(0.50\text{ s}) = 2.38\text{ m}$$

Express and evaluate the ratio of the reaction distance to the total distance:

$$\frac{\Delta x_{\text{react}}}{\Delta x_{\text{tot}}} = \frac{2.38\text{ m}}{4.0\text{ m}} = \boxed{0.60}$$

91 •• **[SSM]** Consider measuring the free-fall motion of a particle (neglect air resistance). Before the advent of computer-driven data-logging software, these experiments typically employed a wax-coated tape placed vertically next to the path of a dropped electrically conductive object. A spark generator would cause an arc to jump between two vertical wires through the falling object and through the tape, thereby marking the tape at fixed time intervals Δt. Show that the change in height during successive time intervals for an object falling from rest follows *Galileo's Rule of Odd Numbers*: $\Delta y_{21} = 3\Delta y_{10}$, $\Delta y_{32} = 5\Delta y_{10}$, . . . , where Δy_{10} is the change in y during the first interval of duration Δt, Δy_{21} is the change in y during the second interval of duration Δt, etc.

Picture the Problem In the absence of air resistance, the particle experiences constant acceleration and we can use constant-acceleration equations to describe its position as a function of time. Choose a coordinate system in which downward is positive, the particle starts from rest ($v_0 = 0$), and the starting height is zero ($y_0 = 0$).

Using a constant-acceleration equation, relate the position of the falling particle to the acceleration and the time. Evaluate the y-position at successive equal time intervals Δt, $2\Delta t$, $3\Delta t$, etc:

$$y_1 = \frac{-g}{2}(\Delta t)^2 = \frac{-g}{2}(\Delta t)^2$$

$$y_2 = \frac{-g}{2}(2\Delta t)^2 = \frac{-g}{2}(4)(\Delta t)^2$$

$$y_3 = \frac{-g}{2}(3\Delta t)^2 = \frac{-g}{2}(9)(\Delta t)^2$$

$$y_4 = \frac{-g}{2}(4\Delta t)^2 = \frac{-g}{2}(16)(\Delta t)^2$$

etc.

Evaluate the changes in those positions in each time interval:

$$\Delta y_{10} = y_1 - 0 = \left(\frac{-g}{2}\right)(\Delta t)^2$$

$$\Delta y_{21} = y_2 - y_1 = 3\left(\frac{-g}{2}\right)(\Delta t)^2 = 3\Delta y_{10}$$

$$\Delta y_{32} = y_3 - y_2 = 5\left(\frac{-g}{2}\right)(\Delta t)^2 = 5\Delta y_{10}$$

$$\Delta y_{43} = y_4 - y_3 = 7\left(\frac{-g}{2}\right)(\Delta t)^2 = 7\Delta y_{10}$$

etc.

93 •• **[SSM]** If it were possible for a spacecraft to maintain a constant acceleration indefinitely, trips to the planets of the Solar System could be undertaken in days or weeks, while voyages to the nearer stars would only take a few years. (*a*) Using data from the tables at the back of the book, find the time it would take for a one-way trip from Earth to Mars (at Mars' closest approach to Earth). Assume that the spacecraft starts from rest, travels along a straight line, accelerates halfway at 1*g*, flips around, and decelerates at 1*g* for the rest of the trip. (*b*) Repeat the calculation for a 4.1×10^{13}-km trip to Proxima Centauri, our nearest stellar neighbor outside of the sun. (See Problem 47.)

Picture the Problem Note: No material body can travel at speeds faster than light. When one is dealing with problems of this sort, the kinematic formulae for displacement, velocity and acceleration are no longer valid, and one must invoke the special theory of relativity to answer questions such as these. For now, ignore such subtleties. Although the formulas you are using (i.e., the constant-acceleration equations) are not quite correct, your answer to Part (*b*) will be wrong by about 1%.

(*a*) Let $t_{1/2}$ represent the time it takes to reach the halfway point. Then the total trip time is:

$$t = 2\,t_{1/2} \qquad (1)$$

Use a constant- acceleration equation to relate the half-distance to Mars Δx to the initial speed, acceleration, and half-trip time $t_{1/2}$:

$$\Delta x = v_0 t + \tfrac{1}{2} a t_{1/2}^2$$

Because $v_0 = 0$ and $a = g$:

$$t_{1/2} = \sqrt{\frac{2\Delta x}{a}}$$

Substitute in equation (1) to obtain:

$$t = 2\sqrt{\frac{2\Delta x}{a}} \qquad (2)$$

The distance from Earth to Mars at closest approach is 7.8×10^{10} m. Substitute numerical values and evaluate t :

$$t_{\text{round trip}} = 2\sqrt{\frac{2(3.9\times10^{10}\ \text{m})}{9.81\,\text{m/s}^2}} = 18\times10^{4}\ \text{s}$$

$$\approx \boxed{2.1\,\text{d}}$$

(*b*) From Problem 47 we have:

$$d_{\text{Proxima Centauri}} = 4.1\times10^{13}\ \text{km}$$

Substitute numerical values in equation (2) to obtain:

$$t_{\text{round trip}} = 2\sqrt{\frac{2(4.1\times10^{13}\ \text{km})}{9.81\,\text{m/s}^2}} = 18\times10^{7}\ \text{s}$$

$$\approx \boxed{5.8\ \text{y}}$$

Remarks: Our result in Part (*a*) seems remarkably short, considering how far Mars is and how low the acceleration is.

99 ••• **[SSM]** A speeder traveling at a constant speed of 125 km/h races past a billboard. A patrol car pursues from rest with constant acceleration of (8.0 km/h)/s until it reaches its maximum speed of 190 km/h, which it maintains until it catches up with the speeder. (*a*) How long does it take the patrol car to catch the speeder if it starts moving just as the speeder passes? (*b*) How far does each car travel? (*c*) Sketch $x(t)$ for each car.

Picture the Problem This is a two-stage constant-acceleration problem. Choose a coordinate system in which the direction of the motion of the cars is the positive direction. The pictorial representation summarizes what we know about the motion of the speeder's car and the patrol car.

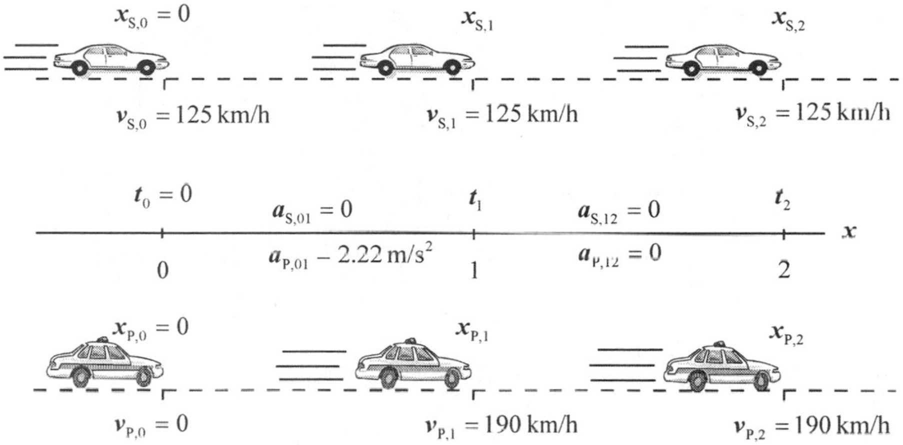

$x_{S,0} = 0$ $\quad x_{S,1}$ $\quad x_{S,2}$

$v_{S,0} = 125\text{ km/h}$ $\quad v_{S,1} = 125\text{ km/h}$ $\quad v_{S,2} = 125\text{ km/h}$

$t_0 = 0$ $\quad a_{S,01} = 0$ $\quad t_1$ $\quad a_{S,12} = 0$ $\quad t_2$ $\quad x$

0 $\quad a_{P,01} - 2.22\text{ m/s}^2$ $\quad$ 1 $\quad a_{P,12} = 0$ $\quad$ 2

$x_{P,0} = 0$ $\quad x_{P,1}$ $\quad x_{P,2}$

$v_{P,0} = 0$ $\quad v_{P,1} = 190\text{ km/h}$ $\quad v_{P,2} = 190\text{ km/h}$

Convert the speeds of the vehicles and the acceleration of the police car into SI units:

$$8.0\frac{\text{km}}{\text{h}\cdot\text{s}} = 8.0\frac{\text{km}}{\text{h}\cdot\text{s}} \times \frac{1\text{ h}}{3600\text{ s}} = 2.2\text{ m/s}^2,$$

$$125\frac{\text{km}}{\text{h}} = 125\frac{\text{km}}{\text{h}} \times \frac{1\text{ h}}{3600\text{ s}} = 34.7\text{ m/s},$$

and

$$190\frac{\text{km}}{\text{h}} = 190\frac{\text{km}}{\text{h}} \times \frac{1\text{ h}}{3600\text{ s}} = 52.8\text{ m/s}$$

(*a*) Express the condition that determines when the police car catches the speeder; that is, that their displacements will be the same:

$$\Delta x_{P,02} = \Delta x_{S,02}$$

Using a constant-acceleration equation, relate the displacement of the patrol car to its displacement while accelerating and its displacement once it reaches its maximum velocity:

$$\Delta x_{P,02} = \Delta x_{P,01} + \Delta x_{P,12} = \Delta x_{P,01} + v_{P,1}(t_2 - t_1)$$

Using a constant-acceleration equation, relate the displacement of the speeder to its constant velocity and the time it takes the patrol car to catch it:

$$\Delta x_{S,02} = v_{S,02}\Delta t_{02} = (34.7\text{ m/s})t_2$$

Calculate the time during which the police car is speeding up:

$$\Delta t_{P,01} = \frac{\Delta v_{P,01}}{a_{P,01}} = \frac{v_{P,1} - v_{P,0}}{a_{P,01}} = \frac{52.8\text{ m/s} - 0}{2.2\text{ m/s}^2} = 24\text{ s}$$

Express the displacement of the patrol car:

$$\Delta x_{P,01} = v_{P,0}\Delta t_{P,01} + \tfrac{1}{2}a_{P,01}\Delta t_{P,01}^2 = 0 + \tfrac{1}{2}(2.2\text{ m/s}^2)(24\text{ s})^2 = 630\text{ m}$$

Equate the displacements of the two vehicles:

$$\Delta x_{P,02} = \Delta x_{P,01} + \Delta x_{P,12} = \Delta x_{P,01} + v_{P,1}(t_2 - t_1) = 630\text{ m} + (52.8\text{ m/s})(t_2 - 24\text{ s})$$

Substitute for $\Delta x_{P,02}$ to obtain:

$$(34.7\text{ m/s})\ t_2 = 630\text{ m} + (52.8\text{ m/s})(t_2 - 24\text{ s})$$

Solving for the time to catch up yields:

$$t_2 = 35.19\text{ s} = \boxed{35\text{ s}}$$

(*b*) The distance traveled is the displacement, $\Delta x_{02,S}$, of the speeder during the catch:

$$\Delta x_{S,02} = v_{S,02}\Delta t_{02} = (35\text{ m/s})(35.19\text{ s}) = \boxed{1.2\text{ km}}$$

(*c*) The graphs of x_S and x_P follow. The straight line (solid) represents $x_S(t)$ and the parabola (dashed) represents $x_P(t)$.

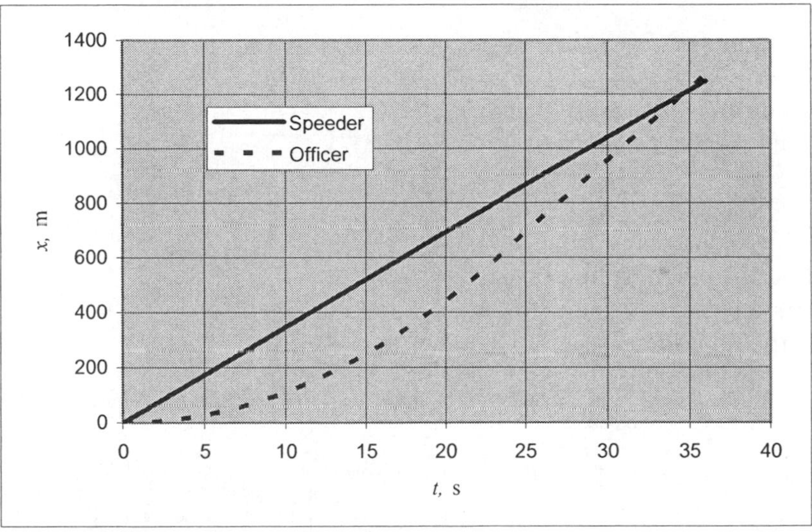

Integration of the Equations of Motion

103 • **[SSM]** The velocity of a particle is given by $v_x(t) = (6.0\ \text{m/s}^2)t + (3.0\ \text{m/s})$. (*a*) Sketch *v* versus *t* and find the area under the curve for the interval $t = 0$ to $t = 5.0$ s. (*b*) Find the position function $x(t)$. Use it to calculate the displacement during the interval $t = 0$ to $t = 5.0$ s.

Picture the Problem The integral of a function is equal to the "area" between the curve for that function and the independent-variable axis.

(*a*) The following graph was plotted using a spreadsheet program:

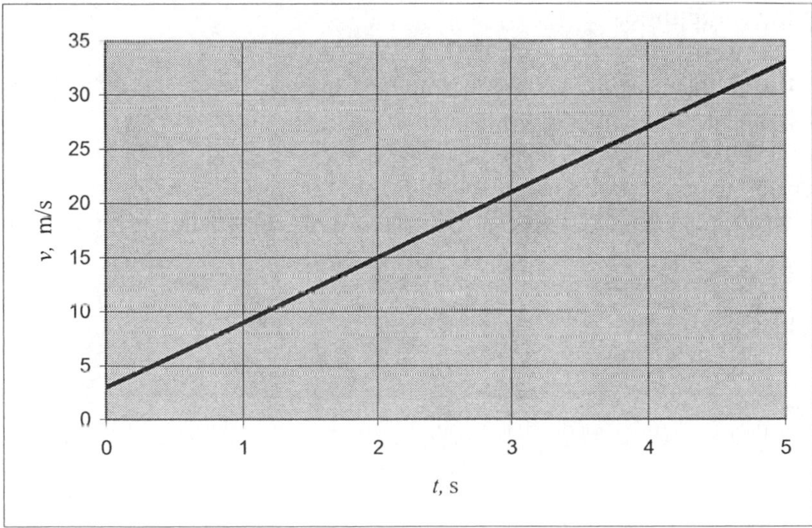

The distance is found by determining the area under the curve. There are approximately 36 blocks each having an area of

$$(5.0\ \text{m/s})(0.5\ \text{s}) = 2.5\ \text{m}.$$

$$A_{\substack{\text{under}\\\text{curve}}} = (36\ \text{blocks})(2.5\ \text{m/block}) = \boxed{90\ \text{m}}$$

You can confirm this result by using the formula for the area of a trapezoid:

$$A = \left(\frac{33\,\text{m/s} + 3\,\text{m/s}}{2}\right)(5.0\,\text{s} - 0\,\text{s}) = 90\,\text{m}$$

(*b*) To find the position function $x(t)$, we integrate the velocity function $v(t)$ over the time interval in question:

$$x(t) = \int_0^t v(t)\,dt = \int_0^t \left[\left(6.0\,\text{m/s}^2\right)t + (3.0\,\text{m/s})\right] dt$$

and

$$x(t) = \boxed{\left(3.0\,\text{m/s}^2\right)t^2 + (3.0\,\text{m/s})t}$$

Now evaluate $x(t)$ at 0 s and 5.0 s respectively and subtract to obtain Δx:

$$\Delta x = x(5.0\,\text{s}) - x(0\,\text{s}) = 90\,\text{m} - 0\,\text{m} = \boxed{90\ \text{m}}$$

109 ••• **[SSM]** Figure 2-45 shows a plot of x versus t for a body moving along a straight line. For this motion, sketch graphs (using the same t axis) of (*a*) v_x as a function of t, and (*b*) a_x as a function of t. (*c*) Use your sketches to compare the times when the object is at its largest distance from the origin to the times when its speed is greatest. Explain why they do *not* occur at the same time. (*d*) Use your sketches to compare the time(s) when the object is moving fastest when the time(s) when its acceleration is the largest. Explain why they do *not* occur at the same time.

Picture the Problem Because the position of the body is not described by a parabolic function, the acceleration is not constant.

(*a*) Select a series of points on the graph of $x(t)$ (e.g., at the extreme values and where the graph crosses the t axis), draw tangent lines at those points, and measure their slopes. In doing this, you are evaluating $v = dx/dt$ at these points. Plot these slopes above the times at which you measured the slopes. Your graph should closely resemble the following graph.

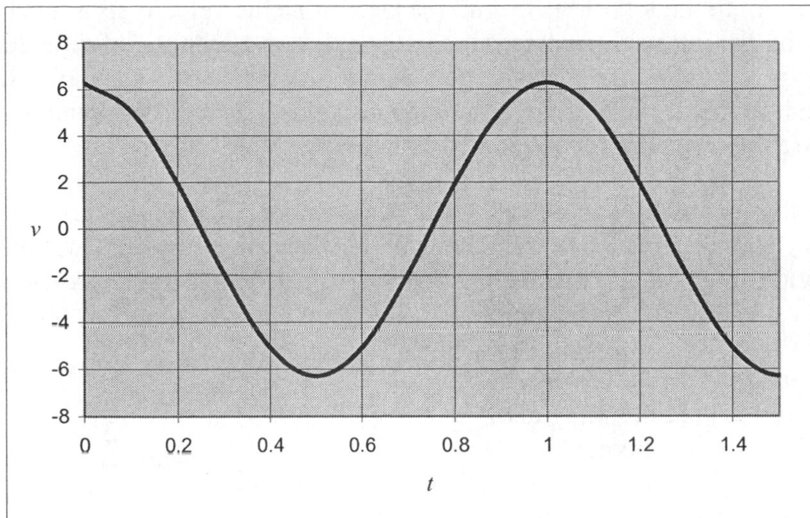

(*b*) Select a series of points on the graph of $v(t)$ (e.g., at the extreme values and where the graph crosses the t axis), draw tangent lines at those points, and measure their slopes. In doing this, you are evaluating $a = dv/dt$ at these points. Plot these slopes above the times at which you measured the slopes. Your graph should closely resemble the following graph.

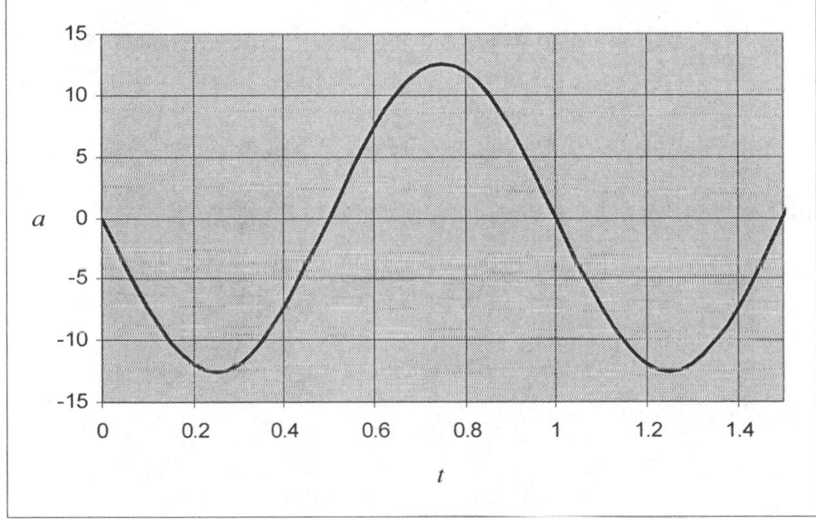

(*c*) The points at the greatest distances from the time axis correspond to turn-around points. The velocity of the body is zero at these points.

(*d*) The body is moving fastest as it goes through the origin. At these times the velocity is not changing and hence the acceleration is zero. The maximum acceleration occurs at the maximum distances where the velocity is zero but changing direction rapidly.

111 ••• **[SSM]** In the time interval from 0.0 s to 10.0 s, the acceleration of a particle traveling in a straight line is given by $a_x = (0.20\ \text{m/s}^3)t$. Let to the right be the $+x$ direction. A particle initially has a velocity to the right of 9.5 m/s and is located 5.0 m to the left of the origin. (*a*) Determine the velocity as a function of time during the interval, (*b*) determine the position as a function of time during the interval, (*c*) determine the average velocity between $t = 0.0$ s and10.0 s, and compare it to the average of the instantaneous velocities at the start and ending times. Are these two averages equal? Explain.

Picture the Problem The acceleration is a function of time; therefore it is not constant. The instantaneous velocity can be determined by integration of the acceleration function and the average velocity from the general expression for the average value of a non-linear function.

(*a*) The ***instantaneous velocity*** function $v(t)$ is the time-integral of the acceleration function:

$$v(t) = \int a_x dt = \int bt\,dt = \frac{b}{2}t^2 + C_1$$

where $b = 0.20\ \text{m/s}^3$

The initial conditions are:

1) $\boldsymbol{x}(0) = -5.0\ \text{m}$
and
2) $\boldsymbol{v}(0) = 9.5\ \text{m/s}$

Use initial condition 2) to obtain:

$$\boldsymbol{v}(0) = 9.5\ \text{m/s} = \boldsymbol{C}_1$$

Substituting in $v(t)$ for b and C_1 yields:

$$\boldsymbol{v}(\boldsymbol{t}) = \boxed{(0.10\ \text{m/s}^3)\boldsymbol{t}^2 + 9.5\ \text{m/s}} \quad (1)$$

(*b*) The ***instantaneous position*** function $x(t)$ is the time-integral of the velocity function:

$$x(t) = \int v(t)dt = \int (ct^2 + C_1)dt$$

$$= \frac{c}{6}t^3 + C_1 t + C_2$$

where $c = 0.10\ \text{m/s}^3$.

Using initial condition 1) yields:

$$\boldsymbol{x}(0) = -5.0\ \text{m} = \boldsymbol{C}_2$$

Substituting in $x(t)$ for C_1 and C_2 yields:

$$x(t) = \boxed{\tfrac{1}{6}(0.20\ \text{m/s}^3)t^3 + (9.5\ \text{m/s})t - 5.0\ \text{m/s}}$$

(*c*) The average value of $v(t)$ over the interval Δt is given by:

$$\bar{v} = \frac{1}{\Delta t}\int_{t_1}^{t_2} v(t)dt$$

Substitute for $v(t)$ and evaluate the integral to obtain:

$$\bar{v} = \frac{1}{\Delta t}\int_{t_1}^{t_2}\left(\frac{b}{2}t^2 + C_1\right)dt = \frac{1}{\Delta t}\left[\frac{b}{6}t^3 + C_1 t\right]_{t_1}^{t_2} = \frac{1}{\Delta t}\left[\frac{b}{6}t_2^3 + C_1 t_2 - \left(\frac{b}{6}t_1^3 + C_1 t_1\right)\right]$$

Simplifying this expression yields:

$$\bar{v} = \frac{1}{\Delta t}\left[\frac{b}{6}\left(t_2^3 - t_1^3\right) + C_1\left(t_2 - t_1\right)\right]$$

Because $t_1 = 0$:

$$\bar{v} = \frac{1}{\Delta t}\left[\frac{b}{6}t_2^3 + C_1 t_2\right]$$

Substitute numerical values and simplify to obtain:

$$\bar{v} = \frac{1}{10.0\text{ s}}\left[\left(\frac{0.20\text{ m/s}^3}{6}\right)(10.0\text{ s})^3 + (9.5\text{ m/s})(10.0\text{ s})\right] = \boxed{13\text{ m/s}}$$

The average of the initial instantaneous and final instantaneous velocities is given by:

$$v_{av} = \frac{v(0) + v(10.0\text{ s})}{2} \qquad (2)$$

Using equation (1), evaluate $v(0)$ and $v(10\text{ s})$:

$$v(0) = 9.5\text{ m/s}$$

and

$$v(10\text{ s}) = \left(0.10\text{ m/s}^3\right)(10.0\text{ s})^2 + 9.5\text{ m/s} = 19.5\text{ m/s}$$

Substitute in equation (2) to obtain:

$$v_{av} = \frac{9.5\text{ m/s} + 19.5\text{ m/s}}{2} = \boxed{15\text{ m/s}}$$

v_{av} is not the same as $\bar{v}$ because the velocity does not change linearly with time. The velocity does not change linearly with time because the acceleration is not constant.

General Problems

115 ••• **[SSM]** Consider an object that is attached to a horizontally oscillating piston. The object moves with a velocity given by $v = B\sin(\omega t)$, where B and ω are constants and ω is in s^{-1}. (*a*) Explain why B is equal to the maximum speed v_{max}. (*b*) Determine the acceleration of the object as a function of time. Is the acceleration constant? (*c*) What is the maximum acceleration (magnitude) in terms of ω and v_{max}. (*d*) At $t = 0$, the object's position is known to be x_0. Determine the position as a function of time in terms of in terms of t, ω, x_0 and v_{max}.

Determine the Concept Because the velocity varies nonlinearly with time, the acceleration of the object is not constant. We can find the acceleration of the object by differentiating its velocity with respect to time and its position function by integrating the velocity function.

(*a*) The maximum value of the sine function (as in $v = v_{max}\sin(\omega t)$) is 1. Hence the coefficient *B* represents the maximum possible speed v_{max}.

(*b*) The acceleration of the object is the derivative of its velocity with respect to time:

$$a = \frac{dv}{dt} = \frac{d}{dt}\left[v_{max}\sin(\omega t)\right] = \boxed{\omega v_{max}\cos(\omega t)}$$

Because *a* varies sinusoidally with time it is *not* constant.

(*c*) Examination of the coefficient of the cosine function in the expression for *a* leads one to the conclusion that $|\mathbf{a}_{max}| = \boxed{\omega v_{max}.}$

(*d*) The position of the object as a function of time is the integral of the velocity function:

$$\int dx = \int v(t)\,dt$$

Integrating the left-hand side of the equation and substituting for *v* on the right-hand side yields:

$$x = \int v_{max}\sin(\omega t)\,dt + C$$

Integrate the right-hand side to obtain:

$$x = \frac{-v_{max}}{\omega}\cos(\omega t) + C \qquad (1)$$

Use the initial condition $x(0) = x_0$ to obtain:

$$x_0 = \frac{-v_{max}}{\omega} + C$$

Solving for *C* yields:

$$C = x_0 + \frac{v_{max}}{\omega}$$

Substitute for *C* in equation (1) to obtain:

$$x = \frac{-v_{max}}{\omega}\cos(\omega t) + x_0 + \frac{v_{max}}{\omega}$$

Solving this equation for *x* yields:

$$x = \boxed{x_0 + \frac{v_{max}}{\omega}\left[1 - \cos(\omega t)\right]}$$

117 ••• **[SSM]** A rock falls through water with a continuously decreasing acceleration. Assume that the rock's acceleration as a function of *velocity* has the form $\mathbf{a}_y = \mathbf{g} - \mathbf{b}\mathbf{v}_y$ where *b* is a positive constant. (The +*y* direction is directly downward.) (*a*) What are the SI units of *b*? (*b*) Prove mathematically that if the rock enters the water at time $t = 0$, the acceleration will depend exponentially on *time* according to $\mathbf{a}_y(t) = \mathbf{g}e^{-bt}$. (*c*) What is the terminal speed for the rock in terms of *g* and *b*? (See Problem 38 for an explanation of the phenomenon of *terminal speed*.)

Picture the Problem Because the acceleration of the rock is a function of its velocity, it is not constant and we will have to integrate the acceleration function in order to find the velocity function. Choose a coordinate system in which downward is positive and the origin is at the point of release of the rock.

(*a*) All three terms in $\boldsymbol{a}_y = \boldsymbol{g} - \boldsymbol{b}\boldsymbol{v}_y$ must have the same units in order for the equation to be valid. Hence the units of bv_y must be acceleration units. Because the SI units of v_y are m/s, b must have units of $\boxed{\text{s}^{-1}.}$

(*b*) Rewrite $a_y = g - bv_y$ explicitly as a differential equation:

$$\frac{dv_y}{dt} = g - bv_y$$

Separate the variables, v_y on the left, t on the right:

$$\frac{dv_y}{g - bv_y} = dt$$

Integrate the left-hand side of this equation from 0 to v_y and the right-hand side from 0 to t:

$$\int_0^{v_y} \frac{dv_y}{g - bv_y} = \int_0^t dt$$

Integrating this equation yields:

$$-\frac{1}{b}\ln\left(\frac{g - bv_y}{g}\right) = t$$

Solve this expression for v_y to obtain:

$$v_y = \frac{g}{b}\left(1 - e^{-bt}\right) \quad (1)$$

Differentiate this expression with respect to time to obtain an expression for the acceleration and complete the proof:

$$a_y = \frac{dv_y}{dt} = \frac{d}{dt}\left(\frac{g}{b}\left(1 - e^{-bt}\right)\right) = \boxed{ge^{-bt}}$$

(*c*) Take the limit, as $t \to \infty$, of both sides of equation (1):

$$\lim_{t\to\infty} v_y = \lim_{t\to\infty}\left[\frac{g}{b}\left(1 - e^{-bt}\right)\right]$$

and

$$v_t = \boxed{\frac{g}{b}}$$

Notice that this result depends only on b (inversely so). Thus b must include things like the shape and cross-sectional area of the falling object, as well as properties of the liquid such as density and temperature.

Chapter 3
Motion in Two and Three Dimensions

Conceptual Problems

1 • **[SSM]** Can the magnitude of the displacement of a particle be less than the distance traveled by the particle along its path? Can its magnitude be more than the distance traveled? Explain.

Determine the Concept The distance traveled along a path can be represented as a sequence of displacements.

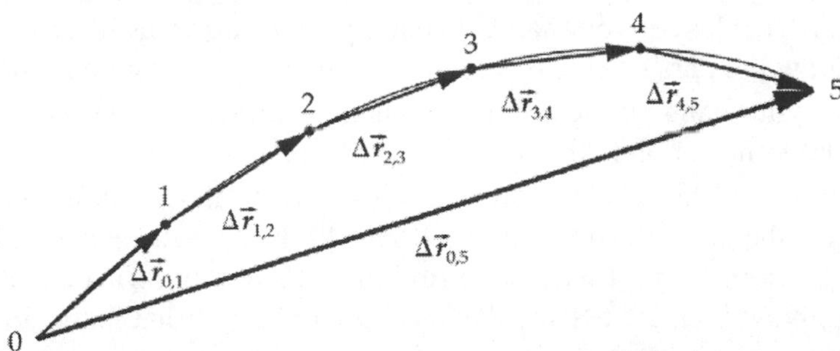

Suppose we take a trip along some path and consider the trip as a sequence of many very small displacements. The net displacement is the vector sum of the very small displacements, and the total distance traveled is the sum of the magnitudes of the very small displacements. That is,

$$\textit{total distance} = \left|\Delta\vec{r}_{0,1}\right| + \left|\Delta\vec{r}_{1,2}\right| + \left|\Delta\vec{r}_{2,3}\right| + \ldots + \left|\Delta\vec{r}_{N-1,N}\right|$$

where N is the number of very small displacements. (For this to be exactly true we have to take the limit as N goes to infinity and each displacement magnitude goes to zero.) Now, using "the shortest distance between two points is a straight line," we have

$$\left|\Delta\vec{r}_{0,N}\right| \le \left|\Delta\vec{r}_{0,1}\right| + \left|\Delta\vec{r}_{1,2}\right| + \left|\Delta\vec{r}_{2,3}\right| + \ldots + \left|\Delta\vec{r}_{N-1,N}\right|,$$

where $\left|\Delta\vec{r}_{0,N}\right|$ is the magnitude of the net displacement.

Hence, we have shown that the magnitude of the displacement of a particle is less than or equal to the distance it travels along its path.

11 • **[SSM]** Give examples of motion in which the directions of the velocity and acceleration vectors are (*a*) opposite, (*b*) the same, and (*c*) mutually perpendicular.

Determine the Concept The velocity vector is defined by $\vec{v} = d\vec{r}/dt$, while the acceleration vector is defined by $\vec{a} = d\vec{v}/dt$.

(*a*) A car moving along a straight road while braking.

(*b*) A car moving along a straight road while speeding up.

(*c*) A particle moving around a circular track at constant speed.

13 • **[SSM]** Imagine throwing a dart straight upward so that it sticks into the ceiling. After it leaves your hand, it steadily slows down as it rises before it sticks. (*a*) Draw the dart's velocity vector at times t_1 and t_2, where t_1 and t_2 occur after it leaves your hand but before it impacts the ceiling, and $\Delta t = t_2 - t_1$ is small. From your drawing find the direction of the change in velocity $\Delta\vec{v} = \vec{v}_2 - \vec{v}_1$, and thus the direction of the acceleration vector. (*b*) After it has stuck in the ceiling for a few seconds, the dart falls down to the floor. As it falls it speeds up, of course, until it hits the floor. Repeat Part (*a*) to find the direction of its acceleration vector as it falls. (*c*) Now imagine tossing the dart horizontally. What is the direction of its acceleration vector after it leaves your hand, but before it strikes the floor?

Determine the Concept The acceleration vector is in the same direction as the *change in velocity vector,* $\Delta\vec{v}$.

(*a*) The sketch for the dart thrown upward is shown to the right. The acceleration vector is in the direction of the *change* in the velocity vector $\Delta\vec{v}$.

(*b*) The sketch for the falling dart is shown to the right. Again, the acceleration vector is in the direction of the *change* in the velocity vector $\Delta\vec{v}$.

(*c*) The acceleration vector is in the direction of the *change* in the velocity vector … and hence is downward as shown to the right:

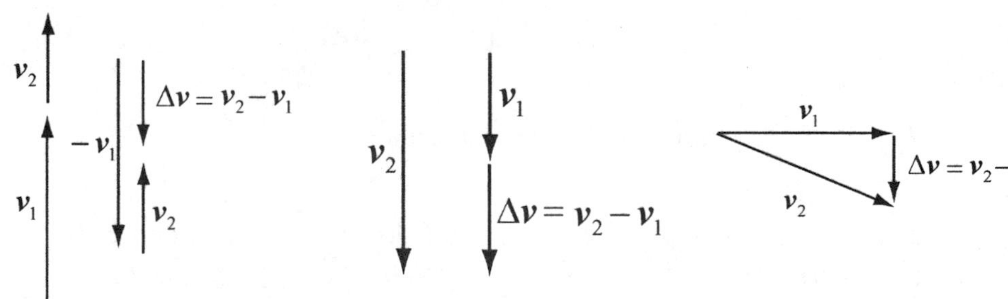

17 • **[SSM]** During a heavy rain, the drops are falling at a constant velocity and at an angle of 10° west of the vertical. You are walking in the rain and notice that only the top surfaces of your clothes are getting wet. In what direction are you walking? Explain.

Determine the Concept You must be walking west to make it appear to you that the rain is exactly vertical.

21 • **[SSM]** A projectile is fired at 35° above the horizontal. Any effects due to air resistance are negligible. The initial velocity of the projectile has a vertical component that is (*a*) less than 20 m/s, (*b*) greater than 20 m/s, (*c*) equal to 20 m/s, (*d*) cannot be determined from the data given.

Determine the Concept $\boxed{(a)}$ is correct. Because the initial horizontal velocity is 20 m/s, and the launch angle is less than 45 degrees, the initial vertical velocity must be less than 20 m/s.

25 • **[SSM]** True or false:

(*a*) If an object's speed is constant, then its acceleration must be zero.
(*b*) If an object's acceleration is zero, then its speed must be constant.
(*c*) If an object's acceleration is zero, its velocity must be constant.
(*d*) If an object's speed is constant, then its velocity must be constant.
(*e*) If an object's velocity is constant, then its speed must be constant.

Determine the Concept Speed is a scalar quantity, whereas acceleration, equal to the *rate of change* of velocity, is a vector quantity.

(*a*) False. Consider a ball on the end of a string. The ball can move with constant *speed* (a scalar) even though its *acceleration* (a vector) is always changing direction.

(*b*) True. From its definition, if the acceleration is zero, the velocity must be constant and so, therefore, must be the speed.

(*c*) True. An object's velocity must change in order for the object to have an acceleration other than zero.

(*d*) False. Consider an object moving at constant speed along a circular path. Its velocity changes continuously along such a path.

(*e*) True. If the velocity of an object is constant, then both its direction and magnitude (speed) must be constant.

33 •• **[SSM]** During your rookie bungee jump, your friend records your fall using a camcorder. By analyzing it frame by frame, he finds that the *y*-component of your velocity is (recorded every 1/20th of a second) as follows:

t (s)	12.05	12.10	12.15	12.20	12.25	12.30	12.35	12.40	12.45
v_y (m/s)	−0.78	−0.69	−0.55	−0.35	−0.10	0.15	0.35	0.49	0.53

(*a*) Draw a motion diagram. Use it to find the direction and relative magnitude of your average acceleration for each of the eight successive 0.050 s time intervals in the table. (*b*) Comment on how the *y* component of your acceleration does or does not vary in sign and magnitude as you reverse your direction of motion.

Determine the Concept (*a*) The motion diagram shown below was constructed using the data in the table shown below the motion diagram. Note that the motion diagram has been rotated 90°. The upward direction is to the left. The column for Δv in the table to the right was calculated using $\Delta v = v_i - v_{i-1}$ and the column for a was calculated using $a = (v_i - v_{i-1})/\Delta t$.

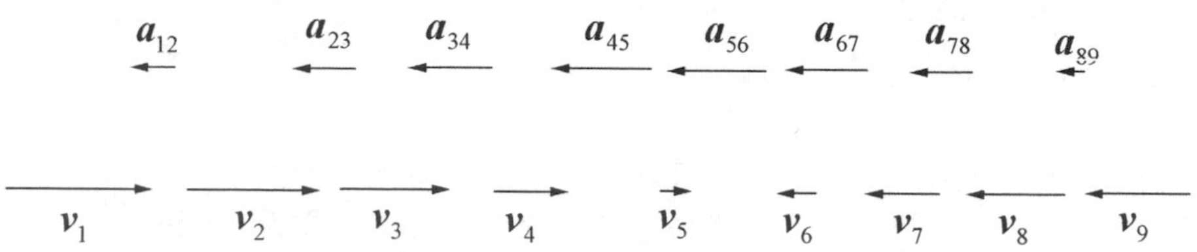

i	v	Δv	a_{ave}
	(m/s)	(m/s)	(m/s^2)
1	−0.78		
2	−0.69	0.09	1.8
3	−0.55	0.14	2.8
4	−0.35	0.20	4.0
5	−0.10	0.25	5.0
6	0.15	0.25	5.0
7	0.35	0.20	4.0
8	0.49	0.14	2.8
9	0.53	0.04	0.8

(*b*) The acceleration vector always points upward and so the sign of its *y* component does not change. The magnitude of the acceleration vector is greatest when the bungee cord has its maximum extension (your speed, the magnitude of your velocity, is least at this time and times near it) and is less than this maximum value when the bungee cord has less extension.

Position, Displacement, Velocity and Acceleration Vectors

39 • [SSM] In Problem 38, find the displacements of the tip of each hand (that is, $\Delta\vec{A}$ and $\Delta\vec{B}$) when the time advances from 3:00 P.M. to 6:00 P.M.

Picture the Problem Let the +*y* direction be straight up, the +*x* direction be to the right, and use the vectors describing the ends of the hour and minute hands in Problem 38 to find the displacements $\Delta\vec{A}$ and $\Delta\vec{B}$.

The displacement of the minute hand as time advances from 3:00 P.M. to 6:00 P.M. is given by:

$$\Delta\vec{B} = \vec{B}_6 - \vec{B}_3$$

From Problem 38:

$$\vec{B}_6 = (0.50\,\text{m})\hat{j} \text{ and } \vec{B}_3 = (0.50\,\text{m})\hat{j}$$

Substitute and simplify to obtain:

$$\Delta\vec{B} = (0.50\,\text{m})\hat{j} - (0.50\,\text{m})\hat{j} = \boxed{0}$$

The displacement of the hour hand as time advances from 3:00 P.M. to 6:00 P.M. is given by:

$$\Delta\vec{A} = \vec{A}_6 - \vec{A}_3$$

From Problem 38:

$$\vec{A}_6 = -(0.25\,\text{m})\hat{j} \text{ and } \vec{A}_3 = (0.25\,\text{m})\hat{i}$$

Substitute and simplify to obtain:

$$\Delta\vec{A} = \boxed{-(0.25\,\text{m})\hat{j} - (0.25\,\text{m})\hat{i}}$$

43 •• **[SSM]** The faces of a cubical storage cabinet in your garage has 3.0-m-long edges that are parallel to the *xyz* coordinate planes. The cube has one corner at the origin. A cockroach, on the hunt for crumbs of food, begins at that corner and walks along three edges until it is at the far corner. (*a*) Write the roach's displacement using the set of $\hat{i}$, $\hat{j}$, and $\hat{k}$ unit vectors, and (*b*) find the magnitude of its displacement.

Picture the Problem While there are several walking routes the cockroach could take to get from the origin to point C, its displacement will be the same for all of them. One possible route is shown in the figure.

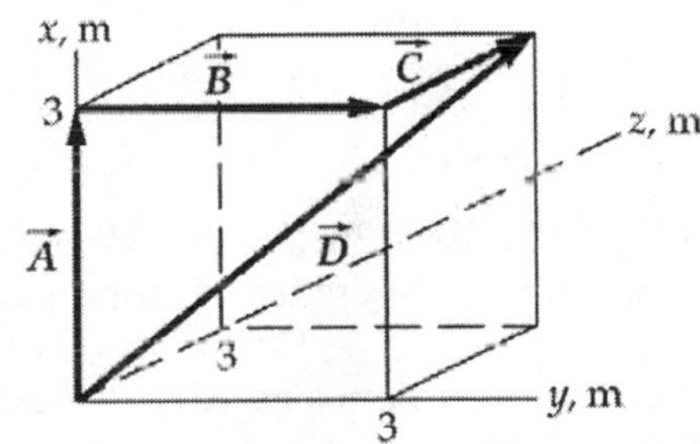

(*a*) The roach's displacement $\vec{D}$ during its trip from the origin to point C is:

$$\vec{D} = \vec{A} + \vec{B} + \vec{C}$$

$$= \boxed{(3.0\,\text{m})\hat{i} + (3.0\,\text{m})\hat{j} + (3.0\,\text{m})\hat{k}}$$

(*b*) The magnitude of the roach's displacement is given by:

$$D = \sqrt{D_x^2 + D_y^2 + D_z^2}$$

Substitute for D_x, D_y, and D_z and evaluate D to obtain:

$$D = \sqrt{(3.0\,\text{m})^2 + (3.0\,\text{m})^2 + (3.0\,\text{m})^2} = \boxed{5.2\,\text{m}}$$

Velocity and Acceleration Vectors

47 • **[SSM]** A particle moving at a velocity of 4.0 m/s in the $+x$ direction is given an acceleration of 3.0 m/s^2 in the $+y$ direction for 2.0 s. Find the final speed of the particle.

Picture the Problem The magnitude of the velocity vector at the end of the 2 s of acceleration will give us its speed at that instant. This is a constant-acceleration problem.

Find the final velocity vector of the particle:

$$\vec{v} = v_x\hat{i} + v_y\hat{j} = v_{x0}\hat{i} + a_y t\hat{j}$$
$$= (4.0\,\text{m/s})\hat{i} + (3.0\,\text{m/s}^2)(2.0\,\text{s})\hat{j}$$
$$= (4.0\,\text{m/s})\hat{i} + (6.0\,\text{m/s})\hat{j}$$

The magnitude of $\vec{v}$ is:

$$|\vec{v}| = \sqrt{v_x^2 + v_y^2}$$

Substitute for v_x and v_y and evaluate $|\vec{v}|$:

$$|\vec{v}| = \sqrt{(4.0\,\text{m/s})^2 + (6.0\,\text{m/s})^2} = \boxed{7.2\text{ m/s}}$$

51 •• **[SSM]** A particle has a position vector given by $\vec{r} = (30t)\hat{i} + (40t - 5t^2)\hat{j}$, where r is in meters and t is in seconds. Find the instantaneous-velocity and instantaneous-acceleration vectors as functions of time t.

Picture the Problem The velocity vector is the time-derivative of the position vector and the acceleration vector is the time-derivative of the velocity vector.

Differentiate $\vec{r}$ with respect to time:

$$\vec{v} = \frac{d\vec{r}}{dt} = \frac{d}{dt}\left[(30t)\hat{i} + (40t - 5t^2)\hat{j}\right]$$
$$= \boxed{30\hat{i} + (40 - 10t)\hat{j}}$$

where $\vec{v}$ has units of m/s if t is in seconds.

Differentiate $\vec{v}$ with respect to time:

$$\vec{a} = \frac{d\vec{v}}{dt} = \frac{d}{dt}\left[30\hat{i} + (40 - 10t)\hat{j}\right]$$
$$= \boxed{(-10\,\text{m/s}^2)\hat{j}}$$

53 •• **[SSM]** Starting from rest at a dock, a motor boat on a lake heads north while gaining speed at a constant 3.0 m/s^2 for 20 s. The boat then heads west and continues for 10 s at the speed that it had at 20 s. (*a*) What is the average velocity of the boat during the 30-s trip? (*b*) What is the average acceleration of the boat during the 30-s trip? (*c*) What is the displacement of the boat during the 30-s trip?

Picture the Problem The displacements of the boat are shown in the figure. Let the +*x* direction be to the east and the +*y* direction be to the north. We need to determine each of the displacements in order to calculate the average velocity of the boat during the 30-s trip.

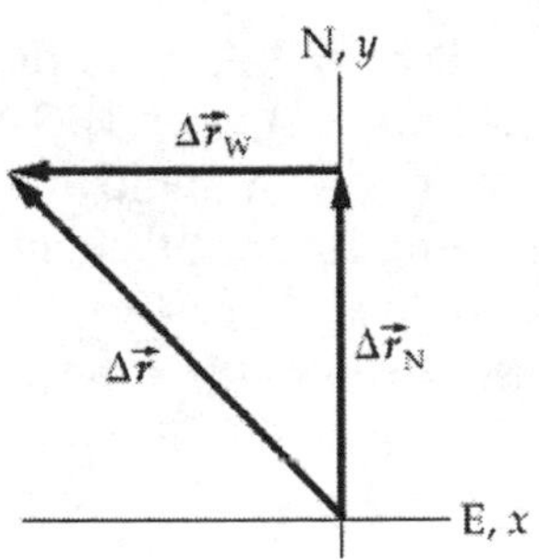

(*a*) The average velocity of the boat is given by:

$$\vec{v}_{av} = \frac{\Delta\vec{r}}{\Delta t} \qquad (1)$$

The total displacement of the boat is given by:

$$\Delta\vec{r} = \Delta\vec{r}_N + \Delta\vec{r}_W = \tfrac{1}{2}a_N\left(\Delta t_N\right)^2\hat{j} + v_W\Delta t_W\left(-\hat{i}\right) \qquad (2)$$

To calculate the displacement we first have to find the speed after the first 20 s:

$$v_W = v_{N,f} = a_N\Delta t_N$$

Substitute numerical values and evaluate v_W:

$$v_W = \left(3.0\ \text{m/s}^2\right)\left(20\ \text{s}\right) = 60\ \text{m/s}$$

Substitute numerical values in equation (2) and evaluate $\Delta\vec{r}(30\ \text{s})$:

$$\Delta\vec{r}(30\ \text{s}) = \tfrac{1}{2}\left(3.0\ \text{m/s}^2\right)\left(20\ \text{s}\right)^2\hat{j} - (60\ \text{m/s})(10\ \text{s})\hat{i} = (600\ \text{m})\hat{j} - (600\ \text{m})\hat{i}$$

Substitute numerical values in equation (1) to find the boat's average velocity:

$$\vec{v}_{av} = \frac{\Delta\vec{r}}{\Delta t} = \frac{(600\ \text{m})\left(-\hat{i}+\hat{j}\right)}{30\ \text{s}} = \boxed{(20\ \text{m/s})\left(-\hat{i}+\hat{j}\right)}$$

(*b*) The average acceleration of the boat is given by:

$$\vec{a}_{av} = \frac{\Delta\vec{v}}{\Delta t} = \frac{\vec{v}_f - \vec{v}_i}{\Delta t}$$

Substitute numerical values and evaluate $\vec{a}_{av}$:

$$\vec{a}_{av} = \frac{(-60\ \text{m/s})\hat{i} - 0}{30\ \text{s}} = \boxed{\left(-2.0\ \text{m/s}^2\right)\hat{i}}$$

(*c*) The displacement of the boat from the dock at the end of the 30-s trip was one of the intermediate results we obtained in Part (*a*).

$$\Delta\vec{r} = (600\text{ m})\hat{j} + (-600\text{ m})\hat{i} = \boxed{(600\text{ m})(-\hat{i} + \hat{j})}$$

Relative Velocity

57 •• **[SSM]** A small plane departs from point A heading for an airport 520 km due north at point B. The airspeed of the plane is 240 km/h and there is a steady wind of 50 km/h blowing directly toward the southeast. Determine the proper heading for the plane and the time of flight.

Picture the Problem Let the velocity of the plane relative to the ground be represented by $\vec{v}_{PG}$; the velocity of the plane relative to the air by $\vec{v}_{PA}$, and the velocity of the air relative to the ground by $\vec{v}_{AG}$. Then

$$\vec{v}_{PG} = \vec{v}_{PA} + \vec{v}_{AG} \quad (1)$$

Choose a coordinate system with the origin at point A, the $+x$ direction to the east, and the $+y$ direction to the north. θ is the angle between north and the direction of the plane's heading. The pilot must head so that the east-west component of $\vec{v}_{PG}$ is zero in order to make the plane fly due north.

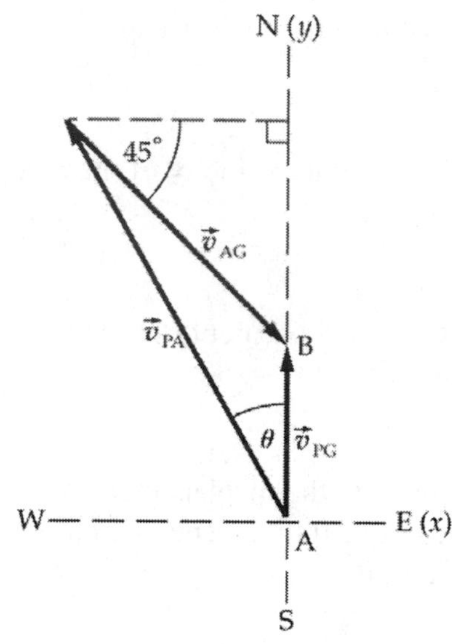

Use the diagram to express the condition relating the eastward component of $\vec{v}_{AG}$ and the westward component of $\vec{v}_{PA}$. This must be satisfied if the plane is to stay on its northerly course. [Note: this is equivalent to equating the x-components of equation (1).]

$$v_{AG}\cos 45° = v_{PA}\sin\theta$$

Now solve for θ to obtain:

$$\theta = \sin^{-1}\left[\frac{v_{AG}\cos 45°}{v_{PA}}\right]$$

Substitute numerical values and evaluate θ:

$$\theta = \sin^{-1}\left[\frac{(50\text{ km/h})\cos 45°}{240\text{ km/h}}\right] = 8.47°$$

$$= \boxed{8.5°}$$

Add the north components of $\vec{v}_{PA}$ and $\vec{v}_{AG}$ to find the velocity of the plane relative to the ground:

$$v_{PG} + v_{AG}\sin 45° = v_{PA}\cos 8.47°$$

Solving for v_{PG} yields:

$$v_{PG} = v_{PA}\cos 8.47° - v_{AG}\sin 45°$$

Substitute numerical values and evaluate v_{PG} to obtain:

$$v_{PG} = (240\text{ km/h})\cos 8.47° - (50\text{ km/h})\sin 45° = 202.0\text{ km/h}$$

The time of flight is given by:

$$t_{flight} = \frac{\text{distance travelled}}{v_{PG}}$$

Substitute numerical values and evaluate t_{flight}:

$$t_{flight} = \frac{520\text{ km}}{202.0\text{ km/h}} = \boxed{2.57\text{ h}}$$

61 •• **[SSM]** Car A is traveling east at 20 m/s toward an intersection. As car A crosses the intersection, car B starts from rest 40 m north of the intersection and moves south, steadily gaining speed at 2.0 m/s^2. Six seconds after A crosses the intersection find (*a*) the position of B relative to A, (*b*) the velocity of B relative to A, and (*c*) the acceleration of B relative to A. (*Hint: Let the unit vectors* $\hat{i}$ *and* $\hat{j}$ *be toward the east and north, respectively, and express your answers using* $\hat{i}$ *and* $\hat{j}$.)

Picture the Problem The position of B relative to A is the vector from A to B; that is,

$$\vec{r}_{AB} = \vec{r}_B - \vec{r}_A$$

The velocity of B relative to A is

$$\vec{v}_{AB} = d\vec{r}_{AB}/dt$$

and the acceleration of B relative to A is

$$\vec{a}_{AB} = d\vec{v}_{AB}/dt$$

Choose a coordinate system with the origin at the intersection, the positive x direction to the east, and the positive y direction to the north.

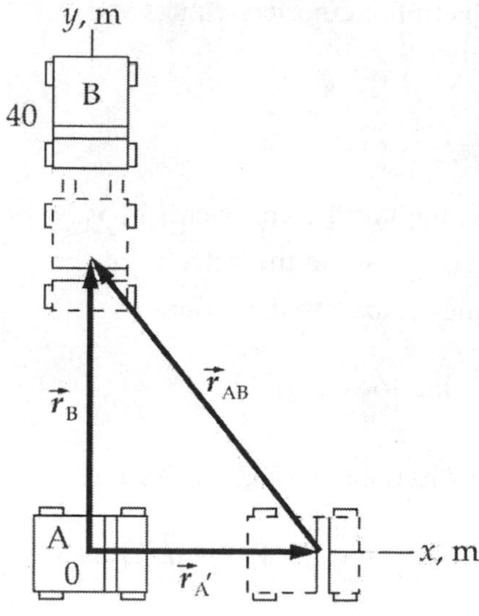

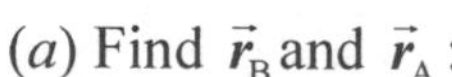

(*a*) Find $\vec{r}_B$ and $\vec{r}_A$:

$$\vec{r}_B = \left[40\text{ m} - \tfrac{1}{2}(2.0\text{ m/s}^2)t^2\right]\hat{j}$$

and

$$\vec{r}_A = [(20\text{ m/s})t]\hat{i}$$

Use $\vec{r}_{AB} = \vec{r}_B - \vec{r}_A$ to find $\vec{r}_{AB}$:

$$\vec{r}_{AB} = [(-20\text{ m/s})t]\hat{i} + \left[40\text{ m} - \tfrac{1}{2}(2.0\text{ m/s}^2)t^2\right]\hat{j}$$

Evaluate $\vec{r}_{AB}$ at $t = 6.0$ s:

$$\begin{aligned}\vec{r}_{AB}(6.0\text{ s}) &= [(-20\text{ m/s})(6.0\text{ s})]\hat{i} + \left[40\text{ m} - \tfrac{1}{2}(2.0\text{ m/s}^2)(6.0\text{ s})^2\right]\hat{j} \\ &= \boxed{(1.2\times10^2\text{ m})\,\hat{i} + (4.0\text{ m})\,\hat{j}}\end{aligned}$$

(*b*) Find $\vec{v}_{AB} = d\vec{r}_{AB}/dt$:

$$\begin{aligned}\vec{v}_{AB} &= \frac{d\vec{r}_{AB}}{dt} = \frac{d}{dt}\left[\{(-20\text{ m/s})t\}\hat{i} + \left\{40\text{ m} - \tfrac{1}{2}(2.0\text{ m/s}^2)t^2\right\}\hat{j}\right] \\ &= (-20\text{ m/s})\,\hat{i} + (-2.0\text{ m/s}^2)\,t\,\hat{j}\end{aligned}$$

Evaluate $\vec{v}_{AB}$ at $t = 6.0$ s:

$$\vec{v}_{AB}(6.0\text{ s}) = (-20\text{ m/s})\,\hat{i} + (-2.0\text{ m/s}^2)(6.0\text{ s})\,\hat{j} = \boxed{(-20\text{ m/s})\hat{i} - (12\text{ m/s})\hat{j}}$$

(*c*) Find $\vec{a}_{AB} = d\vec{v}_{AB}/dt$:

$$\vec{a}_{AB} = \frac{d}{dt}\left[(-20\text{ m/s})\,\hat{i} + (-2.0\text{ m/s}^2)\,t\,\hat{j}\right] = \boxed{\left(-2.0\text{ m/s}^2\right)\hat{j}}$$

Note that $\vec{a}_{AB}$ is independent of time.

63 •• **[SSM]** Ben and Jack are shopping in a department store. Ben leaves Jack at the bottom of the escalator and walks east at a speed of 2.4 m/s. Jack then stands on the escalator, which is inclined at an angle of 37° above the horizontal and travels eastward and upward at a speed of 2.0 m/s. (*a*) What is the velocity of Ben relative to Jack? (*b*) At what speed should Jack walk up the escalator so that he is always directly above Ben (until he reaches the top)?

Picture the Problem The velocity of Ben relative to the velocity of Jack $\vec{v}_{rel}$ is the difference between the vectors $\vec{v}_{Ben}$ and $\vec{v}_{escalator}$. Choose the coordinate system shown and express these vectors using the unit vectors $\hat{i}$ and $\hat{j}$.

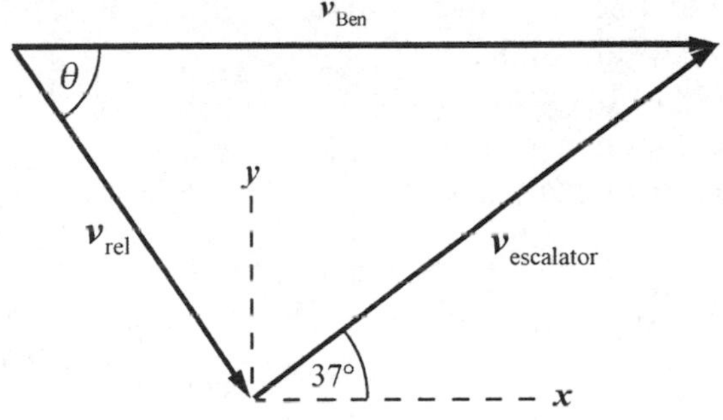

(*a*) The velocity of Ben relative to Jack is given by:

$$\vec{v}_{rel} = \vec{v}_{Ben} - \vec{v}_{escalator}$$

The velocities of the floor walker and the escalator are:

$$\vec{v}_{Ben} = (2.4\text{ m/s})\,\hat{i}$$

and

$$\vec{v}_{escalator} = (2.0\text{ m/s})\cos 37°\,\hat{i} + (2.0\text{ m/s})\sin 37°\,\hat{j}$$

Substitute for $\vec{v}_{Ben}$ and $\vec{v}_{escalator}$ and simplify to obtain:

$$\begin{aligned}\vec{v}_{rel} &= (2.4\text{ m/s})\,\hat{i} - \left[(2.0\text{ m/s})\cos 37°\,\hat{i} + (2.0\text{ m/s})\sin 37°\,\hat{j}\right] \\ &= (0.803\text{ m/s})\,\hat{i} - (1.20\text{ m/s})\,\hat{j} \\ &= \boxed{(0.80\text{ m/s})\,\hat{i} - (1.2\text{ m/s})\,\hat{j}}\end{aligned}$$

The magnitude and direction of $\vec{v}_{rel}$ are given by:

$$|\vec{v}_{rel}| = \sqrt{v_{rel,x}^2 + v_{rel,y}^2}$$

and

$$\theta = \tan^{-1}\left(\frac{v_{rel,y}}{v_{rel,x}}\right)$$

Substitute numerical values and evaluate $|\vec{v}_{rel}|$ and θ:

$$|\vec{v}_{rel}| = \sqrt{(0.803\text{ m/s})^2 + (-1.20\text{ m/s})^2} = \boxed{1.4\text{ m/s}}$$

and

$$\theta = \tan^{-1}\left(\frac{-1.20\text{ m/s}}{0.803\text{ m/s}}\right) = -56.2°$$

$$= \boxed{56° \text{ below the horizontal}}$$

(*b*) If Jack walks so that he is always directly above Ben, the sum of the horizontal component of his velocity and the horizontal component of the escalator's velocity must equal the velocity of Ben:

$$(v_{Jack} + v_{escalator})\cos 37° = v_{Ben}$$

or, solving for v_{Jack},

$$v_{Jack} = \frac{v_{Ben}}{\cos 37°} - v_{escalator}$$

Substitute numerical values and evaluate v_{Jack}:

$$v_{Jack} = \frac{2.4\text{ m/s}}{\cos 37°} - 2.0\text{ m/s} = \boxed{1.0\text{ m/s}}$$

Circular Motion and Centripetal Acceleration

67 ••• **[SSM]** Earth rotates on its axis once every 24 hours, so that objects on its surface that are stationary with respect to the surface execute uniform circular motion about the axis with a period of 24 hours. Consider only the effect of this rotation on the person on the surface. (Ignore Earth's orbital motion about the Sun.) (*a*) What is the speed and what is the magnitude of the acceleration of a person standing on the equator? (Express the magnitude of this acceleration as a percentage of *g*.) (*b*) What is the direction of the acceleration vector? (*c*) What is the speed and what is the magnitude of the centripetal acceleration of a person standing on the surface at 35°N latitude? (*d*) What is the angle between the direction of the acceleration of the person at 35°N latitude and the direction of the acceleration of the person at the equator if both persons are at the same longitude?

Picture the Problem The radius of Earth is 6370 km. Thus at the equator, a person undergoes circular motion with radius equal to Earth's radius, and a period of 24 h = 86400 s. At 35° N latitude, the person undergoes circular motion having radius $r\cos 35° = 5220$ km, and the same period.

The centripetal acceleration experienced by a person traveling with a speed v in a circular path of radius r is given by:

$$a = \frac{v^2}{r} \qquad (1)$$

The speed of the person is the distance the person travels in one revolution divided by the elapsed time (the period T):

$$v = \frac{2\pi r}{T}$$

Substitute for v in equation (1) to obtain:

$$a = \frac{\left(\frac{2\pi r}{T}\right)^2}{r} = \frac{4\pi^2 r}{T^2}$$

(*a*) Substitute numerical values and evaluate v for the person at the equator:

$$v = \frac{2\pi(6370\ \text{km})}{86400\ \text{s}} = \boxed{463\ \text{m/s}}$$

Substitute numerical values and evaluate a for the person at the equator:

$$a = \frac{4\pi^2(6370\ \text{km})}{(86400\ \text{s})^2} = 3.369\times10^{-2}\ \text{m/s}^2$$
$$= \boxed{3.37\ \text{cm/s}^2}$$

The ratio of a to g is:

$$\frac{a}{g} = \frac{3.369\times10^{-2}\ \text{m/s}^2}{9.81\ \text{m/s}^2} = \boxed{0.343\%}$$

(*b*) The acceleration vector points directly at the center of Earth.

(*c*) Substitute numerical values and evaluate v for the person at 35°N latitude:

$$v = \frac{2\pi(5220\ \text{km})}{86400\ \text{s}} = \boxed{380\ \text{m/s}}$$

Substitute numerical values and evaluate v for the person at 35°N latitude:

$$a = \frac{4\pi^2(5220\ \text{km})}{(86400\ \text{s})^2} = \boxed{2.76\ \text{cm/s}^2}$$

(*d*) The plane of the person's path is parallel to the plane of the equator – the acceleration vector is in the plane – so that it is perpendicular to Earth's axis, pointing at the center of the person's revolution, rather than the center of Earth.

Projectile Motion and Projectile Range

77 •• **[SSM]** A ball launched from ground level lands 2.44 s later 40.0-m away from the launch point. Find the magnitude of the initial velocity vector and the angle it is above the horizontal. (Ignore any effects due to air resistance.)

Picture the Problem In the absence of air resistance, the motion of the ball is uniformly accelerated and its horizontal and vertical motions are independent of each other. Choose the coordinate system shown in the figure to the right and use constant-acceleration equations to relate the x and y components of the ball's initial velocity.

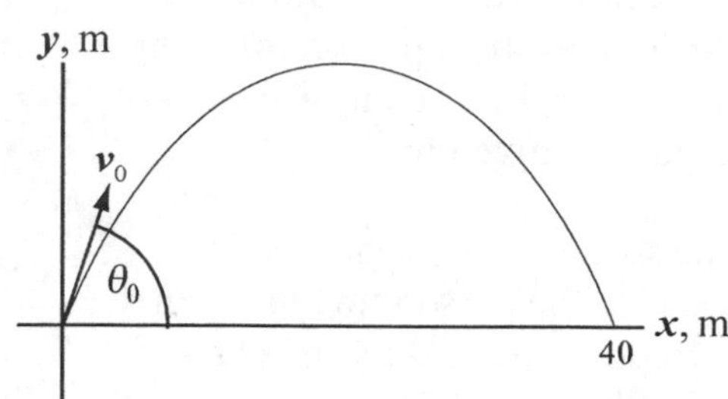

Express θ_0 in terms of v_{0x} and v_{0y}:

$$\theta_0 = \tan^{-1}\left(\frac{v_{0y}}{v_{0x}}\right) \quad (1)$$

Use the Pythagorean relationship between the velocity and its components to express v_0:

$$v_0 = \sqrt{v_{0x}^2 + v_{0y}^2} \quad (2)$$

Using a constant-acceleration equation, express the vertical speed of the projectile as a function of its initial upward speed and time into the flight:

$$v_y = v_{0y} + a_y t \Rightarrow v_{0y} = v_y - a_y t$$

Because $v_y = 0$ halfway through the flight (at maximum elevation) and $a_y = -g$:

$$v_{0y} = g t_{\text{max elevation}}$$

Substitute numerical values and evaluate v_{0y}:

$$v_{0y} = (9.81\text{ m/s}^2)(1.22\text{ s}) = 11.97\text{ m/s}$$

Because there is no acceleration in the horizontal direction, v_{0x} can be found from:

$$v_{0x} = \frac{\Delta x}{\Delta t} = \frac{40.0\text{ m}}{2.44\text{ s}} = 16.39\text{ m/s}$$

Substitute for v_{0x} and v_{0y} in equation (2) and evaluate v_0:

$$v_0 = \sqrt{(16.39\text{ m/s})^2 + (11.97\text{ m/s})^2} = \boxed{20.3\text{ m/s}}$$

Substitute for v_{0x} and v_{0y} in equation (1) and evaluate θ_0 :

$$\theta_0 = \tan^{-1}\left(\frac{11.97\text{ m/s}}{16.39\text{ m/s}}\right) = \boxed{36.1°}$$

81 •• **[SSM]** Wile E. Coyote (*Carnivorous hungribilous*) is chasing the Roadrunner (*Speedibus cantcatchmi*) yet again. While running down the road, they come to a deep gorge, 15.0 m straight across and 100 m deep. The Roadrunner launches himself across the gorge at a launch angle of 15° above the horizontal, and lands with 1.5 m to spare. (*a*) What was the Roadrunner's launch speed? (*b*) Wile E. Coyote launches himself across the gorge with the same initial speed, but at a different launch angle. To his horror, he is short the other lip by 0.50 m. What was his launch angle? (Assume that it was less than 15°.)

Picture the Problem In the absence of air resistance, the accelerations of both Wiley Coyote and the Roadrunner are constant and we can use constant-acceleration equations to express their coordinates at any time during their leaps across the gorge. By eliminating the parameter t between these equations, we can obtain an expression that relates their y coordinates to their x coordinates and that we can solve for their launch angles.

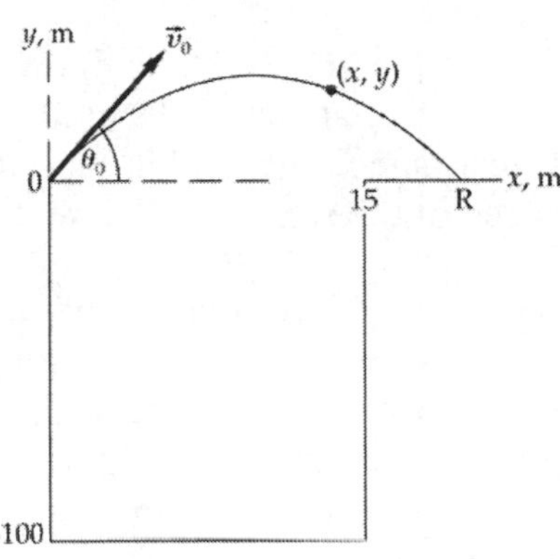

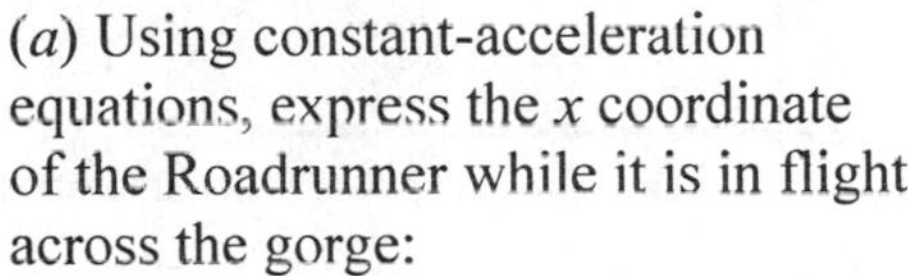

(*a*) Using constant-acceleration equations, express the x coordinate of the Roadrunner while it is in flight across the gorge:

$$x = x_0 + v_{0x}t + \tfrac{1}{2}a_x t^2$$

or, because $x_0 = 0$, $a_x = 0$ and $v_{0x} = v_0 \cos\theta_0$,

$$x = (v_0 \cos\theta_0)t \qquad (1)$$

Using constant-acceleration equations, express the y coordinate of the Roadrunner while it is in flight across the gorge:

$$y = y_0 + v_{0y}t + \tfrac{1}{2}a_y t^2$$

or, because $y_0 = 0$, $a_y = -g$ and $v_{0y} = v_0 \sin\theta_0$,

$$y = (v_0 \sin\theta_0)t - \tfrac{1}{2}gt^2 \qquad (2)$$

Eliminate the variable t between equations (1) and (2) to obtain:

$$y = (\tan\theta_0)x - \frac{g}{2v_0^2\cos^2\theta_0}x^2 \qquad (3)$$

When $y = 0$, $x = R$ and equation (3) becomes:

$$0 = (\tan\theta_0)R - \frac{g}{2v_0^2\cos^2\theta_0}R^2$$

Using the trigonometric identity $\sin 2\theta = 2\sin\theta\cos\theta$, solve for v_0:

$$v_0 = \sqrt{\frac{Rg}{\sin 2\theta_0}}$$

Substitute numerical values and evaluate v_0:

$$v_0 = \sqrt{\frac{(16.5\,\text{m})(9.81\,\text{m/s}^2)}{\sin 30°}} = \boxed{18\,\text{m/s}}$$

(*b*) Letting R represent Wiley's range, solve equation (1) for his launch angle:

$$\theta_0 = \frac{1}{2}\sin^{-1}\left(\frac{Rg}{v_0^2}\right)$$

Substitute numerical values and evaluate θ_0:

$$\theta_0 = \frac{1}{2}\sin^{-1}\left[\frac{(14.5\,\text{m})(9.81\,\text{m/s}^2)}{(18.0\,\text{m/s})^2}\right] = \boxed{13°}$$

83 •• **[SSM]** A stone thrown horizontally from the top of a 24-m tower hits the ground at a point 18 m from the base of the tower. (Ignore any effects due to air resistance.) (*a*) Find the speed with which the stone was thrown. (*b*) Find the speed of the stone just before it hits the ground.

Picture the Problem Choose a coordinate system in which the origin is at the base of the tower and the x- and y-axes are as shown in the figure to the right. In the absence of air resistance, the horizontal speed of the stone will remain constant during its fall and a constant-acceleration equation can be used to determine the time of fall. The final velocity of the stone will be the vector sum of its x and y components.

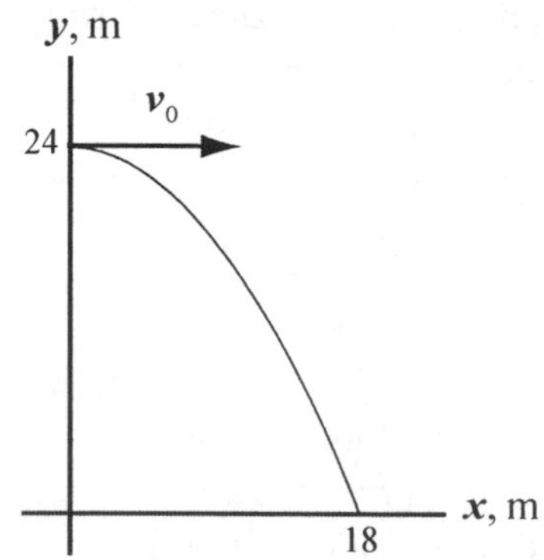

Because the stone is thrown horizontally:

$$v_x = v_{0x} = \frac{\Delta x}{\Delta t} \qquad (1)$$

(*a*) Using a constant-acceleration equation, express the vertical displacement of the stone as a function of the fall time:

$$\Delta y = v_{0y}\Delta t + \tfrac{1}{2}a_y(\Delta t)^2$$

or, because $v_{0y} = 0$ and $a = -g$,

$$\Delta y = -\tfrac{1}{2}g(\Delta t)^2 \Rightarrow \Delta t = \sqrt{-\frac{2\Delta y}{g}}$$

Substituting for Δt in equation (1) and simplifying yields:

$$v_x = \Delta x\sqrt{\frac{g}{-2\Delta y}}$$

Substitute numerical values and evaluate v_x:

$$v_x = (18\,\text{m})\sqrt{\frac{9.81\,\text{m/s}^2}{-2(-24\,\text{m})}} = \boxed{8.1\,\text{m}}$$

(*b*) The speed with which the stone hits the ground is related to the x and y components of its speed:

$$v = \sqrt{v_x^2 + v_y^2} \qquad (2)$$

The y component of the stone's velocity at time t is:

$$v_y = v_{0y} - gt$$

or, because $v_{0y} = 0$,

$$v_y = -gt$$

Substitute for v_x and v_y in equation (2) and simplify to obtain:

$$v = \sqrt{\left(\Delta x\sqrt{\frac{g}{-2\Delta y}}\right)^2 + (-gt)^2} = \sqrt{\frac{g(\Delta x)^2}{-2\Delta y} + g^2t^2}$$

Substitute numerical values and evaluate v:

$$v = \sqrt{\frac{(9.81\text{ m/s}^2)(18\text{ m})^2}{-2(-24\text{ m})} + (9.81\text{ m/s}^2)^2(2.21\text{ s})^2} = \boxed{23\text{ m/s}}$$

87 •• **[SSM]** You are trying out for the position of place-kicker on a professional football team. With the ball teed up 50.0 m from the goalposts with a crossbar 3.05 m off the ground, you kick the ball at 25.0 m/s and 30° above the horizontal. (*a*) Is the field goal attempt good? (*b*) If so, by how much does it clear the bar? If not, by how much does it go under the bar? (*c*) How far behind the plane of the goalposts does the ball land?

Picture the Problem We can use constant-acceleration equations to express the x and y coordinates of the ball along its flight path. Eliminating t between the equations will leave us with an equation for y as a function of x that we can use to find the height of the ball when it has reached the cross bar. We can use this same equation to find the range of the ball and, hence, how far behind the plane of the goalposts the ball lands.

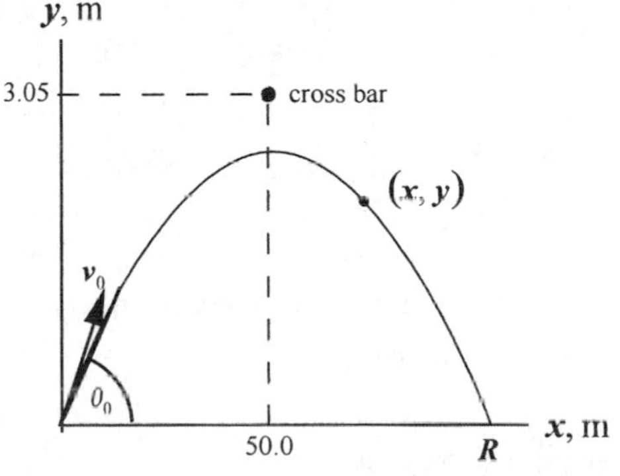

(*a*) Use a constant-acceleration equation to express the x coordinate of the ball as a function of time:

$$x(t) = x_0 + v_{0x}t + \tfrac{1}{2}a_xt^2$$

or, because $x_0 = 0$, $v_{0x} = v_0\cos\theta_0$, and $a_x = 0$,

$$x(t) = (v_0\cos\theta_0)t \Rightarrow t = \frac{x(t)}{v_0\cos\theta_0}$$

Use a constant-acceleration equation to express the y coordinate of the ball as a function of time:

$$y(t) = y_0 + v_{0y}t + \tfrac{1}{2}a_yt^2$$

or, because $y_0 = 0$, $v_{0y} = v_0\sin\theta_0$, and $a_y = -g$,

$$y(t) = (v_0\sin\theta_0)t - \tfrac{1}{2}gt^2$$

Substituting for t yields:

$$y(x)=(v_0\sin\theta_0)\frac{x(t)}{v_0\cos\theta_0}-\tfrac{1}{2}g\left(\frac{x(t)}{v_0\cos\theta_0}\right)^2$$

Simplify to obtain:

$$y(x)=(\tan\theta_0)x(t)-\frac{g}{2v_0^2\cos^2\theta_0}(x(t))^2$$

Substitute numerical values and evaluate $y(50.0\text{ m})$:

$$y(50.0\text{ m})=(\tan 30°)(50.0\text{ m})-\frac{9.81\text{ m/s}^2}{2(25.0\text{ m/s})^2\cos^2 30°}(50.0\text{ m})^2=2.71\text{ m}$$

Because 2.71 m < 3.05 m, the ball goes under the crossbar and the kick is no good.

(*b*) The ball goes under the bar by:

$$d_{\text{under}}=3.05\text{ m}-2.71\text{ m}=\boxed{0.34\text{ m}}$$

(*c*) The distance the ball lands behind the goalposts is given by:

$$d_{\text{behind the goal posts}}=R-50.0\text{ m}\qquad(3)$$

Evaluate the equation derived in (*a*) for $y = 0$ and $x(t) = R$:

$$0=(\tan\theta_0)R-\frac{g}{2v_0^2\cos^2\theta_0}R^2$$

Solving for v_0 yields:

$$R=\frac{v_0^2\sin 2\theta_0}{g}$$

Substitute for R in equation (3) to obtain:

$$d_{\substack{\text{behind the}\\\text{goal posts}}}=\frac{v_0^2\sin 2\theta_0}{g}-50.0\text{ m}$$

Substitute numerical values and evaluate $d_{\substack{\text{behind the}\\\text{goal posts}}}$:

$$d_{\substack{\text{behind the}\\\text{goal posts}}}=\frac{(25.0\text{ m/s})^2\sin 2(30°)}{9.81\text{ m/s}^2}-50.0\text{ m}=\boxed{5.2\text{ m}}$$

95 ••• **[SSM]** In the text, we calculated the range for a projectile that lands at the same elevation from which it is fired as $R=(v_0^2\,/\,g)\sin 2\theta_0$ if the effects of air resistance are negligible. (*a*) Show that for the same conditions the change in the range for a small change in launch speed, and the same initial angle and free-fall acceleration, is given by $\Delta R/R = 2\Delta v_0/v_0$. (*b*) Suppose a projectile's range was 200 m. Use the formula in Part (*a*) to estimate its increase in range if the launch speed were increased by 20.0%. (*c*) Compare your answer in (*b*) to the increase in range by calculating the increase in range directly from $R = (v_0^2/g)\sin 2\theta_0$. If the results for Parts (*b*) and (*c*) are different, is the estimate too large or too small?

Picture the Problem We can show that $\Delta R/R = 2\Delta v_0/v_0$ by differentiating R with respect to v_0 and then using a differential approximation.

(*a*) Differentiate the range equation with respect to v_0:

$$\frac{dR}{dv_0} = \frac{d}{dv_0}\left(\frac{v_0^2}{g}\sin 2\theta_0\right) = \frac{2v_0}{g}\sin 2\theta_0$$
$$= 2\frac{R}{v_0}$$

Approximate dR/dv_0 by $\Delta R/\Delta v_0$:

$$\frac{\Delta R}{\Delta v_0} = 2\frac{R}{v_0}$$

Separate the variables to obtain:

$$\frac{\Delta R}{R} = \boxed{2\frac{\Delta v_0}{v_0}}$$

That is, for small changes in the launch velocity ($v_0 \approx v_0 \pm \Delta v_0$), the fractional change in R is twice the fractional change in v_0.

(*b*) Solve the proportion derived in (*a*) for ΔR to obtain:

$$\Delta R = 2R\frac{\Delta v_0}{v_0}$$

Substitute numerical values and evaluate ΔR:

$$\Delta R = 2(200\text{ m})\frac{0.20v_0}{v_0} = \boxed{80\text{ m}}$$

(*c*) The same-elevation equation is:

$$R = \frac{v_0^2}{g}\sin 2\theta_0 \qquad (1)$$

The longer range R' resulting from a 20.0% increase in the launch speed is given by:

$$R' = \frac{(v_0 + 0.20v_0)^2}{g}\sin 2\theta_0$$
$$= \frac{(1.20v_0)^2}{g}\sin 2\theta_0 \qquad (2)$$

Divide equation (2) by equation (1) and simplify to obtain:

$$\frac{R'}{R} = \frac{\frac{(1.20v_0)^2}{g}\sin 2\theta_0}{\frac{v_0^2}{g}\sin 2\theta_0} = (1.20)^2 = 1.44$$

Solving for R' yields:

$$R' = 1.44R$$

Substitute the numerical value of R and evaluate R':

$$R' = 1.44(200\text{ m}) = \boxed{288\text{ m}}$$

The approximate solution is larger. The estimate ignores higher-order terms and they are important when the differences are not small.

Remarks: The result in (*a*) tells us that as launch velocity increases, the range will increase twice as fast, and vice versa.

97 ••• **[SSM]** A projectile is launched over level ground at an initial elevation angle of θ. An observer standing at the launch site sees the projectile at the point of its highest elevation, and measures the angle ϕ shown in Figure 3-38. Show that
$\tan\phi = \frac{1}{2}\tan\theta$. (Ignore any effects due to air resistance.)

Picture the Problem We can use trigonometry to relate the maximum height of the projectile to its range and the sighting angle at maximum elevation and the range equation to express the range as a function of the launch speed and angle. We can use a constant-acceleration equation to express the maximum height reached by the projectile in terms of its launch angle and speed. Combining these relationships will allow us to conclude that $\tan\phi = \frac{1}{2}\tan\theta$.

Referring to the figure, relate the maximum height of the projectile to its range and the sighting angle ϕ:

$$\tan\phi = \frac{h}{\frac{1}{2}R}$$

Express the range of the rocket and use the trigonometric identity $\sin 2\theta = 2\sin\theta\cos\theta$ to rewrite the expression as:

$$R = \frac{v_0^2}{g}\sin(2\theta) = 2\frac{v_0^2}{g}\sin\theta\cos\theta$$

Using a constant-acceleration equation, relate the maximum height h of a projectile to the vertical component of its launch speed:

$$v_y^2 = v_{0y}^2 - 2gh$$

or, because $v_y = 0$ and $v_{0y} = v_0\sin\theta$,

$$v_0^2\sin^2\theta = 2gh \Rightarrow h = \frac{v_0^2}{2g}\sin^2\theta$$

Substitute for R and h and simplify to obtain:

$$\tan\phi = \frac{2\left(\dfrac{v_0^2}{2g}\sin^2\theta\right)}{2\left(\dfrac{v_0^2}{g}\sin\theta\cos\theta\right)} = \boxed{\tfrac{1}{2}\tan\theta}$$

Hitting Targets and Related Problems

105 ••• **[SSM]** If a bullet that leaves the muzzle of a gun at 250 m/s is to hit a target 100 m away at the level of the muzzle (1.7 m above the level ground), the gun must be aimed at a point above the target. (*a*) How far above the target is that point? (*b*) How far behind the target will the bullet strike the ground? Ignore any effects due to air resistance.

Picture the Problem In the absence of air resistance, the bullet experiences constant acceleration along its parabolic trajectory. Choose a coordinate system with the origin at the end of the barrel and the coordinate axes oriented as shown in the figure and use constant-acceleration equations to express the x and y coordinates of the bullet as functions of time along its flight path.

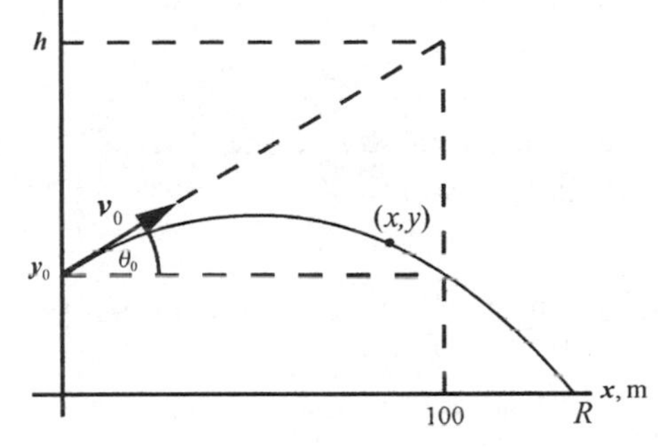

(*a*) Use a constant-acceleration equation to express the bullet's horizontal position as a function of time:

$$x = x_0 + v_{0x}t + \tfrac{1}{2}a_xt^2$$

or, because $x_0 = 0$, $v_{0x} = v_0\cos\theta_0$, and $a_x = 0$,

$$\boldsymbol{x = (v_0\cos\theta_0)t}$$

Use a constant-acceleration equation to express the bullet's vertical position as a function of time:

$$y = y_0 + v_{0y}t + \tfrac{1}{2}a_yt^2$$

or, because $v_{0y} = v_0\sin\theta_0$, and $a_y = -g$,

$$\boldsymbol{y = y_0 + (v_0\sin\theta_0)t - \tfrac{1}{2}gt^2}$$

Eliminate t between the two equations to obtain:

$$\boldsymbol{y = y_0 + (\tan\theta_0)x - \frac{g}{2v_0^2\cos^2\theta_0}x^2}$$

Because $y = y_0$ when the bullet hits the target:

$$\boldsymbol{0 = (\tan\theta_0)x - \frac{g}{2v_0^2\cos^2\theta_0}x^2}$$

Solve for the angle above the horizontal that the rifle must be fired to hit the target:

$$\boldsymbol{\theta_0 = \tfrac{1}{2}\sin^{-1}\left(\frac{xg}{v_0^2}\right)}$$

Substitute numerical values and evaluate θ_0:

$$\theta_0 = \tfrac{1}{2}\sin^{-1}\left[\frac{(100\,\text{m})(9.81\,\text{m/s}^2)}{(250\,\text{m/s})^2}\right]$$

$$= 0.450°$$

Note: A second value for θ_0, 89.6° is physically unreasonable.

Referring to the diagram, relate h to θ_0:

$$\tan\theta_0 = \frac{h}{100\,\text{m}} \Rightarrow h = (100\,\text{m})\tan\theta_0$$

Substitute numerical values and evaluate h:

$$h = (100\,\text{m})\tan(0.450°) = \boxed{0.785\,\text{m}}$$

(*b*) The distance Δx behind the target where the bullet will strike the ground is given by:

$$\Delta x = R - 100\,\text{m} \qquad (1)$$

where R is the range of the bullet.

When the bullet strikes the ground, $y = 0$ and $x = R$:

$$0 = y_0 + (\tan\theta_0)R - \frac{g}{2v_0^2\cos^2\theta_0}R^2$$

Substitute numerical values and simplify to obtain:

$$\frac{9.81\,\text{m/s}^2}{2(250\,\text{m/s})^2\cos^2 0.450°}R^2 - (\tan 0.450°)R - 1.7\,\text{m} = 0$$

Use the quadratic formula or your graphing calculator to obtain:

$$R = 206\,\text{m}$$

Substitute for R in equation (1) to obtain:

$$\Delta x = 206\,\text{m} - 100\,\text{m} = \boxed{105\,\text{m}}$$

General Problems

111 •• **[SSM]** Plane A is headed due east, flying at an air speed of 400 mph. Directly below, at a distance of 4000 ft, plane B is headed due north, flying at an air speed of 700 mph. Find the velocity vector of plane B relative to A.

Picture the Problem The velocity of plane B relative to plane A is independent of the reference frame in which the calculation is done. The solution that follows uses the reference frame of the air. Choose a coordinate system in which the +*x* axis is to the east and the +*y* axis is to the north and write the velocity vectors for the two airplanes using unit vector notation. We can then use this vector to express the relative velocity in speed and heading form.

The velocity of plane B relative to plane A is given by:

$$\vec{v}_{\text{BA}} = \vec{v}_{\text{Ba}} - \vec{v}_{\text{Aa}} \qquad (1)$$

where the subscript a refers to the air.

Using unit vector notation, write expressions for $\vec{v}_{\text{Bg}}$ and $\vec{v}_{\text{Ag}}$:

$$\vec{v}_{\text{Ba}} = (700\ \text{mi/h})\hat{j}$$

and

$$\vec{v}_{\text{Aa}} = (400\ \text{mi/h})\hat{i}$$

Substitute for $\vec{v}_{\text{Ba}}$ and $\vec{v}_{\text{Aa}}$ in equation (1) to obtain:

$$\vec{v}_{\text{BA}} = (700\ \text{mi/h})\hat{j} - (400\ \text{mi/h})\hat{i}$$
$$= -(400\ \text{mi/h})\hat{i} + (700\ \text{mi/h})\hat{j}$$

The speed of plane B relative to plane is the magnitude of $\vec{v}_{\text{BA}}$:

$$|\vec{v}_{\text{BA}}| = \sqrt{(-400\ \text{mi/h})^2 + (700\ \text{mi/h})^2}$$
$$= \boxed{806\ \text{mi/h}}$$

The heading of plane B relative to plane A is:

$$\theta_{\text{BA}} = \tan^{-1}\left(\frac{700\ \text{mi/h}}{-400\ \text{mi/h}}\right)$$
$$= \boxed{60.3^\circ\ \text{north of west}}$$

113 •• **[SSM]** A small steel ball is projected horizontally off the top of a long flight of stairs. The initial speed of the ball is 3.0 m/s. Each step is 0.18 m high and 0.30 m wide. Which step does the ball strike first?

Picture the Problem In the absence of air resistance; the steel ball will experience constant acceleration. Choose a coordinate system with its origin at the initial position of the ball, the x direction to the right, and the y direction downward. In this coordinate system $y_0 = 0$ and $a = g$. Letting (x, y) be a point on the path of the ball, we can use constant-acceleration equations to express both x and y as functions of time and, using the geometry of the staircase, find an expression for the time of flight of the ball. Knowing its time of flight, we can find its range and identify the step it strikes first.

The angle of the steps, with respect to the horizontal, is:

$$\theta = \tan^{-1}\left(\frac{0.18\,\text{m}}{0.30\,\text{m}}\right) = 31.0^\circ$$

Using a constant-acceleration equation, express the x coordinate of the steel ball in its flight:

$$x = x_0 + v_{0x}t + \tfrac{1}{2}a_x t^2$$

or, because $x_{0x} = 0$ and $a_x = 0$,

$$x = v_{0x}t = v_0 t \qquad (1)$$

Using a constant-acceleration equation, express the y coordinate of the steel ball in its flight:

$$y = y_0 + v_{0y}t + \tfrac{1}{2}a_y t^2$$

or, because $y_0 = 0$, $v_{0y} = 0$, and $a_y = g$,

$$y = \tfrac{1}{2}gt^2$$

The equation of the dashed line in the figure is:

$$\frac{y}{x} = \tan\theta = \frac{gt}{2v_0} \Rightarrow t = \frac{2v_0}{g}\tan\theta$$

Substitute for t in equation (1) to find the x coordinate of the landing position:

$$\boldsymbol{x} = \boldsymbol{v}_0\boldsymbol{t} = \boldsymbol{v}_0\left(\frac{2\boldsymbol{v}_0}{\boldsymbol{g}}\tan\boldsymbol{\theta}\right) = \frac{2\boldsymbol{v}_0^2}{\boldsymbol{g}}\tan\boldsymbol{\theta}$$

Substitute numerical values and evaluate x:

$$\boldsymbol{x} = \frac{2(3.0\,\text{m/s})^2}{9.81\,\text{m/s}^2}\tan 31° = 1.1\,\text{m}$$

The first step with $x > 1.1$ m is the 4th step.

117 •• **[SSM]** Galileo showed that, if any effects due to air resistance are neglected, the ranges for projectiles (on a level field) whose angles of projection exceed or fall short of 45° by the same amount are equal. Prove Galileo's result.

Picture the Problem In the absence of air resistance, the acceleration of the projectile is constant and the equation of a projectile for equal initial and final elevations, which was derived from the constant-acceleration equations, is applicable. We can use the equation giving the range of a projectile for equal initial and final elevations to evaluate the ranges of launches that exceed or fall short of 45° by the same amount.

Express the range of the projectile as a function of its initial speed and angle of launch:

$$R = \frac{v_0^2}{g}\sin 2\theta_0$$

Letting $\theta_0 = 45° \pm \theta$ yields:

$$R = \frac{v_0^2}{g}\sin(90° \pm 2\theta) = \frac{v_0^2}{g}\cos(\pm 2\theta)$$

Because $\cos(-\theta) = \cos(+\theta)$ (the cosine function is an *even* function):

$$\boxed{R(45° + \theta) = R(45° - \theta)}$$

Chapter 4
Newton's Laws

Conceptual Problems

3 • **[SSM]** You are riding in a limousine that has opaque windows which do not allow you to see outside. The car can accelerate by speeding up, slowing down, or turning. Equipped with just a small heavy object on the end of a string, how can you use it to determine if the limousine is changing either speed or direction? Can you determine the limousine's velocity?

Determine the Concept The sum of the external forces on the object is always proportional to its acceleration relative to an inertial reference frame. Any reference frame that maintains a zero acceleration relative to an inertial reference frame is itself an inertial reference frame, and vice versa. The ground is an inertial reference frame. If the limo does not accelerate (that is if it does not change direction or speed) relative to the ground, the pendulum will dangle straight down so that the net force on the bob is zero (no acceleration). In that case, the limo is an inertial reference frame. Just with this apparatus and not looking outside you cannot tell the limo's velocity; all you know is that it is constant.

9 •• **[SSM]** You are riding in an elevator. Describe two situations in which your apparent weight is greater than your true weight.

Determine the Concept Your apparent weight is the reading of a scale. If the acceleration of the elevator (and you) is directed upward, the normal force exerted by the scale on you is greater than your weight. You could be moving down but slowing or moving up and speeding up. In both cases your acceleration is upward.

13 •• **[SSM]** Suppose a block of mass m_1 rests on a block of mass m_2 and the combination rests on a table as shown in Figure 4-33. Tell the name of the force and its category (contact versus action-at-a-distance) for each of the following forces; (*a*) force exerted by m_1 on m_2, (*b*) force exerted by m_2 on m_1, (*c*) force exerted by m_2 on the table, (*d*) force exerted by the table on m_2, (*e*) force exerted by the earth on m_2. Which, if any, of these forces constitute a Newton's third law pair of forces?

Determine the Concept

(*a*) The force exerted by m_1 on m_2. Normal force, contact type.

(*b*) The force exerted by m_2 on m_1. Normal force, contact type.

(*c*) The force exerted by m_2 on the table. Normal force, contact type.

(*d*) The force exerted by the table on m_2. Normal force, contact type.

(*e*) The force exerted by the earth on m_2. Gravitational force, action-at-a-distance type.

The Newton's 3rd law force pairs are the two normal forces between the two blocks and the normal force between the table and the bottom block. The gravitational force has a 3rd law force pair, that acts on Earth and, so, is not in the question set.

21 •• **[SSM]** A 2.5-kg object hangs at rest from a string attached to the ceiling. (*a*) Draw a free body diagram of the object, indicate the reaction force to each force drawn, and tell what object the reaction force acts on. (*b*) Draw a free body diagram of the string, indicate the reaction force to each force drawn, and tell what object each reaction force acts on. Do not neglect the mass of the string.

Determine the Concept The force diagrams will need to include forces exerted by the ceiling, on the string, on the object, and forces exerted by the earth.

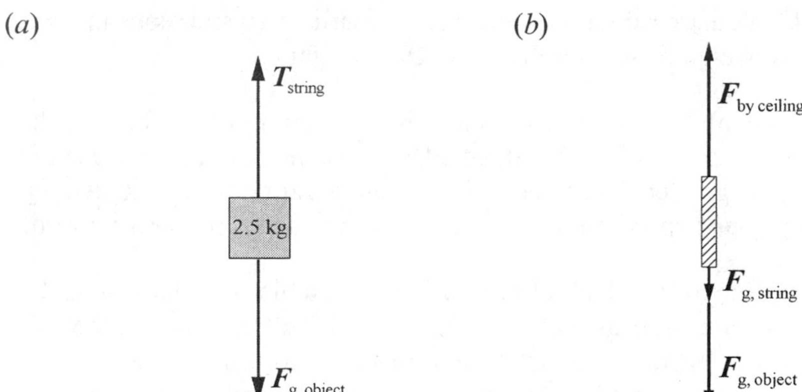

(*a*) The forces acting on the 2.5-kg object are the gravitational force $\vec{F}_{\text{g, object}}$ and the tension $\vec{T}_{\text{string}}$ in the string. The reaction to $\vec{T}_{\text{string}}$ is the force the object exerts downward on the string. The reaction to $\vec{F}_{\text{g, object}}$ is the force the object exerts upward on the earth.

(*b*) The forces acting on the string are its weight $\vec{F}_{\text{g, string}}$, the weight of the object $\vec{F}_{\text{g, object}}$, and $\vec{F}_{\text{by ceiling}}$, the force exerted by the ceiling. The reaction to $\vec{F}_{\text{g, string}}$ is the force the string exerts upward on the earth. The reaction to $\vec{F}_{\text{by ceiling}}$ is a downward force the string exerts on the ceiling. The reaction to $\vec{F}_{\text{g, object}}$ is the force the object exerts upward on the earth.

Estimation and Approximation

27 •• **[SSM]** Estimate the force exerted on the goalie's glove by the puck when he catches a hard slap shot for a save.

Picture the Problem Suppose the goalie's glove slows the puck from 60 m/s to zero as it recoils a distance of 10 cm. Further, assume that the puck's mass is 200 g. Because the force the puck exerts on the goalie's glove and the force the goalie's glove exerts on the puck are action-and-reaction forces, they are equal in magnitude. Hence, if we use a constant-acceleration equation to find the puck's acceleration and Newton's 2nd law to find the force the glove exerts on the puck, we'll have the magnitude of the force exerted on the goalie's glove.

Apply Newton's 2nd law to the puck as it is slowed by the goalie's glove to express the magnitude of the force the glove exerts on the puck:

$$F_{\text{glove on puck}} = \left|m_{\text{puck}} a_{\text{puck}}\right| \qquad (1)$$

Use a constant-acceleration equation to relate the initial and final speeds of the puck to its acceleration and stopping distance:

$$v^2 = v_0^2 + 2a_{\text{puck}}(\Delta x)_{\text{puck}}$$

Solving for a_{puck} yields:

$$a_{\text{puck}} = \frac{v^2 - v_0^2}{2(\Delta x)_{\text{puck}}}$$

Substitute for a_{puck} in equation (1) to obtain:

$$F_{\text{glove on puck}} = \left|\frac{m_{\text{puck}}\left(v^2 - v_0^2\right)}{2(\Delta x)_{\text{puck}}}\right|$$

Substitute numerical values and evaluate $F_{\text{glove on puck}}$:

$$F_{\text{glove on puck}} = \left|\frac{(0.200\text{ kg})\left(0 - (60\text{ m/s})^2\right)}{2(0.10\text{ m})}\right| \approx \boxed{3.6\text{ kN}}$$

Remarks: The force on the puck is about 1800 times its weight.

Newton's First and Second Laws: Mass, Inertia, and Force

35 •• **[SSM]** A bullet of mass 1.80×10^{-3} kg moving at 500 m/s impacts a tree stump and penetrates 6.00 cm into the wood before coming to rest. (*a*) Assuming that the acceleration of the bullet is constant, find the force (including direction) exerted by the wood on the bullet. (*b*) If the same force acted on the bullet and it had the same speed but half the mass, how far would it penetrate into the wood?

Picture the Problem Choose a coordinate system in which the $+x$ direction is in the direction of the motion of the bullet and use Newton's 2nd law and a constant-acceleration equation to express the relationship between F_{stopping} and the mass of the bullet and its displacement as it is brought to rest in the block of wood.

(*a*) Apply Newton's 2nd law to the bullet to obtain:

$$\sum F_x = F_{\text{stopping}} = ma_x \qquad (1)$$

Use a constant-acceleration equation to relate the bullet's initial and final speeds, acceleration, and stopping distance:

$$v_{\text{fx}}^2 = v_{\text{ix}}^2 + 2a_x\Delta x$$

or, because $v_{\text{fx}} = 0$,

$$0 = v_{\text{ix}}^2 + 2a_x\Delta x \Rightarrow a_x = \frac{-v_{\text{ix}}^2}{2\Delta x}$$

Substitute for a_x in equation (1) to obtain:

$$F_{\text{stopping}} = -m\frac{v_{\text{ix}}^2}{2\Delta x} \qquad (2)$$

Substitute numerical values and evaluate F_{stopping}:

$$F_{\text{stopping}} = -\left(1.80\times10^{-3}\ \text{kg}\right)\frac{(500\ \text{m/s})^2}{2(6.00\ \text{cm})} = \boxed{-3.8\ \text{kN}}$$

where the minus sign indicates that F_{stopping} opposes the motion of the bullet.

(*b*) Solving equation (2) for Δx yields:

$$\Delta x = -m\frac{v_{\text{ix}}^2}{2F_{\text{stopping}}} \qquad (3)$$

For $m = m'$ and $\Delta x = \Delta x'$:

$$\Delta x' = -m'\frac{v_{\text{ix}}^2}{2F_{\text{stopping}}}$$

Evaluate this expression for $m' = \frac{1}{2}m$ to obtain:

$$\Delta x' = -m\frac{v_{\text{ix}}^2}{4F_{\text{stopping}}} \qquad (4)$$

Dividing equation (4) by equation (3) yields:

$$\frac{\Delta x'}{\Delta x} = \frac{-m\dfrac{v_{\text{ix}}^2}{4F_{\text{stopping}}}}{-m\dfrac{v_{\text{ix}}^2}{2F_{\text{stopping}}}} = \frac{1}{2}$$

or

$$\Delta x' = \tfrac{1}{2}\Delta x$$

Substitute numerical values and evaluate $\Delta x'$:

$$\Delta x' = \tfrac{1}{2}(6.00\ \text{cm}) = \boxed{3.00\ \text{cm}}$$

Mass and Weight

45 •• **[SSM]** To train astronauts to work on the moon, where the acceleration due to gravity is only about 1/6 of that on earth, NASA submerges them in a tank of water. If an astronaut, who is carrying a backpack, air conditioning unit, oxygen supply, and other equipment, has a total mass of 250 kg, determine the following quantities. (*a*) her weight on Earth, (*b*) her weight on the moon, (*c*) the required upward buoyancy force of the water during her training for the moon's environment on Earth.

Picture the Problem We can use the relationship between weight (gravitational force) and mass, together with the given information about the acceleration due to gravity on the moon, to find the astronaut's weight on Earth and on the moon.

(*a*) Her weight on Earth is the product of her mass and the gravitational field at the surface of the earth:

$$w_{\text{earth}} = mg$$

Substitute numerical values and evaluate *w*:

$$w_{\text{earth}} = (250\text{ kg})(9.81\text{ m/s}^2) = 2.453\text{ kN} = \boxed{2.45\text{ kN}}$$

(*b*) Her weight on the moon is the product of her mass and the gravitational field at the surface of the moon:

$$w_{\text{moon}} = mg_{\text{moon}} = \tfrac{1}{6}mg = \tfrac{1}{6}w_{\text{earth}}$$

Substitute for her weight on Earth and evaluate her weight on the moon:

$$w_{\text{moon}} = \tfrac{1}{6}(2.453\text{ kN}) = \boxed{409\text{ N}}$$

(*c*) The required upward buoyancy force of the water equals her weight of earth:

$$w_{\text{buoyancy}} = w_{\text{earth}} = \boxed{2.45\text{ kN}}$$

Free-Body Diagrams: Static Equilibrium

49 •• **[SSM]** In Figure 4-38*a*, a 0.500-kg block is suspended at the midpoint of a 1.25-m-long string. The ends of the string are attached to the ceiling at points separated by 1.00 m. (*a*) What angle does the string make with the ceiling? (*b*) What is the tension in the string? (*c*) The 0.500-kg block is removed and two 0.250-kg blocks are attached to the string such that the lengths of the three string segments are equal (Figure 4-38*b*). What is the tension in each segment of the string?

Picture the Problem The free-body diagrams for parts (*a*), (*b*), and (*c*) are shown below. In both cases, the block is in equilibrium under the influence of the forces and we can use Newton's 2nd law of motion and geometry and trigonometry to obtain relationships between θ and the tensions.

(*a*) and (*b*)

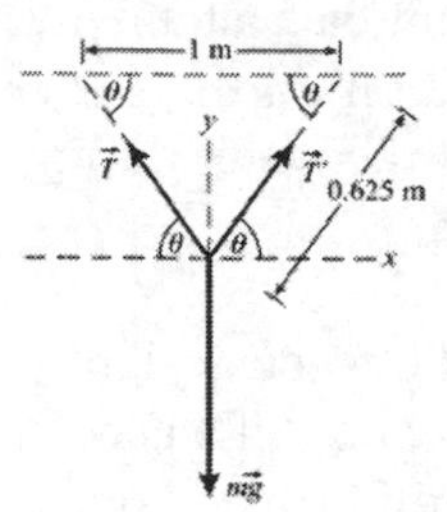

(*c*)

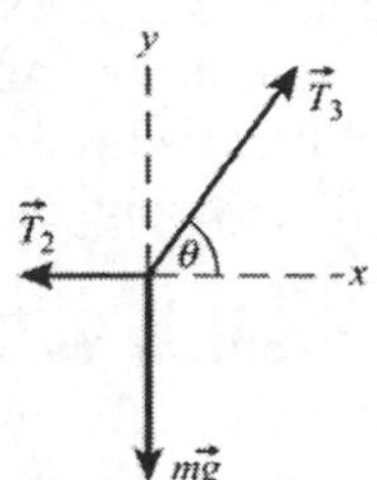

(*a*) Referring to the free-body diagram for part (*a*), use trigonometry to determine θ:

$$\theta = \cos^{-1}\left(\frac{0.50\,\text{m}}{0.625\,\text{m}}\right) = 36.9° = \boxed{37°}$$

(*b*) Noting that $T = T'$, apply $\sum F_y = ma_y$ to the 0.500-kg block and solve for the tension *T*:

$$2T\sin\theta - mg = 0 \text{ because } a = 0$$

and

$$T = \frac{mg}{2\sin\theta}$$

Substitute numerical values and evaluate *T*:

$$T = \frac{(0.500\,\text{kg})(9.81\,\text{m/s}^2)}{2\sin 36.9°} = \boxed{4.1\,\text{N}}$$

(*c*) The length of each segment is:

$$\frac{1.25\,\text{m}}{3} = 0.41667\,\text{m}$$

Find the distance *d*:

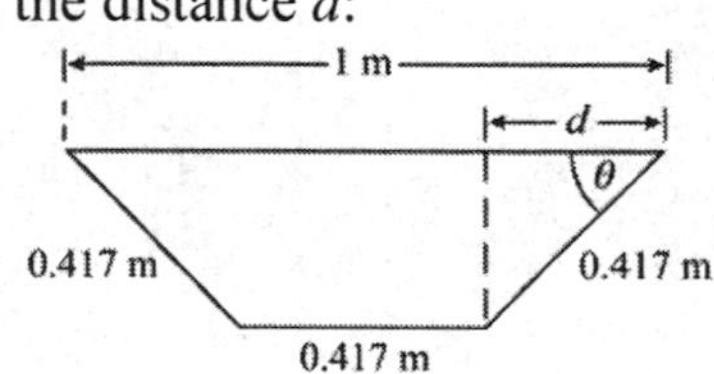

$$d = \frac{1.00\,\text{m} - 0.41667\text{m}}{2}$$
$$= 0.29167\,\text{m}$$

Express θ in terms of *d* and solve for its value:

$$\theta = \cos^{-1}\left(\frac{d}{0.417\,\text{m}}\right)$$
$$= \cos^{-1}\left(\frac{0.2917\,\text{m}}{0.4167\,\text{m}}\right) = 45.57°$$

Apply $\sum F_y = ma_y$ to the 0.250-kg block:

$$T_3\sin\theta - mg = 0 \Rightarrow T_3 = \frac{mg}{\sin\theta}$$

Substitute numerical values and evaluate T_3:

$$T_3 = \frac{(0.250\,\text{kg})(9.81\,\text{m/s}^2)}{\sin 45.57°} = 3.434\,\text{N} = \boxed{3.4\,\text{N}}$$

Apply $\sum F_x = ma_x$ to the 0.250-kg block and solve for the tension T_2:

$T_3 \cos\theta - T_2 = 0$ since $a = 0$.
and
$T_2 = T_3 \cos\theta$

Substitute numerical values and evaluate T_2:

$$T_2 = (3.434\,\text{N})\cos 45.57° = \boxed{2.4\,\text{N}}$$

By symmetry:

$$T_1 = T_3 = \boxed{3.4\,\text{N}}$$

51 •• **[SSM]** A 10-kg object on a frictionless table is subjected to two horizontal forces, $\vec{F}_1$ and $\vec{F}_2$, with magnitudes $F_1 = 20$ N and $F_2 = 30$ N, as shown in Figure 4-40. Find the third force $\vec{F}_3$ that must be applied so that the object is in static equilibrium.

Picture the Problem The acceleration of *any* object is directly proportional to the *net* force acting on it. Choose a coordinate system in which the positive *x* direction is the same as that of $\vec{F}_1$ and the positive *y* direction is to the right. Add the two forces to determine the net force and then use Newton's 2nd law to find the acceleration of the object. If $\vec{F}_3$ brings the system into equilibrium, it must be true that $\vec{F}_3 + \vec{F}_1 + \vec{F}_2 = 0$.

Express $\vec{F}_3$ in terms of $\vec{F}_1$ and $\vec{F}_2$:

$$\vec{F}_3 = -\vec{F}_1 - \vec{F}_2 \qquad (1)$$

Express $\vec{F}_1$ and $\vec{F}_2$ in unit vector notation:

$$\vec{F}_1 = (20\,\text{N})\hat{i}$$

and

$$\vec{F}_2 = \{(-30\,\text{N})\sin 30°\}\hat{i} + \{(30\,\text{N})\cos 30°\}\hat{j} = (-15\,\text{N})\hat{i} + (26\,\text{N})\hat{j}$$

Substitute for $\vec{F}_1$ and $\vec{F}_2$ in equation (1) and simplify to obtain:

$$\vec{F}_3 = -(20\,\text{N})\hat{i} - \left[(-15\,\text{N})\hat{i} + (26\,\text{N})\hat{j}\right] = \boxed{(-5.0\,\text{N})\hat{i} + (-26\,\text{N})\hat{j}}$$

Free-Body Diagrams: Inclined Planes and the Normal Force

61 •• **[SSM]** A 65-kg student weighs himself by standing on a scale mounted on a skateboard that is rolling down an incline, as shown in Figure 4-47. Assume there is no friction so that the force exerted by the incline on the skateboard is normal to the incline. What is the reading on the scale if $\theta = 30°$?

Picture the Problem The scale reading (the boy's apparent weight) is the force the scale exerts on the boy. Draw a free-body diagram for the boy, choosing a coordinate system in which the positive x-axis is parallel to and down the inclined plane and the positive y-axis is in the direction of the normal force the incline exerts on the boy. Apply Newton's 2nd law of motion in the y direction.

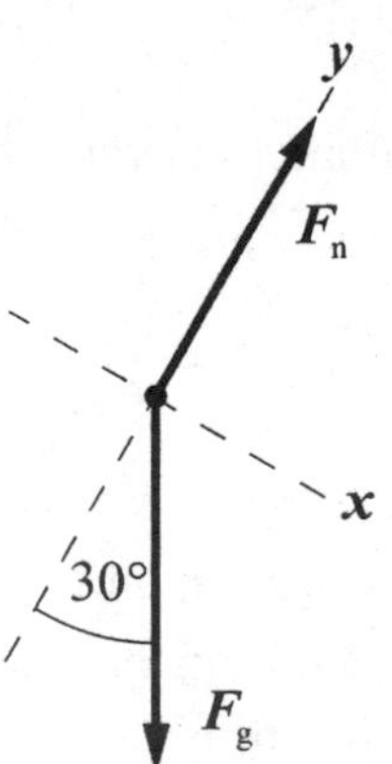

Apply $\sum F_y = ma_y$ to the boy to find F_n. Remember that there is no acceleration in the y direction:

$$\boldsymbol{F}_n - \boldsymbol{F}_g \cos 30° = 0$$

or, because $F_g = mg$,

$$\boldsymbol{F}_n - \boldsymbol{mg} \cos 30° = 0$$

Solving for F_n yields:

$$F_n = mg \cos 30°$$

Substitute numerical values and evaluate F_n:

$$F_n = (65\,\text{kg})(9.81\,\text{m/s}^2)\cos 30° = \boxed{0.55\,\text{kN}}$$

Free-Body Diagrams: Elevators

63 • **[SSM]** (*a*) Draw the free body diagram (with accurate relative force magnitudes) for an object that is hung by a rope from the ceiling of an elevator that is ascending but slowing. (*b*) Repeat Part (*a*) but for the situation in which the elevator is descending and speeding up. (*c*) Can you tell the difference between the two diagrams? Comment. Explain why the diagram does not tell anything about the object's velocity.

Picture the Problem
(a) The free body diagram for an object that is hung by a rope from the ceiling of an ascending elevator that is slowing down is shown to the right. Note that because $F_g > T$, the net force acting on the object is downward; as it must be if the object is slowing down as it is moving upward.

(b) The free body diagram for an object that is hung by a rope from the ceiling of an elevator that is descending and speeding up is shown to the right. Note that because $F_g > T$, the net force acting on the object is downward; as it must be if the object is speeding up as it descends.

(c) No. In both cases the acceleration is downward. You can only tell the direction of the acceleration, not the direction of the velocity.

Free-Body Diagrams: Several Objects and Newton's Third Law

71 •• **[SSM]** A block of mass m is being lifted vertically by a rope of mass M and length L. The rope is being pulled upward by a force applied at its top end, and the rope and block are accelerating upward with acceleration a. The distribution of mass in the rope is uniform. Show that the tension in the rope at a distance x (where $x < L$) above the block is $(a + g)[m + (x/L)M]$.

Picture the Problem Because the distribution of mass in the rope is uniform, we can express the mass m' of a length x of the rope in terms of the total mass of the rope M and its length L. We can then express the total mass that the rope must support at a distance x above the block and use Newton's 2nd law to find the tension as a function of x.

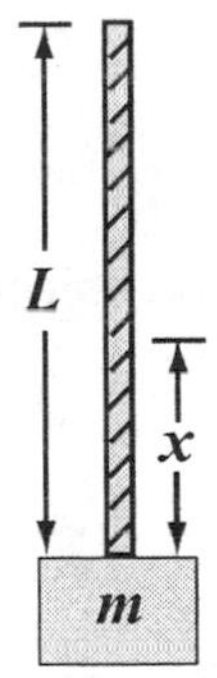

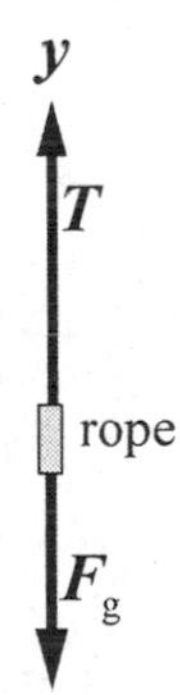

Set up a proportion expressing the mass m' of a length x of the rope as a function of M and L and solve for m':

$$\frac{m'}{x} = \frac{M}{L} \Rightarrow m' = \frac{M}{L}x$$

Express the total mass that the rope must support at a distance x above the block:

$$m + m' = m + \frac{M}{L}x$$

Apply $\sum F_y = ma_y$ to the block and a length x of the rope:

$$T - F_g = T - (m + m')g = (m + m')a_y$$

Substituting for $m + m'$ yields:

$$T - \left(m + \frac{M}{L}x\right)g = \left(m + \frac{M}{L}x\right)a_y$$

Solve for T and simplify to obtain:

$$\boxed{T = (a_y + g)\left(m + \frac{M}{L}x\right)}$$

73 •• **[SSM]** A 40.0-kg object supported by a vertical rope is initially at rest. The object is then accelerated upward from rest so that it attains a speed of 3.50 m/s in 0.700 s. (*a*) Draw the object's free body diagram with the relative lengths of the vectors showing the relative magnitudes of the forces. (*b*) Use the free body diagram and Newton's laws to determine the tension in the rope.

Picture the Problem A *net* force is required to accelerate the object. In this problem the net force is the difference between $\vec{T}$ and $\vec{F}_g (= m\vec{g})$.

(*a*) The free-body diagram of the object is shown to the right. A coordinate system has been chosen in which the upward direction is positive. The magnitude of $\vec{T}$ is approximately 1.5 times the length of $\vec{F}_g$.

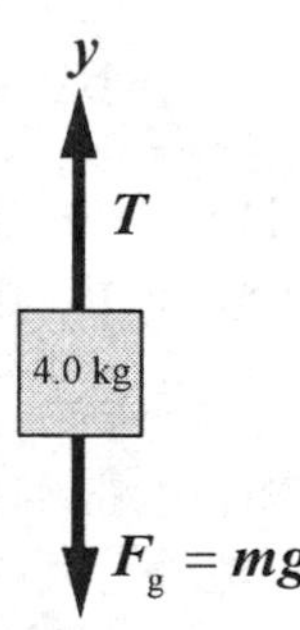

(*b*) Apply $\sum F = ma_y$ to the object to obtain:

$$T - mg = ma_y$$

Solving for T yields:

$$T = ma + mg = m(a_y + g)$$

Using its definition, substitute for a_y to obtain:

$$T = m\left(\frac{\Delta v_y}{\Delta t} + g\right)$$

Substitute numerical values and evaluate T:

$$T = (40.0\text{ kg})\left(\frac{3.50\text{ m/s}}{0.700\text{ s}} + 9.81\text{ m/s}^2\right)$$

$$= \boxed{592\text{ N}}$$

79 •• **[SSM]** A 60-kg housepainter stands on a 15-kg aluminum platform. The platform is attached to a rope that passes through an overhead pulley, which allows the painter to raise herself and the platform (Figure 4-57). (*a*) To accelerate herself and the platform at a rate of 0.80 m/s^2, with what force F must she pull down on the rope? (*b*) When her speed reaches 1.0 m/s, she pulls in such a way that she and the platform go up at a constant speed. What force is she exerting on the rope now? (Ignore the mass of the rope.)

Picture the Problem Choose a coordinate system in which the upward direction is the positive y direction. Note that $\vec{F}$ is the force exerted by the painter on the rope and that $\vec{T}$ is the resulting tension in the rope. Hence the net upward force on the painter-plus-platform is $2\vec{T}$.

(*a*) Letting $m_{\text{tot}} = m_{\text{frame}} + m_{\text{painter}}$, apply $\sum F_y = ma_y$ to the frame-plus-painter:

$$2T - m_{\text{tot}}g = m_{\text{tot}}a_y$$

Solving for T yields:

$$T = \frac{m_{\text{tot}}(a_y + g)}{2}$$

Substitute numerical values and evaluate T:

$$T = \frac{(75\,\text{kg})(0.80\,\text{m/s}^2 + 9.81\,\text{m/s}^2)}{2} = 398\,\text{N}$$

Because $F = T$:

$$F = 398\,\text{N} = \boxed{0.40\,\text{kN}}$$

(*b*) Apply $\sum F_y = 0$ to obtain:

$$2T - m_{\text{tot}}g = 0 \Rightarrow T = \tfrac{1}{2}m_{\text{tot}}g$$

Substitute numerical values and evaluate T:

$$T = \tfrac{1}{2}(75\,\text{kg})(9.81\,\text{m/s}^2) = \boxed{0.37\,\text{kN}}$$

General Problems

87 •• **[SSM]** The mast of a sailboat is supported at bow and stern by stainless steel wires, the forestay and backstay, anchored 10 m apart (Figure 4-61). The 12.0-m-long mast weighs 800 N and stands vertically on the deck of the boat. The mast is positioned 3.60 m behind where the forestay is attached. The tension in the forestay is 500 N. Find the tension in the backstay and the force that the mast exerts on the deck.

Picture the Problem The free-body diagram shows the forces acting at the top of the mast. Choose the coordinate system shown and use Newton's 2nd and 3rd laws of motion to analyze the forces acting on the deck of the sailboat.

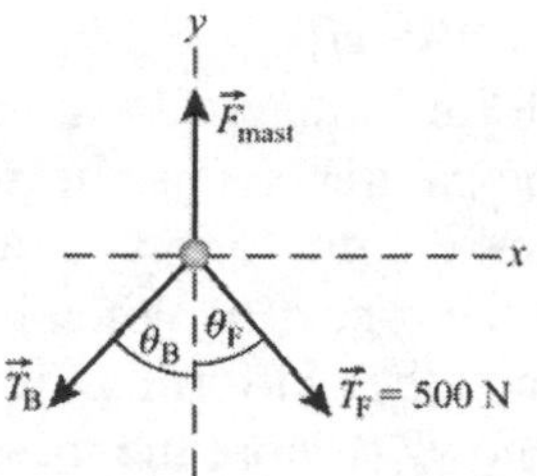

Apply $\sum F_x = ma_x$ to the top of the mast:

$$T_F \sin\theta_F - T_B \sin\theta_B = 0$$

Find the angles that the forestay and backstay make with the vertical:

$$\theta_F = \tan^{-1}\left(\frac{3.60\,\text{m}}{12.0\,\text{m}}\right) = 16.7°$$

and

$$\theta_B = \tan^{-1}\left(\frac{6.40\,\text{m}}{12.0\,\text{m}}\right) = 28.1°$$

Solving the x-direction equation for T_B yields:

$$T_B = T_F \frac{\sin\theta_F}{\sin\theta_B}$$

Substitute numerical values and evaluate T_B:

$$T_B = (500\,\text{N})\frac{\sin 16.7°}{\sin 28.1°} = \boxed{305\,\text{N}}$$

Apply $\sum F_y = 0$ to the mast:

$$\sum F_y = F_{\text{mast}} - T_F\cos\theta_F - T_B\cos\theta_B = 0$$

Solve for F_{mast} to obtain:

$$F_{\text{mast}} = T_F\cos\theta_F + T_B\cos\theta_B$$

Substitute numerical values and evaluate F_{mast}:

$$F_{\text{mast}} = (500\,\text{N})\cos 16.7° + (305\,\text{N})\cos 28.1° = 748\,\text{N}$$

The force that the mast exerts on the deck is the sum of its weight and the downward forces exerted on it by the forestay and backstay:

$$F_{\text{mast on the deck}} = 748\,\text{N} + 800\,\text{N} = \boxed{1.55\,\text{kN}}$$

93 ••• **[SSM]** The masses attached to each side of an ideal Atwood's machine consist of a stack of five washers, each of mass m, as shown in Figure 4-65. The tension in the string is T_0. When one of the washers is removed from the left side, the remaining washers accelerate and the tension decreases by 0.300 N. (*a*) Find m. (*b*) Find the new tension and the acceleration of each mass when a second washer is removed from the left side.

Picture the Problem Because the system is initially in equilibrium, it follows that $T_0 = 5mg$. When one washer is moved from the left side to the right side, the remaining washers on the left side will accelerate upward (and those on the right side downward) in response to the net force that results. The free-body diagrams show the forces under this unbalanced condition. Applying Newton's 2nd law to each collection of washers will allow us to determine both the acceleration of the system and the mass of a single washer.

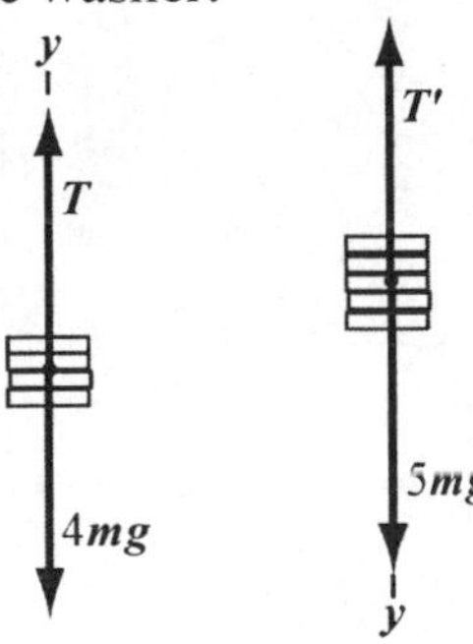

(*a*) Apply $\sum F_y = ma_y$ to the rising washers: $$T - 4mg = (4m)a_y \quad (1)$$

Noting that $T = T'$, apply $\sum F_y = ma_y$ to the descending masses: $$5mg - T = (5m)a_y \quad (2)$$

Eliminate T between these equations to obtain: $$a_y = \tfrac{1}{9}g$$

Use this acceleration in equation (1) or equation (2) to obtain: $$T = \frac{40}{9}mg$$

Expressing the difference ΔT between T_0 and T yields: $$\Delta T = 5mg - \frac{40}{9}mg \Rightarrow m = \frac{\tfrac{9}{5}\Delta T}{g}$$

Substitute numerical values and evaluate m: $$m = \frac{\tfrac{9}{5}(0.300\,\text{N})}{9.81\,\text{m/s}^2} = \boxed{55.0\,\text{g}}$$

(*b*) Proceed as in (*a*) to obtain: $$T - 3mg = 3ma_y \text{ and } 5mg - T = 5ma_y$$

Add these equations to eliminate T and solve for a_y to obtain: $$a_y = \tfrac{1}{4}g$$

Substitute numerical values and evaluate a_y: $$a_y = \tfrac{1}{4}(9.81\,\text{m/s}^2) = \boxed{2.45\,\text{m/s}^2}$$

Eliminate a_y in either of the motion equations and solve for T to obtain:

$$\boldsymbol{T} = \tfrac{15}{4}\boldsymbol{mg}$$

Substitute numerical values and evaluate T:

$$\boldsymbol{T} = \tfrac{15}{4}(0.05505\,\text{kg})(9.81\,\text{m/s}^2) = \boxed{2.03\,\text{N}}$$

Chapter 5
Additional Applications of Newton's Laws

Conceptual Problems

1 • **[SSM]** Various objects lie on the bed of a truck that is moving along a straight horizontal road. If the truck gradually speeds up, what force acts on the objects to cause them to speed up too? Explain why some of the objects might stay stationary on the floor while others might slip backward on the floor.

Determine the Concept The forces acting on the objects are the normal and frictional forces exerted by the truck bed and the gravitational force exerted by Earth., and t The static (if the objects do not slip) frictional forces exerted by the floor of the truck bed cause them to speed up. Because the objects are speeding up (accelerating), there must be a net force acting on them. Of these forces, the only one that acts in the direction of the acceleration is the $\boxed{\text{static friction force}}$. The maximum acceleration is determined not by the mass of the objects but instead by the value of the coefficient of static friction. This will vary from object to object depending on its material and surface characteristics.

11 ••• **[SSM]** The following question is an excellent "braintwister" invented by Boris Korsunsky. Two identical blocks are attached by a massless string running over a pulley as shown in Figure 5-58. The rope initially runs over the pulley at the rope's midpoint, and the surface that block 1 rests on is frictionless. Blocks 1 and 2 are initially at rest when block 2 is released with the string taut and horizontal. Will block 1 hit the pulley before or after block 2 hits the wall? (Assume that the initial distance from block 1 to the pulley is the same as the initial distance from block 2 to the wall.) There is a very simple solution.

Picture the Problem The following free-body diagrams show the forces acting on the two objects some time after block 2 is dropped. Note that, while $\vec{T}_1 \neq \vec{T}_2$, $T_1 = T_2$. The only force pulling block 2 to the left is the horizontal component of the tension $\vec{T}_2$. Because this force is smaller than the magnitude of the tension, the acceleration of block 1, which is identical to block 2, to the right ($T_1 = T_2$) will always be greater than the acceleration of block 2 to the left.

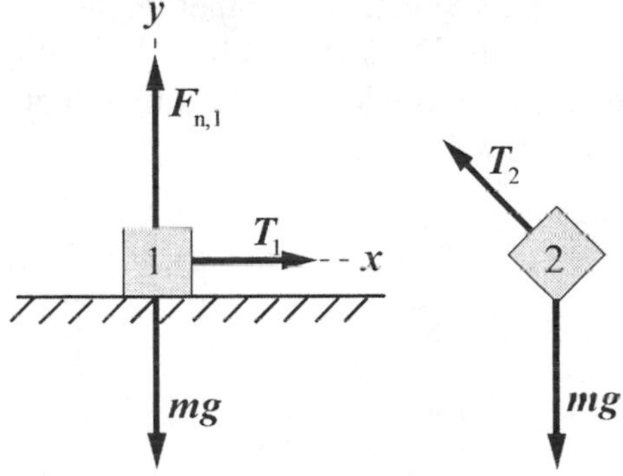

Because the initial distance from block 1 to the pulley is the same as the initial distance of block 2 to the wall, block 1 will hit the pulley before block 2 hits the wall.

13 •• **[SSM]** Jim decides to attempt to set a record for terminal speed in skydiving. Using the knowledge he has gained from a physics course, he makes the following plans. He will be dropped from as high an altitude as possible (equipping himself with oxygen), on a warm day and go into a "knife" position in which his body is pointed vertically down and his hands are pointed ahead. He will outfit himself with a special sleek helmet and rounded protective clothing. Explain how each of these factors helps Jim attain the record.

Determine the Concept On a warm day the air is less dense. The air is also less dense at high altitudes. Pointing his hands results in less area being presented to air drag forces and, hence, reduces them. Rounded and sleek clothing has the same effect as pointing his hands. All are attempts to maximize his acceleration to near g for a good part of the drop by minimizing air drag forces.

15 • **[SSM]** The mass of the moon is only about 1% of that of Earth. Therefore, the force that keeps the moon in its orbit around Earth (*a*) is much smaller than the gravitational force exerted on the moon by Earth, (*b*) is much greater than the gravitational force exerted on the moon by Earth, (*c*) is the gravitational force exerted on the moon by Earth, (*d*) cannot be answered yet, because we have not yet studied Newton's law of gravity.

Determine the Concept The centripetal force that keeps the moon in its orbit around the earth is provided by the gravitational force the earth exerts on the moon. As described by Newton's 3rd law, this force is equal in magnitude to the force the moon exerts on the earth. $\boxed{(c)}$ is correct.

17 •• **[SSM]** (*a*) A pebble and a feather held at the same height above the ground are simultaneously dropped. During the first few milliseconds following release the drag force on the pebble is less than that on the feather, but later on during the fall the *opposite* is true. Explain. (*b*) In light of this result, explain how the pebble's acceleration can be so obviously larger than that of the feather. (*Hint*: *Draw a free-body diagram of each object.*)

Determine the Concept The drag force acting on the objects is given by $F_d = \frac{1}{2}CA\rho v^2$, where A is the projected surface area, v is the object's speed, ρ is the density of air, and C is a dimensionless coefficient. We'll assume that, over the height of the fall, the density of air ρ is constant. The free-body diagrams for a feather and a pebble several milliseconds into their fall are shown to the right. The forces acting on both objects are the downward gravitational force $\vec{F}_g$ and an upward drag force $\vec{F}_d$.

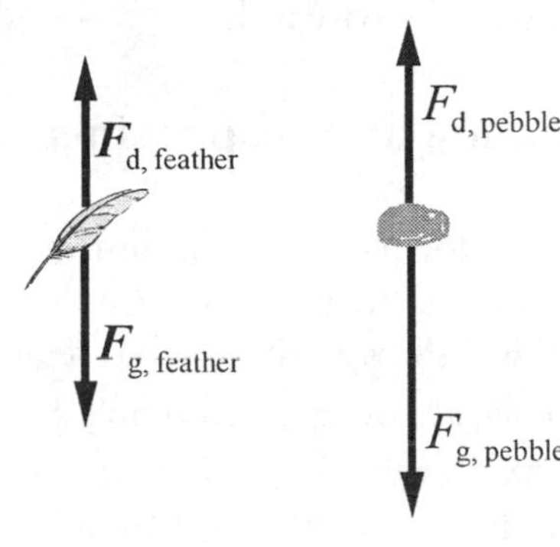

(*a*) The drag force on an object is proportional to some power of its speed. For a millisecond or two following release, the speeds of both the pebble and the feather are negligible, so the drag forces are negligible and they both fall with the same free-fall acceleration *g*. During this brief period their speeds remain equal, so the object that presents the greater area has the greater drag force. It is the feather that presents the greater area, so during this brief period the drag force on the feather is greater than that on the pebble.

A short time after the initial period the feather reaches terminal speed, after which the drag force on it remains equal to the gravitational force on it. However, the gravitational force on the pebble is much greater than that on the feather, so the pebble continues to gain speed long after the feather reaches terminal speed. As the pebble continues to gain speed, the drag force on it continues to increase. As a result, the drag force on the pebble eventually exceeds the drag force on the feather.

(*b*) The acceleration of the feather rapidly decreases because the drag force on it approaches the gravitational force on it shortly after release. However, the drag force on the pebble does not approach the gravitational force on it until much higher speeds are attained, which means the acceleration of the pebble remains high for a longer period of time.

23 •• **[SSM]** When you are standing upright, your center of mass is located within the volume of your body. However, as you bend over (say to pick up a package), its location changes. Approximately where is it when you are bent over at right angles and what change in your body caused the center of mass location to change? Explain.

Determine the Concept Relative to the ground, your center of mass moves downward. This is because some of your mass (hips) moved backward, some of your mass (your head and shoulders) moved forward, and the top half of your body moved downward.

Estimation and Approximation

27 •• **[SSM]** Using dimensional analysis, determine the units and dimensions of the constant b in the retarding force bv^n if (*a*) $n = 1$ and (*b*) $n = 2$. (*c*) Newton showed that the air resistance of a falling object with a circular cross section should be approximately $\frac{1}{2}\rho\pi r^2v^2$, where $\rho = 1.20\text{ kg/m}^3$, the density of air. Show that this is consistent with your dimensional analysis for part (*b*). (*d*) Find the terminal speed for a 56.0-kg skydiver; approximate his cross-sectional area as a disk of radius 0.30 m. The density of air near the surface of Earth is 1.20 kg/m^3. (*e*) The density of the atmosphere decreases with height above the surface of Earth; at a height of 8.0 km, the density is only 0.514 kg/m^3. What is the terminal velocity at this height?

Picture the Problem We can use the dimensions of force and velocity to determine the dimensions of the constant b and the dimensions of ρ, r, and v to show that, for $n = 2$, Newton's expression is consistent dimensionally with our result from part (*b*). In Parts (*d*) and (*e*), we can apply Newton's 2nd law under terminal velocity conditions to find the terminal velocity of the sky diver near the surface of the earth and at a height of 8 km. Assume that $g = 9.81\text{ m/s}^2$ remains constant. (Note: At 8 km, $g = 9.78\text{ m/s}^2$. However, it will not affect the result in Part (*e*).)

(*a*) Solve the drag force equation for b with $n = 1$:

$$b = \frac{F_{\text{d}}}{v}$$

Substitute the dimensions of F_{d} and v and simplify to obtain:

$$[b] = \frac{\frac{\text{ML}}{\text{T}^2}}{\frac{\text{L}}{\text{T}}} = \boxed{\frac{\text{M}}{\text{T}}}$$

and the units of b are $\boxed{\text{kg/s}}$

(*b*) Solve the drag force equation for b with $n = 2$:

$$b = \frac{F_{\text{d}}}{v^2}$$

Substitute the dimensions of F_d and v and simplify to obtain:

$$[b]=\frac{\frac{\text{ML}}{\text{T}^2}}{\left(\frac{\text{L}}{\text{T}}\right)^2}=\boxed{\frac{\text{M}}{\text{L}}}$$

and the units of b are $\boxed{\text{kg/m}}$

(c) Express the dimensions of Newton's expression:

$$[F_\text{d}]=\left[\tfrac{1}{2}\rho\pi r^2v^2\right]=\left(\frac{\text{M}}{\text{L}^3}\right)(\text{L})^2\left(\frac{\text{L}}{\text{T}}\right)^2$$

$$=\boxed{\frac{\text{ML}}{\text{T}^2}}$$

From Part (b) we have:

$$[F_\text{d}]=\left[bv^2\right]=\left(\frac{\text{M}}{\text{L}}\right)\left(\frac{\text{L}}{\text{T}}\right)^2=\boxed{\frac{\text{ML}}{\text{T}^2}}$$

(d) Letting the downward direction be the positive y direction, apply $\sum F_y=ma_y$ to the sky diver:

$$mg-\tfrac{1}{2}\rho\pi r^2v_\text{t}^2=0\Rightarrow v_\text{t}=\sqrt{\frac{2mg}{\rho\pi r^2}}$$

Substitute numerical values and evaluate v_t:

$$v_\text{t}=\sqrt{\frac{2(56\,\text{kg})(9.81\,\text{m/s}^2)}{\pi(1.2\,\text{kg/m}^3)(0.30\,\text{m})^2}}=\boxed{57\,\text{m/s}}$$

(e) Evaluate v_t at a height of 8 km:

$$v_\text{t}=\sqrt{\frac{2(56\,\text{kg})(9.81\,\text{m/s}^2)}{\pi(0.514\,\text{kg/m}^3)(0.30\,\text{m})^2}}$$

$$=\boxed{87\,\text{m/s}}$$

Friction

31 • **[SSM]** A block of mass m slides at constant speed down a plane inclined at an angle of θ with the horizontal. It follows that (a) $\mu_\text{k}=mg\sin\theta$, ($b$) $\mu_\text{k}=\tan\theta$, (c) $\mu_\text{k}=1-\cos\theta$, ($d$) $\mu_\text{k}=\cos\theta-\sin\theta$.

Picture the Problem The block is in equilibrium under the influence of $\vec{F}_\text{n}$, $m\vec{g}$, and $\vec{f}_\text{k}$; that is, $\vec{F}_\text{n}+m\vec{g}+\vec{f}_\text{k}=0$. We can apply Newton's 2nd law to determine the relationship between f_k, θ, and mg.

A pictorial representation showing the forces acting on the sliding block is shown to the right.

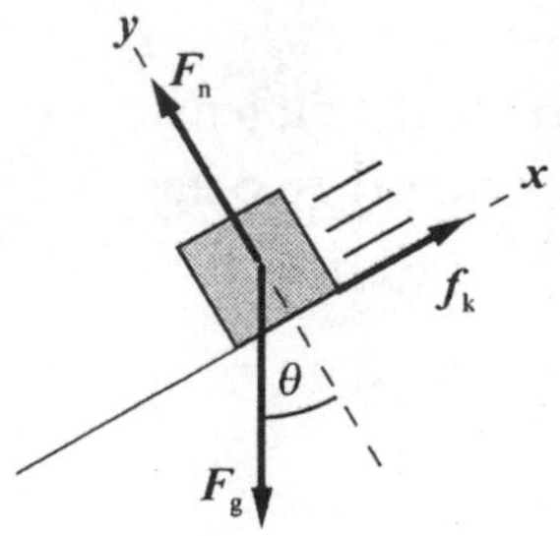

Using its definition, express the coefficient of kinetic friction:

$$\mu_k = \frac{f_k}{F_n} \quad (1)$$

Apply $\sum F_x = ma_x$ to the block:

$f_k - mg\sin\theta = ma_x$
or, because $a_x = 0$,
$f_k = mg\sin\theta$

Apply $\sum F_y = ma_y$ to the block:

$F_n - mg\cos\theta = ma_y$
or, because $a_y = 0$,
$F_n = mg\cos\theta$

Substitute for f_k and F_n in equation (1) and simplify to obtain:

$$\mu_k = \frac{mg\sin\theta}{mg\cos\theta} = \tan\theta$$

and $\boxed{(b)}$ is correct.

33 • **[SSM]** A block weighing 20-N rests on a horizontal surface. The coefficients of static and kinetic friction between the surface and the block are $\mu_s = 0.80$ and $\mu_k = 0.60$. A horizontal string is then attached to the block and a constant tension T is maintained in the string. What is the subsequent force of friction acting on the block if (*a*) $T = 15$ N or (*b*) $T = 20$ N?

Picture the Problem Whether the friction force is that due to static friction or kinetic friction depends on whether the applied tension is greater than the maximum static friction force. We can apply the definition of the maximum static friction to decide whether $f_{s,max}$ or T is greater.

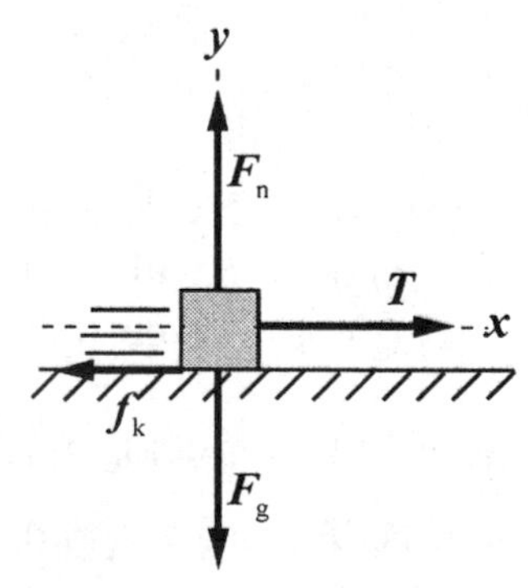

Noting that $F_n = F_g$, calculate the maximum static friction force:

$$f_{s,max} = \mu_s F_n = \mu_s F_g = (0.80)(20\text{ N})$$
$$= 16\text{ N}$$

(*a*) Because $f_{s,max} > T$:

$$f = f_s = T = \boxed{15\ \text{N}}$$

(*b*) Because $T > f_{s,max}$:

$$f = f_k = \mu_k F_n = \mu_k F_g = (0.60)(20\ \text{N}) = \boxed{12\ \text{N}}$$

35 • **[SSM]** A 100-kg crate rests on a thick-pile carpet. A weary worker then pushes on the crate with a horizontal force of 500 N. The coefficients of static and kinetic friction between the crate and the carpet are 0.600 and 0.400, respectively. Find the subsequent frictional force exerted by the carpet on the crate.

Picture the Problem Whether the friction force is that due to static friction or kinetic friction depends on whether the applied tension is greater than the maximum static friction force. If it is, then the box moves and the friction force is the force of kinetic friction. If it is less, the box does not move.

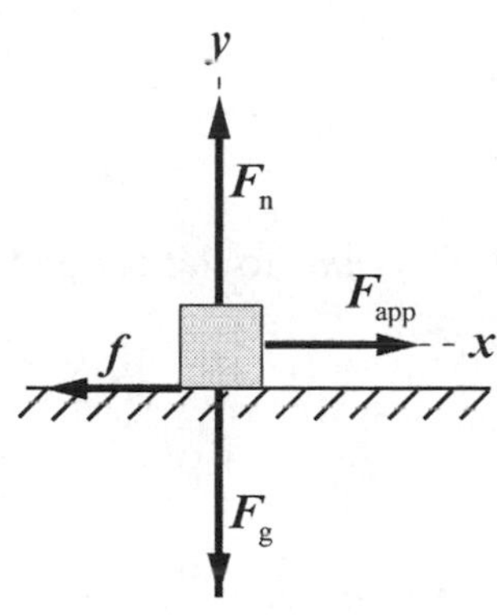

The maximum static friction force is given by:

$$f_{s,max} = \mu_s F_n$$

or, because $F_n = F_g = mg$,

$$f_{s,max} = \mu_s mg$$

Substitute numerical values and evaluate $f_{s,max}$:

$$f_{s,max} = (0.600)(100\ \text{kg})(9.81\ \text{m/s}^2) = 589\ \text{N}$$

Because $f_{s,\max} > F_{app}$, the box does not move and :

$$F_{app} = f_s = \boxed{500\ \text{N}}$$

37 • **[SSM]** The coefficient of static friction between the tires of a car and a horizontal road is 0.60. Neglecting air resistance and rolling friction, (*a*) what is the magnitude of the maximum acceleration of the car when it is braked? (*b*) What is the shortest distance in which the car can stop if it is initially traveling at 30 m/s?

Picture the Problem Assume that the car is traveling to the right and let the positive x direction also be to the right. We can use Newton's 2nd law of motion and the definition of μ_s to determine the maximum acceleration of the car. Once we know the car's maximum acceleration, we can use a constant-acceleration equation to determine the least stopping distance.

(*a*) A pictorial representation showing the forces acting on the car is shown to the right.

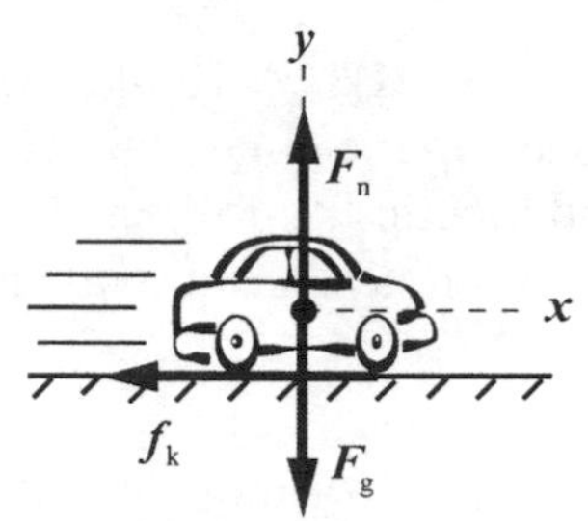

Apply $\sum F_x = ma_x$ to the car:

$$-f_{s,max} = -\mu_s F_n = ma_x \qquad (1)$$

Apply $\sum F_y = ma_y$ to the car and solve for F_n:

$$F_n - F_g = ma_y$$

$$F_n - w = ma_y = 0$$

or, because $a_y = 0$ and $F_g = mg$,

$$F_n = mg \qquad (2)$$

Substitute for F_n in equation (1) to obtain:

$$-f_{s,max} = -\mu_s mg = ma_x$$

Solving for $a_{x,max}$ yields:

$$a_{x,max} = \mu_s g$$

Substitute numerical values and evaluate $a_{x,max}$:

$$\boldsymbol{a}_{x,max} = (0.60)(9.81\,\text{m/s}^2) = 5.89\,\text{m/s}^2$$

$$= \boxed{5.9\,\text{m/s}^2}$$

(b) Using a constant-acceleration equation, relate the stopping distance of the car to its initial velocity and its acceleration:

$$v_x^2 = v_{0x}^2 + 2a_x\Delta x$$

or, because $v_x = 0$,

$$0 = v_{0x}^2 + 2a_x\Delta x \Rightarrow \Delta x = \frac{-v_{0x}^2}{2a_x}$$

Using ax = –5.89 m/s2 because the acceleration of the car is to the left, substitute numerical values and evaluate Δx:

$$\boldsymbol{\Delta x} = \frac{-(30\,\text{m/s})^2}{2(-5.89\,\text{m/s}^2)} = \boxed{76\,\text{m}}$$

43 •• **[SSM]** A block of mass $m_1 = 250$ g is at rest on a plane that makes an angle of $\theta = 30°$ with the horizontal. The coefficient of kinetic friction between the block and the plane is 0.100. The block is attached to a second block of mass

$m_2 = 200$ g that hangs freely by a string that passes over a frictionless, massless pulley (Figure 5-62). When the second block has fallen 30.0 cm, what will be its speed?

Picture the Problem Choose a coordinate system in which the $+x$ direction is up the incline for the block whose mass is m_1 and downward for the block whose mass is m_2. We can find the speed of the system when it has moved a given distance by using a constant-acceleration equation. We'll assume that the string is massless and that it does not stretch. Under the influence of the forces shown in the free-body diagrams, the blocks will have a common acceleration a. The application of Newton's 2nd law to each block, followed by the elimination of the tension T and the use of the definition of f_k, will allow us to determine the acceleration of the system.

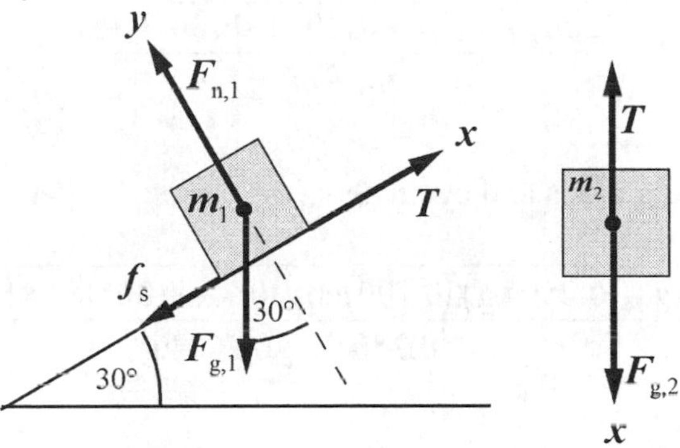

Using a constant-acceleration equation, relate the speed of the system to its acceleration and displacement:

$$v_x^2 = v_{0x}^2 + 2a_x\Delta x$$

and, because $v_{0x} = 0$,

$$v_x^2 = 2a_x\Delta x \Rightarrow v_x = \sqrt{2a_x\Delta x} \qquad (1)$$

Apply $\sum \vec{F} = m\vec{a}$ to the block whose mass is m_1:

$$\sum F_x = T - f_k - F_{g,1}\sin 30° = m_1 a_x \qquad (2)$$

and

$$\sum F_y = F_{n,1} - F_{g,1}\cos 30° = 0 \qquad (3)$$

Because $F_{g,1} = m_1 g$, equations (2) and (3) can be written as:

$$T - f_k - m_1 g\sin 30° = m_1 a_x \qquad (4)$$

and

$$F_{n,1} = m_1 g\cos 30° \qquad (5)$$

Using $f_k = \mu_k F_{n,1}$, substitute equation (5) in equation (4) to obtain:

$$T - \mu_k m_1 g\cos 30° - m_1 g\sin 30° = m_1 a_x \qquad (6)$$

Apply $\sum F_x = ma_x$ to the block whose mass is m_2:

$$F_{g,2} - T = m_2a_x$$

or, because $F_{g,2} = m_2g$,

$$m_2g - T = m_2a_x \qquad (7)$$

Add equations (6) and (7) to eliminate T and solve for a_x to obtain:

$$a_x = \frac{(m_2 - \mu_k m_1 \cos 30° - m_1 \sin 30°)g}{m_1 + m_2}$$

Substituting for a_x in equation (1) and simplifying yields:

$$v_x = \sqrt{\frac{2[m_2 - m_1(\mu_k \cos 30° + \sin 30°)]g\Delta x}{m_1 + m_2}}$$

Substitute numerical values and evaluate v_x:

$$v_x = \sqrt{\frac{2[0.200\text{ kg} - (0.250\text{ kg})((0.100)\cos 30° + \sin 30°)](9.81\text{ m/s}^2)(0.300\text{ m})}{0.250\text{ kg} + 0.200\text{ kg}}}$$

$$= \boxed{84\text{ m/s}}$$

47 •• **[SSM]** A 150-g block is projected up a ramp with an initial speed of 7.0 m/s. The coefficient of kinetic friction between the ramp and the block is 0.23. (*a*) If the ramp is inclined 25° with the horizontal, how far along the surface of the ramp does the block slide before coming to a stop? (*b*) The block then slides back down the ramp. What is the minimum coefficient of static friction between the block and the ramp?

Picture the Problem The force diagram shows the forces acting on the block as it slides up the ramp. Note that the block is accelerated by $\vec{f}_k$ and the x component of $\vec{F}_g$. We can use a constant-acceleration equation to express the displacement of the block up the ramp as a function of its acceleration and Newton's 2nd law to find the acceleration of the block as it slides up the ramp.

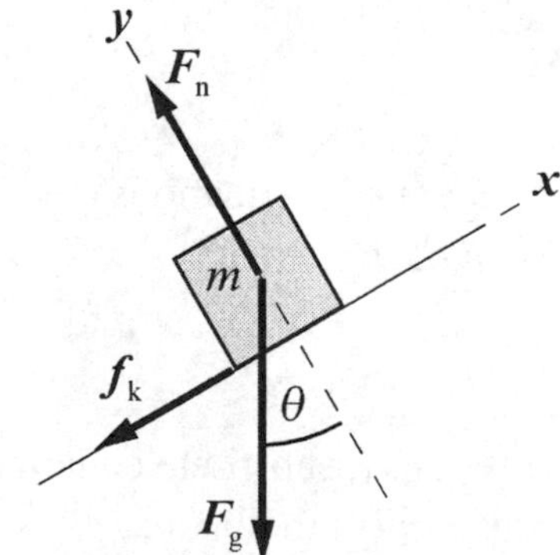

(*a*) Use a constant-acceleration equation to relate the distance the block slides up the incline to its initial speed and acceleration:

$$v_x^2 = v_{0x}^2 + 2a_x\Delta x$$

or, because $v_x = 0$,

$$0 = v_{0x}^2 + 2a_x\Delta x \Rightarrow \Delta x = \frac{-v_{0x}^2}{2a_x} \quad (1)$$

Apply $\sum\vec{F} = m\vec{a}$ to the block:

$$\sum F_x = -f_k - F_g\sin\theta = ma_x \quad (2)$$

and

$$\sum F_y = F_n - F_g\cos\theta = 0 \quad (3)$$

Substituting $f_k = \mu_k F_n$ and $F_g = mg$ in equations (2) and (3) yields:

$$-\mu_k F_n - mg\sin\theta = ma_x \quad (4)$$

and

$$F_n - mg\cos\theta = 0 \quad (5)$$

Eliminate F_n between equations (4) and (5) to obtain:

$$-\mu_k mg\cos\theta - mg\sin\theta = ma_x$$

Solving for a_x yields:

$$a_x = -(\mu_k\cos\theta + \sin\theta)g$$

Substitute for a in equation (1) to obtain:

$$\Delta x = \frac{v_{0x}^2}{2(\mu_k\cos\theta + \sin\theta)g}$$

Substitute numerical values and evaluate Δx:

$$\Delta x = \frac{(7.0\ \text{m/s})^2}{2[(0.23)\cos 25° + \sin 25°](9.81\ \text{m/s}^2)} = 3.957\ \text{m} = \boxed{4.0\ \text{m}}$$

(*b*) At the point at which the block is instantaneously at rest, static friction becomes operative and, if the static friction coefficient is too high, the block will not resume motion, but will remain at the high point. We can determine the minimum value of the coefficient of static friction for which this occurs by considering the equality of the static friction force and the component of the weight of the block down the ramp.

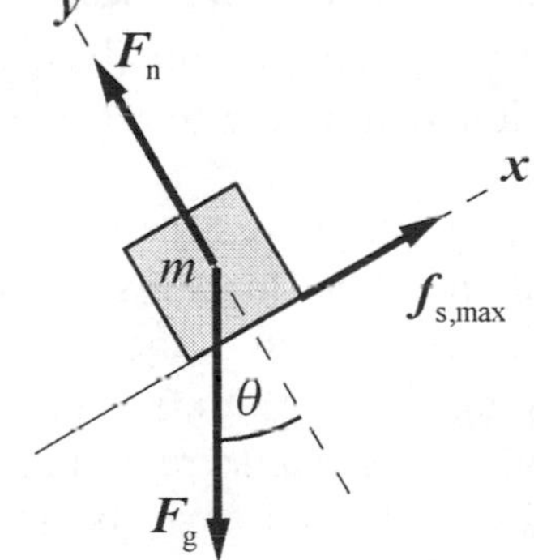

Apply $\sum\vec{F} = m\vec{a}$ to the block when it is in equilibrium at the point at which it is momentarily at rest:

$$\sum F_x = f_{\text{s,max}} - F_{\text{g}}\sin\theta = 0 \qquad (5)$$

and

$$\sum F_y = F_{\text{n}} - F_{\text{g}}\cos\theta = 0 \qquad (6)$$

Solving equation (6) for F_{n} yields:

$$F_{\text{n}} = F_{\text{g}}\cos\theta$$

Because $f_{\text{s,max}} = \mu_{\text{s}}F_{\text{n}}$, equation (5) becomes:

$$\mu_{\text{s}}F_{\text{g}}\cos\theta - F_{\text{g}}\sin\theta = 0$$

or

$$\mu_{\text{s}}\cos\theta - \sin\theta = 0 \Rightarrow \mu_{\text{s}} = \tan\theta$$

Substitute the numerical value of θ and evaluate μ_{s}:

$$\mu_{\text{s}} = \tan 25° = \boxed{0.47}$$

53 •• **[SSM]** A block of mass *m* rests on a horizontal table (Figure 5-65). The block is pulled by a massless rope with a force $\vec{F}$ at an angle θ. The coefficient of static friction is 0.60. The minimum value of the force needed to move the block depends on the angle θ. (*a*) Discuss qualitatively how you would expect this force to depend on θ. (*b*) Compute the force for the angles $\theta = 0°$, 10°, 20°, 30°, 40°, 50°, and 60°, and make a plot of *F* versus θ for $mg = 400$ N. From your plot, at what angle is it most efficient to apply the force to move the block?

Picture the Problem The vertical component of $\vec{F}$ reduces the normal force; hence, the static friction force between the surface and the block. The horizontal component is responsible for any tendency to move and equals the static friction force until it exceeds its maximum value. We can apply Newton's 2nd law to the box, under equilibrium conditions, to relate *F* to θ.

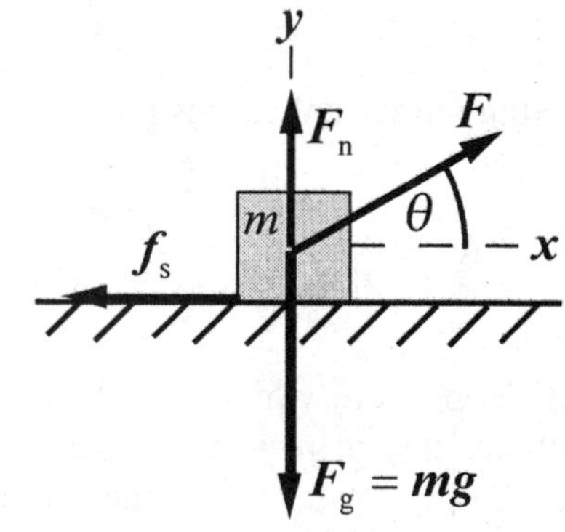

(*a*) The static-frictional force opposes the motion of the object, and the maximum value of the static-frictional force is proportional to the normal force F_{N}. The normal force is equal to the weight minus the vertical component F_{V} of the force *F*. Keeping the magnitude *F* constant while increasing θ from zero results in an increase in F_{V} and a decrease in F_{n}; thus decreasing the maximum static-frictional force f_{max}. The object will begin to move if the horizontal component F_{H} of the force *F* exceeds f_{max}. An increase in θ results in a decrease in F_{H}. As θ increases from 0, the decrease in F_{N} is larger than the decrease in F_{H}, so the object is more and more likely to slip. However, as θ approaches 90°, F_{H} approaches zero and no movement will be initiated. If *F* is large enough and if θ increases from 0, then at some value of θ the block will start to move.

(*b*) Apply $\sum\vec{F} = m\vec{a}$ to the block:

$$\sum F_x = F\cos\theta - f_s = 0 \quad (1)$$

and

$$\sum F_y = F_n + F\sin\theta - mg = 0 \quad (2)$$

Assuming that $f_s = f_{s,max}$, eliminate f_s and F_n between equations (1) and (2) and solve for F:

$$F = \frac{\mu_s mg}{\cos\theta + \mu_s\sin\theta}$$

Use this function with $mg = 400$ N to generate the following table:

θ (deg)	0	10	20	30	40	50	60
F (N)	240	220	210	206	208	218	235

The following graph of $F(\theta)$ was plotted using a spreadsheet program.

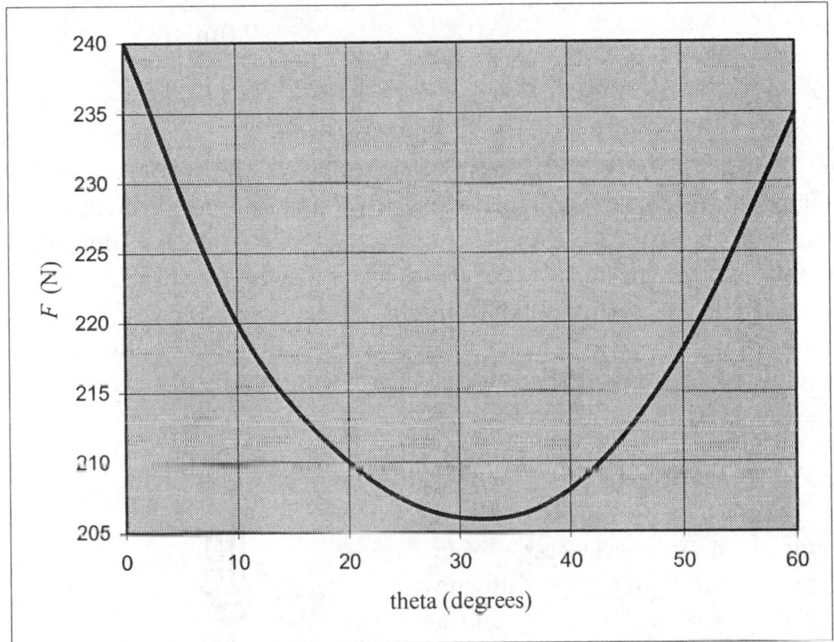

From the graph, we can see that the minimum value for F occurs when $\theta \approx 32°$.

Remarks: An alternative to manually plotting F as a function of θ or using a spreadsheet program is to use a graphing calculator to enter and graph the function.

Drag Forces

67 • **[SSM]** A Ping-Pong ball has a mass of 2.3 g and a terminal speed of 9.0 m/s. The drag force is of the form bv^2. What is the value of b?

Picture the Problem The ping-pong ball experiences a downward gravitational force exerted by the earth and an upward drag force exerted by the air. We can apply Newton's 2nd law to the Ping-Pong ball to obtain its equation of motion. Applying terminal speed conditions will yield an expression for b that we can evaluate using the given numerical values. Let the downward direction be the $+y$ direction.

Apply $\sum F_y = ma_y$ to the Ping-Pong ball:

$$mg - bv^2 = ma_y$$

When the Ping-Pong ball reaches its terminal speed $v = v_t$ and $a_y = 0$:

$$mg - bv_t^2 = 0 \Rightarrow b = \frac{mg}{v_t^2}$$

Substitute numerical values and evaluate b:

$$b = \frac{(2.3\times10^{-3}\,\text{kg})(9.81\,\text{m/s}^2)}{(9.0\,\text{m/s})^2} = \boxed{2.8\times10^{-4}\,\text{kg/m}}$$

69 •• **[SSM]** A common classroom demonstration involves dropping basket-shaped coffee filters and measuring the time required for them to fall a given distance. A professor drops a single basket-shaped coffee filter from a height h above the floor, and records the time for the fall as Δt. How far will a stacked set of n identical filters fall during the same time interval Δt? Consider the filters to be so light that they instantaneously reach their terminal velocities. Assume a drag force that varies as the square of the speed and assume the filters are released oriented right-side up.

Picture the Problem The force diagram shows n coffee filters experiencing an upward drag force exerted by the air that varies with the square of their terminal velocity and a downward gravitational force exerted by the earth. We can use the definition of average velocity and Newton's 2nd law to establish the dependence of the distance of fall on n.

Express the distance fallen by 1 coffee filter, falling at its terminal speed, in time Δt:

$$d_{1\,\text{filter}} = v_{t,1\,\text{filter}}\Delta t \qquad (1)$$

Express the distance fallen by n coffee filters, falling at their terminal speed, in time Δt:

$$d_{n\text{ filters}} = v_{\text{t},\,n\text{ filters}}\Delta t \qquad (2)$$

Divide equation (2) by equation (1) to obtain:

$$\frac{d_{n\text{ filters}}}{d_{1\text{ filter}}} = \frac{v_{\text{t},\,n\text{ filters}}\Delta t}{v_{\text{t},\,1\text{ filter}}\Delta t} = \frac{v_{\text{t},\,n\text{ filters}}}{v_{\text{t},\,1\text{ filter}}}$$

Solving for $d_{n\text{ filters}}$ yields:

$$d_{n\text{ filters}} = \left(\frac{v_{\text{t},\,n\text{ filters}}}{v_{\text{t},\,1\text{ filter}}}\right)d_{1\text{ filter}} \qquad (3)$$

Apply $\sum F_y = ma_y$ to the coffee filters:

$$nmg - Cv_\text{t}^2 = ma_y$$

or, because $a_y = 0$,

$$nmg - Cv_\text{t}^2 = 0$$

Solving for v_t yields:

$$v_{\text{t},\,n\text{ filters}} = \sqrt{\frac{nmg}{C}} = \sqrt{\frac{mg}{C}}\sqrt{n}$$

or, because $v_{\text{t},\,1\text{ filter}} = \sqrt{\dfrac{mg}{C}}$,

$$v_{\text{t},\,n\text{ filters}} = v_{\text{t},\,1\text{ filter}}\sqrt{n}$$

Substitute for $v_{\text{t},\,n\text{ filters}}$ in equation (3) to obtain:

$$d_{n\text{ filters}} = \left(\frac{v_{\text{t},\,1\text{ filter}}\sqrt{n}}{v_{\text{t},\,1\text{ filter}}}\right)d_{1\text{ filter}} = \boxed{\sqrt{n}\,d_{1\text{ filter}}}$$

This result tells us that n filters will fall farther, in the same amount of time, than 1 filter by a factor of $\sqrt{n}$.

73 ••• **[SSM]** You are on an environmental chemistry internship, and in charge of a sample of air containing pollution particles of the size and density given in Problem 72. You capture the sample in an 8.0-cm-long test tube. You then place the test tube in a centrifuge with the midpoint of the test tube 12 cm from the center of the centrifuge. You set the centrifuge to spin at 800 revolutions per minute. (*a*) Estimate the time you have to wait so that nearly all of the pollution particles settle to the end of the test tube. (*b*) Compare this to the time required for a pollution particle to fall 8.0 cm under the action of gravity and subject to the drag force given in Problem 72.

Picture the Problem The motion of the centrifuge will cause the pollution particles to migrate to the end of the test tube. We can apply Newton's 2nd law and Stokes' law to derive an expression for the terminal speed of the sedimentation particles. We can then use this terminal speed to calculate the sedimentation time. We'll use the 12 cm distance from the center of the centrifuge as the average radius of the pollution particles as they settle in the test tube. Let R represent the radius of a particle and r the radius of the particle's circular path in the centrifuge.

(*a*) Express the sedimentation time in terms of the sedimentation speed v_t:

$$\Delta t_{\text{sediment}} = \frac{\Delta x}{v_t} \qquad (1)$$

Apply $\sum F_{\text{radial}} = ma_{\text{radial}}$ to a pollution particle:

$$6\pi\eta R v_t = ma_c$$

Express the mass of the particle in terms of its radius R and density ρ:

$$m = \rho V = \tfrac{4}{3}\pi R^3 \rho$$

Express the acceleration of the pollution particles due to the motion of the centrifuge in terms of their orbital radius r and period T:

$$a_c = \frac{v^2}{r} = \frac{\left(\frac{2\pi r}{T}\right)^2}{r} = \frac{4\pi^2 r}{T^2}$$

Substitute for m and a_c and simplify to obtain:

$$6\pi\eta R v_t = \tfrac{4}{3}\pi R^3 \rho\left(\frac{4\pi^2 r}{T^2}\right) = \frac{16\pi^3 \rho r R^3}{3T^2}$$

Solving for v_t yields:

$$v_t = \frac{8\pi^2 \rho r R^2}{9\eta T^2}$$

Substitute for v_t in equation (1) and simplify to obtain:

$$\Delta t_{\text{sediment}} = \frac{\Delta x}{\frac{8\pi^2 \rho r R^2}{9\eta T^2}} = \frac{9\eta T^2 \Delta x}{8\pi^2 \rho r R^2}$$

Substitute numerical values and evaluate $\Delta t_{\text{sediment}}$:

$$\Delta t_{\text{sediment}} = \frac{9\left(1.8\times10^{-5}\,\frac{\text{N}\cdot\text{s}}{\text{m}^2}\right)\left(\frac{1}{800\frac{\text{rev}}{\text{min}}\cdot\frac{1\,\text{min}}{60\,\text{s}}}\right)^2(8.0\,\text{cm})}{8\pi^2\left(2000\frac{\text{kg}}{\text{m}^3}\right)(0.12\,\text{m})\left(10^{-5}\,\text{m}\right)^2} = 38.47\,\text{ms} = \boxed{38\,\text{ms}}$$

(*b*) In Problem 72 it was shown that the rate of fall of the particles in air is 2.42 cm/s. Find the time required to fall 8.0 cm in air under the influence of gravity:

$$\Delta t_{air} = \frac{\Delta x}{v} = \frac{8.0\,\text{cm}}{2.42\,\text{cm/s}} = 3.31\,\text{s}$$

Find the ratio of the two times:

$$\frac{\Delta t_{air}}{\Delta t_{sediment}} = \frac{3.31\,\text{s}}{38.47\,\text{ms}} = 86$$

With the drag force in Problem 72 it takes about 86 times longer than it does using the centrifuge.

Motion Along a Curved Path

75 • **[SSM]** A 95-g stone is whirled in a horizontal circle on the end of an 85-cm-long string. The stone takes 1.2 s to make one complete revolution. Determine the angle that the string makes with the horizontal.

Picture the Problem The only forces acting on the stone are the tension in the string and the gravitational force. The centripetal force required to maintain the circular motion is a component of the tension. We'll solve the problem for the general case in which the angle with the horizontal is θ by applying Newton's 2nd law of motion to the forces acting on the stone.

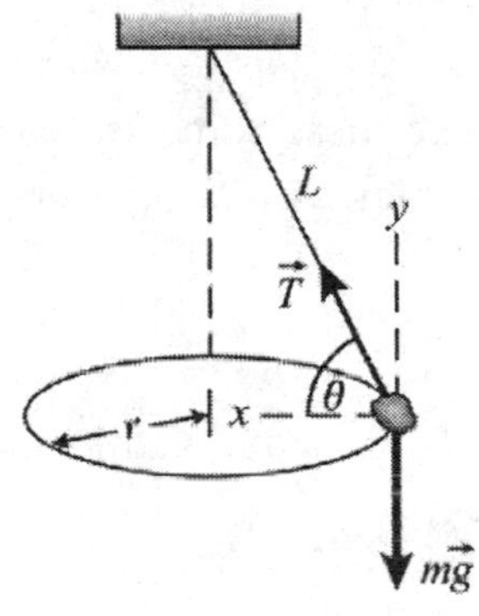

Apply $\sum \vec{F} = m\vec{a}$ to the stone:

$$\sum F_x = T\cos\theta = ma_c = m\frac{v^2}{r} \quad (1)$$

and

$$\sum F_y = T\sin\theta - mg = 0 \quad (2)$$

Use the right triangle in the diagram to relate r, L, and θ:

$$r = L\cos\theta \quad (3)$$

Eliminate T and r between equations (1), (2) and (3) and solve for v_2:

$$v^2 = gL\cot\theta\cos\theta \quad (4)$$

Express the speed of the stone in terms of its period:

$$v = \frac{2\pi r}{t_{1\,\text{rev}}} \quad (5)$$

Eliminate v between equations (4) and (5) and solve for θ:

$$\theta = \sin^{-1}\left(\frac{g t_{1\,\text{rev}}^2}{4\pi^2 L}\right)$$

Substitute numerical values and evaluate θ:

$$\theta = \sin^{-1}\left[\frac{(9.81\,\text{m/s}^2)(1.2\,\text{s})^2}{4\pi^2(0.85\,\text{m})}\right] = \boxed{25°}$$

81 •• **[SSM]** A block of mass m_1 is attached to a cord of length L_1, which is fixed at one end. The block moves in a horizontal circle on a frictionless tabletop. A second block of mass m_2 is attached to the first by a cord of length L_2 and also moves in a circle on the same frictionless tabletop, as shown in Figure 5-73. If the period of the motion is T, find the tension in each cord in terms of the given symbols.

Picture the Problem The free-body diagrams show the forces acting on each block. We can use Newton's 2nd law to relate these forces to each other and to the masses and accelerations of the blocks.

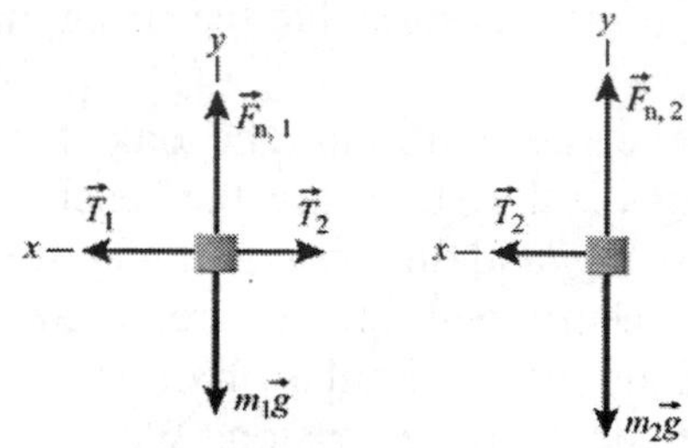

Apply $\sum F_x = ma_x$ to the block whose mass is m_1:

$$T_1 - T_2 = m_1\frac{v_1^2}{L_1}$$

Apply $\sum F_x = ma_x$ to the block whose mass is m_2:

$$T_2 = m_2\frac{v_2^2}{L_1 + L_2}$$

Relate the speeds of each block to their common period T and their distance from the center of the circle:

$$v_1 = \frac{2\pi L_1}{T} \text{ and } v_2 = \frac{2\pi(L_1 + L_2)}{T}$$

In the second force equation, substitute for v_2, and simplify to obtain:

$$T_2 = \boxed{[m_2(L_1 + L_2)]\left(\frac{2\pi}{T}\right)^2}$$

Substitute for T_2 and v_1 in the first force equation to obtain:

$$T_1 = \boxed{[m_2(L_1 + L_2) + m_1 L_1]\left(\frac{2\pi}{T}\right)^2}$$

Centripetal Force

89 • **[SSM]** The radius of curvature of the track at the top of a loop-the-loop on a roller-coaster ride is 12.0 m. At the top of the loop, the force that the seat exerts on a passenger of mass *m* is 0.40*mg*. How fast is the roller-coaster car moving as it moves through the highest point of the loop.

Picture the Problem The speed of the roller coaster is embedded in the expression for its radial acceleration. The radial acceleration is determined by the net radial force acting on the passenger. We can use Newton's 2nd law to relate the net force on the passenger to the speed of the roller coaster.

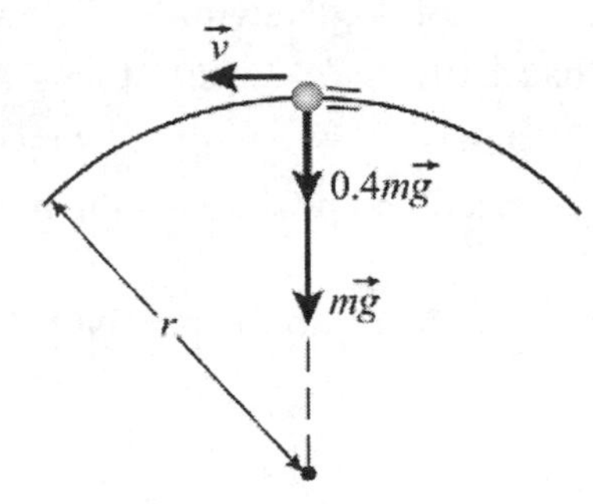

Apply $\sum F_{\text{radial}} = ma_{\text{radial}}$ to the passenger:

$$mg + 0.40mg = m\frac{v^2}{r} \Rightarrow v = \sqrt{1.40gr}$$

Substitute numerical values and evaluate *v*:

$$v = \sqrt{(1.40)(9.81\,\text{m/s}^2)(12.0\,\text{m})} = \boxed{12.8\,\text{m/s}}$$

Euler's Method

99 •• **[SSM]** You throw a baseball straight up with an initial speed of 150 km/h. The ball's terminal speed when falling is also 150 km/h. (*a*) Use Euler's method (**spreadsheet**) to estimate its height 3.50 s after release. (*b*) What is the maximum height it reaches? (*c*) How long after release does it reach its maximum height? (*d*) How much later does it return to the ground? (*e*) Is the time the ball spends on the way up less than, the same as, or greater than the time it spends on the way down?

Picture the Problem The free-body diagram shows the forces acting on the baseball after it has left your hand. In order to use Euler's method, we'll need to determine how the acceleration of the ball varies with its speed. We can do this by applying Newton's 2nd law to the baseball. We can then use $v_{n+1} = v_n + a_n\Delta t$ and $x_{n+1} = x_n + v_n\Delta t$ to find the speed and position of the ball.

y

$\boldsymbol{F}_\text{d} = -bv^2\hat{\boldsymbol{j}}$

$\boldsymbol{F}_\text{g} = -mg\hat{\boldsymbol{j}}$

Apply $\sum F_y = ma_y$ to the baseball:

$$-bv|v| - mg = m\frac{dv}{dt}$$

where $|v| = v$ for the upward part of the flight of the ball and $|v| = -v$ for the downward part of the flight.

Solve for dv/dt to obtain:

$$\frac{dv}{dt} = -g - \frac{b}{m}v|v|$$

Under terminal speed conditions ($|v| = -v_\text{t}$):

$$0 = -g + \frac{b}{m}v_\text{t}^2 \text{ and } \frac{b}{m} = \frac{g}{v_\text{t}^2}$$

Substituting for b/m yields:

$$\frac{dv}{dt} = -g - \frac{g}{v_\text{t}^2}v|v| = -g\left(1 + \frac{v|v|}{v_\text{t}^2}\right)$$

Letting a_n be the acceleration of the ball at time t_n, express its position and speed when $t = t_n + 1$:

$$y_{n+1} = y_n + \tfrac{1}{2}(v_n + v_{n-1})\Delta t$$

and

$$v_{n+1} = v_n + a_n\Delta t$$

where $a_n = -g\left(1 + \frac{v_n|v_n|}{v_\text{t}^2}\right)$ and Δt is an arbitrarily small interval of time.

A spreadsheet solution is shown below. The formulas used to calculate the quantities in the columns are as follows:

Cell	Formula/Content	Algebraic Form
D11	D10+$B$6	$t + \Delta t$
E10	41.7	v_0
E11	E10−$B$4* (1+E10*ABS(E10)/($B$5^2))*$B$6	$v_{n+1} = v_n + a_n\Delta t$

F10	0	y_0
F11	F10+0.5*(E10+E11)*$B$6	$y_{n+1} = y_n + \frac{1}{2}(v_n + v_{n-1})\Delta t$
G10	0	y_0
G11	$E$10*D11−0.5*$B$4*D11^2	$v_0 t - \frac{1}{2} g t^2$

	A	B	C	D	E	F	G
4	g=	9.81	m/s^2				
5	v_t=	41.7	m/s				
6	Δt=	0.1	s				
7							
8							
9				t	v	y	$y_{\text{no drag}}$
10				0.00	41.70	0.00	0.00
11				0.10	39.74	4.07	4.12
12				0.20	37.87	7.95	8.14
40				3.00	3.01	60.13	81.00
41				3.10	2.03	60.39	82.18
42				3.20	1.05	60.54	83.26
43				3.30	0.07	60.60	84.25
44				3.40	−0.91	60.55	85.14
45				3.50	−1.89	60.41	85.93
46				3.60	−2.87	60.17	86.62
78				6.80	−28.34	6.26	56.98
79				6.90	−28.86	3.41	54.44
80				7.00	−29.37	0.49	51.80
81				7.10	−29.87	−2.47	49.06

(*a*) When $t = 3.50$ s, the height of the ball is about $\boxed{60.4\,\text{m}}$.

(*b*) The maximum height reached by the ball is $\boxed{60.6\,\text{m}}$.

(*c*) The time the ball takes to reach its maximum height is about $\boxed{3.0\,\text{s}}$.

(*d*) The ball hits the ground at about $t = \boxed{7.0\,\text{s}}$

(*e*) Because the time the ball takes to reach its maximum height is less than half its time of flight, the time the ball spends on the way up less than the time it spends on the way down

Finding the Center of Mass

105 •• **[SSM]** Find the center of mass of the uniform sheet of plywood in Figure 5-79. Consider this as a system of effectively two sheets, letting one have a "negative mass" to account for the cutout. Thus, one is a square sheet of 3.0-m edge length and mass m_1 and the second is a rectangular sheet measuring 1.0 m × 2.0 m with a mass of $-m_2$. Let the coordinate origin be at the lower left corner of the sheet.

Picture the Problem Let the subscript 1 refer to the 3.0-m by 3.0-m sheet of plywood before the 2.0-m by 1.0-m piece has been cut from it. Let the subscript 2 refer to 2.0-m by 1.0-m piece that has been removed and let σ be the area density of the sheet. We can find the center-of-mass of these two regions; treating the missing region as though it had negative mass, and then finding the center-of-mass of the U-shaped region by applying its definition.

Express the coordinates of the center of mass of the sheet of plywood:

$$x_{\text{cm}} = \frac{m_1 x_{\text{cm},1} - m_2 x_{\text{cm},2}}{m_1 - m_2}$$

and

$$y_{\text{cm}} = \frac{m_1 y_{\text{cm},1} - m_2 y_{\text{cm},2}}{m_1 - m_2}$$

Use symmetry to find $x_{\text{cm},1}$, $y_{\text{cm},1}$, $x_{\text{cm},2}$, and $y_{\text{cm},2}$:

$$x_{\text{cm},1} = 1.5\,\text{m},\ y_{\text{cm},1} = 1.5\,\text{m}$$

and

$$x_{\text{cm},2} = 1.5\,\text{m},\ y_{\text{cm},2} = 2.0\,\text{m}$$

Determine m_1 and m_2:

$$m_1 = \sigma A_1 = 9\sigma\,\text{kg}$$

and

$$m_2 = \sigma A_2 = 2\sigma\,kg$$

Substitute numerical values and evaluate x_{cm}:

$$x_{\text{cm}} = \frac{(9\sigma\,\text{kg})(1.5\,\text{m}) - (2\sigma\,\text{kg})(1.5\,\text{kg})}{9\sigma\,\text{kg} - 2\sigma\,\text{kg}} = 1.5\,\text{m}$$

Substitute numerical values and evaluate y_{cm}:

$$y_{\text{cm}} = \frac{(9\sigma\,\text{kg})(1.5\,\text{m}) - (2\sigma\,\text{kg})(2.0\,\text{m})}{9\sigma\,\text{kg} - 2\sigma\,\text{kg}} = 1.4\,\text{m}$$

The center of mass of the U-shaped sheet of plywood is at $\boxed{(1.5\,\text{m}, 1.4\,\text{m})}$.

107 •• **[SSM]** Two identical uniform rods each of length L are glued together so that the angle at the joint is 90°. Determine the location of the center of mass (in terms of L) of this configuration relative to the origin taken to be at the joint. (*Hint: You do not need the mass of the rods (why?), but you should start by assuming a mass m and see that it cancels out.*)

Picture the Problem A pictorial representation of the system is shown to the right. The x and y coordinates of the two rods are

$$\left(x_{1,\text{cm}}, y_{1,\text{cm}}\right) = \left(0, \tfrac{1}{2}L\right)$$

and

$$\left(x_{2,\text{cm}}, y_{2,\text{cm}}\right) = \left(\tfrac{1}{2}L, 0\right).$$

We can use the definition of the center of mass to find the coordinates $\left(x_{\text{cm}}, y_{\text{cm}}\right)$.

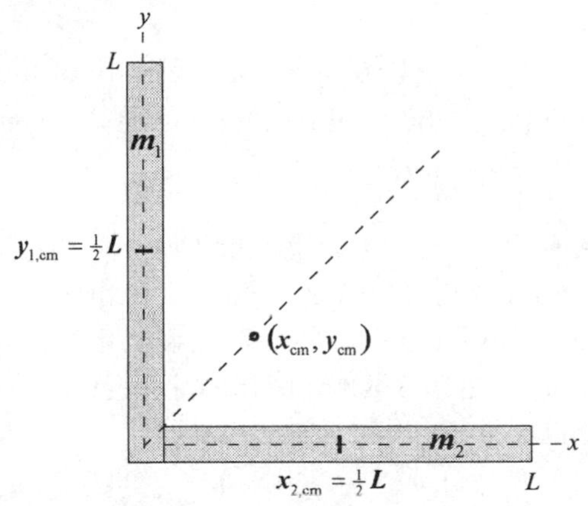

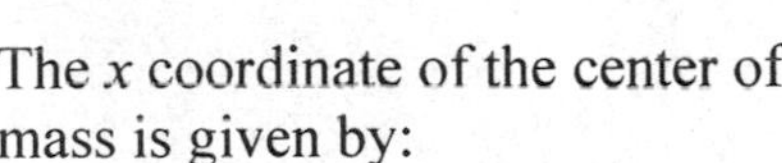

The x coordinate of the center of mass is given by:

$$x_{\text{cm}} = \frac{x_{1,\text{cm}}m_1 + x_{2,\text{cm}}m_2}{m_1 + m_2}$$

Substitute numerical values and evaluate x_{cm}:

$$x_{\text{cm}} = \frac{(0)m_1 + \left(\frac{1}{2}L\right)m_2}{m_1 + m_2}$$

or, because $m_1 = m_2 = m$,

$$x_{\text{cm}} = \frac{(0)m + \left(\frac{1}{2}L\right)m}{m + m} = \tfrac{1}{4}L$$

The y coordinate of the center of mass is given by:

$$y_{\text{cm}} = \frac{y_{1,\text{cm}}m_1 + y_{2,\text{cm}}m_2}{m_1 + m_2}$$

Substitute numerical values and evaluate y_{cm}:

$$y_{\text{cm}} = \frac{\left(\frac{1}{2}L\right)m_1 + (0)m_2}{m_1 + m_2}$$

or, because $m_1 = m_2 = m$,

$$y_{\text{cm}} = \frac{\left(\frac{1}{2}L\right)m + (0)m}{m + m} = \tfrac{1}{4}L$$

The center of mass of this system is located at $\boxed{\left(\tfrac{1}{4}L, \tfrac{1}{4}L\right)}$

Remarks: Note that the center of mass is located at a distance $d = \frac{1}{2}L\cos 45°$ from the vertex on the axis of symmetry bisecting the two arms.

Motion of the Center of Mass

113 • **[SSM]** Two 3.0-kg particles have velocities $\vec{v}_1 = (2.0\text{ m/s})\,\hat{i} + (3.0\text{ m/s})\,\hat{j}$ and $\vec{v}_2 = (4.0\text{ m/s})\,\hat{i} - (6.0\text{ m/s})\,\hat{j}$. Find the velocity of the center of mass of the system.

Picture the Problem The velocity of the center of mass of a system of particles is related to the total momentum of the system through $\vec{P} = \sum_i m_i\vec{v}_i = M\vec{v}_{cm}$.

Use the expression for the total momentum of a system to relate the velocity of the center of mass of the two-particle system to the momenta of the individual particles:

$$\vec{v}_{cm} = \frac{\sum_i m_i\vec{v}_i}{M} = \frac{m_1\vec{v}_1 + m_2\vec{v}_2}{m_1 + m_2}$$

Substitute numerical values and evaluate $\vec{v}_{cm}$:

$$\vec{v}_{cm} = \frac{(3.0\,\text{kg})\left[(2.0\text{ m/s})\hat{i} + (3.0\text{ m/s})\hat{j}\right] + (3.0\,\text{kg})\left[(4.0\text{ m/s})\hat{i} - (6.0\text{ m/s})\hat{j}\right]}{3.0\,\text{kg} + 3.0\,\text{kg}}$$

$$= \boxed{(3.0\,\text{m/s})\hat{i} - (1.5\,\text{m/s})\hat{j}}$$

117 •• **[SSM]** The bottom end of a massless, vertical spring of force constant k rests on a scale and the top end is attached to a massless cup, as in Figure 5-81. Place a ball of mass m_b gently into the cup and ease it down into an equilibrium position where it sits at rest in the cup. (*a*) Draw the separate free-body diagrams for the ball and the spring. (*b*) Show that in this situation, the spring compression d is given by $d = m_b g/k$. (*c*) What is the scale reading under these conditions?

Picture the Problem (*b*) We can apply Newton's 2nd law to the ball to find an expression for the spring's compression when the ball is at rest in the cup. (*c*) The scale reading is the force exerted by the spring on the scale and can be found from the application of Newton's 2nd law to the cup (considered as part of the spring).

(*a*) The free-body diagrams for the ball and spring follow. Note that, because the ball has been eased down into the cup, both its speed and acceleration are zero.

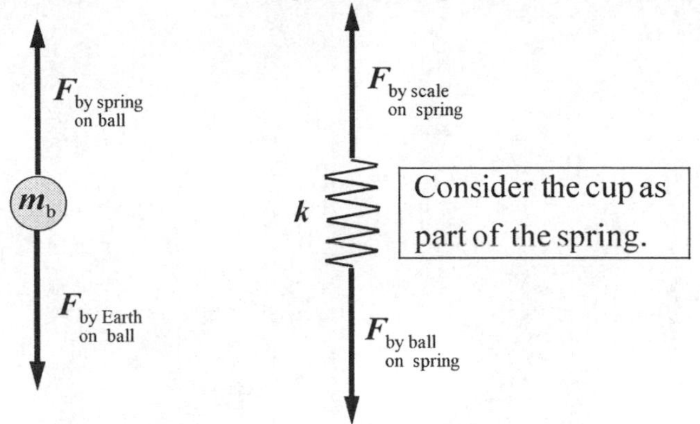

(*b*) Letting the upward direction be the positive *y* direction, apply $\sum F_y = ma_y$ to the ball when it is at rest in the cup and the spring has been compressed a distance *d*:

$$F_{\substack{\text{by spring}\\\text{on ball}}} - F_{\substack{\text{by Earth}\\\text{on ball}}} = m_b a_y$$

or, because $a_y = 0$,

$$F_{\substack{\text{by spring}\\\text{on ball}}} - F_{\substack{\text{by Earth}\\\text{on ball}}} = 0 \qquad (1)$$

Because $F_{\substack{\text{by spring}\\\text{on ball}}} = kd$ and $F_{\substack{\text{by Earth}\\\text{on ball}}} = m_b g$:

$$kd - m_b g = 0 \Rightarrow d = \boxed{\frac{m_b g}{k}}$$

(*c*) Apply $\sum F_y = ma_y$ to the spring:

$$F_{\substack{\text{by scale}\\\text{on spring}}} - F_{\substack{\text{by ball}\\\text{on spring}}} = m_{\text{spring}} a_y \qquad (2)$$

or, because $a_y = 0$,

$$F_{\substack{\text{by scale}\\\text{on spring}}} - F_{\substack{\text{by ball}\\\text{on spring}}} = 0$$

Because $F_{\substack{\text{by spring}\\\text{on ball}}} = F_{\substack{\text{by ball}\\\text{on spring}}}$, adding equations (1) and (2) yields:

$$F_{\substack{\text{by scale}\\\text{on spring}}} - F_{\substack{\text{by Earth}\\\text{on ball}}} = 0$$

Solving for $F_{\substack{\text{by scale}\\\text{on spring}}}$ yields:

$$F_{\substack{\text{by scale}\\\text{on spring}}} = F_{\substack{\text{by Earth}\\\text{on ball}}} = \boxed{m_b g}$$

Chapter 6
Work and Kinetic Energy

Conceptual Problems

7 • **[SSM]** How does the work required to stretch a spring 2.0 cm from its unstressed length compare with the work required to stretch it 1.0 cm from its unstressed length?

Determine the Concept The work required to stretch or compress a spring a distance x is given by $W = \frac{1}{2}kx^2$ where k is the spring's stiffness constant. Because $W \propto x^2$, doubling the distance the spring is stretched will require four times as much work.

13 •• **[SSM]** True or false: (*a*) The scalar product cannot have units. (*b*) If the scalar product of two nonzero vectors is zero, then they are parallel. (*c*) If the scalar product of two nonzero vectors is equal to the product of their magnitudes, then the two vectors are parallel. (*d*) As an object slides up an incline, the sign of the scalar product of the force of gravity on it and its displacement is negative.

(*a*) False. Work is the scalar product of force and displacement.

(*b*) False. Because $\vec{A} \cdot \vec{B} = AB\cos\theta$, where θ is the angle between $\vec{A}$ and $\vec{B}$, if the scalar product of the vectors is zero, then θ must be 90° (or some odd multiple of 90°) and the vectors are perpendicular.

(*c*) True. Because $\vec{A} \cdot \vec{B} = AB\cos\theta$, where θ is the angle between $\vec{A}$ and $\vec{B}$, if the scalar product of the vectors is equal to the product of their magnitudes, then θ must be 0° and the vectors are parallel.

(*d*) True. Because the angle between the gravitational force and the displacement of the object is greater than 90°, its cosine is negative and, hence, the scalar product is negative.

17 •• **[SSM]** You are driving a car that accelerates from rest on a level road without spinning its wheels. Use the center-of-mass work–translational-kinetic-energy relation and free-body diagrams to clearly explain which force (or forces) is (are) directly responsible for the gain in translational kinetic energy of both you and the car. *Hint: The relation refers to external forces only, so the car's engine is not the answer. Pick your "system" correctly for each case.*

Determine the Concept The car shown in the free-body diagram is accelerating in the positive x direction, as are you (shown to the right). The net external force (neglecting air resistance) acting on the car (and on you) is the static friction force

$\vec{f}_s$ exerted by the road and acting on the car tires. The positive center of mass work this friction force does is translated into a gain of kinetic energy.

Work, Kinetic Energy and Applications

27 •• **[SSM]** A 3.0-kg particle moving along the x axis has a velocity of +2.0 m/s as it passes through the origin. It is subjected to a single force, F_x, that varies with position, as shown in Figure 6-30. (*a*) What is the kinetic energy of the particle as it passes through the origin? (*b*) How much work is done by the force as the particle moves from x = 0.0 m to x = 4.0 m? (*c*) What is the speed of the particle when it is at x = 4.0 m?

Picture the Problem The pictorial representation shows the particle as it moves along the positive x axis. The particle's kinetic energy increases because work is done on it. We can calculate the work done on it from the graph of F_x vs. x and relate its kinetic energy when it is at x = 4.0 m to its kinetic energy when it was at the origin and the work done on it by using the work-kinetic energy theorem.

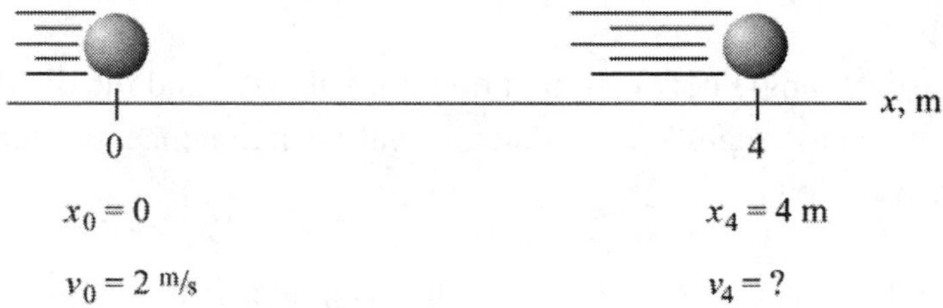

(*a*) Calculate the kinetic energy of the particle when it is at x = 0:

$$K_0 = \tfrac{1}{2}mv^2 = \tfrac{1}{2}(3.0\,\text{kg})(2.0\,\text{m/s})^2 = \boxed{6.0\,\text{J}}$$

(*b*) Because the force and displacement are parallel, the work done is the area under the curve. Use the formula for the area of a triangle to calculate the area under the $F(x)$ graph:

$$W_{0\to 4} = \tfrac{1}{2}(\text{base})(\text{altitude}) = \tfrac{1}{2}(4.0\,\text{m})(6.0\,\text{N}) = \boxed{12\,\text{J}}$$

(*c*) Express the kinetic energy of the particle at $x = 4.0$ m in terms of its speed and mass:

$$K_{4\,\text{s}} = \tfrac{1}{2}mv_{4\,\text{s}}^2 \Rightarrow v_{4\,\text{s}} = \sqrt{\frac{2K_{4\,\text{s}}}{m}} \quad (1)$$

Using the work-kinetic energy theorem, relate the work done on the particle to its *change* in kinetic energy:

$$W_{0\to 4.0\,\text{m}} = K_{4\,\text{s}} - K_0$$

Solve for the particle's kinetic energy at $x = 4.0$ m:

$$K_{4\,\text{s}} = K_0 + W_{0\to 4}$$

Substitute numerical values and evaluate $K_{4\,\text{s}}$:

$$K_{4\,\text{s}} = 6.0\,\text{J} + 12\,\text{J} = 18\,\text{J}$$

Substitute numerical values in equation (1) and evaluate v_4:

$$v_4 = \sqrt{\frac{2(18\,\text{J})}{3.0\,\text{kg}}} = \boxed{3.5\,\text{m/s}}$$

33 •• **[SSM]** You are designing a jungle-vine–swinging sequence for the latest Tarzan movie. To determine his speed at the low point of the swing and to make sure it does not exceed mandatory safety limits, you decide to model the system of Tarzan + vine as a pendulum. Assume your model consists of a particle (Tarzan, mass 100 kg) hanging from a light string (the vine) of length ℓ attached to a support. The angle between the vertical and the string is written as ϕ. (*a*) Draw a free-body diagram for the object on the end of the string (Tarzan on the vine). (*b*) An infinitesimal distance along the arc (along which the object travels) is $\ell d\phi$. Write an expression for the total work dW_{total} done on the particle as it traverses that distance for an arbitrary angle ϕ. (*c*) If the $\ell = 7.0$ m, and if the particle starts from rest at an angle 50°, determine the particle's kinetic energy and speed at the low point of the swing using the work–kinetic-energy theorem.

Picture the Problem Because Tarzan's displacement is always perpendicular to the tension force exerted by the rope, this force can do no work on him. The net work done on Tarzan is the work done by the gravitational force acting on him. We can use the definition of work and the work-kinetic energy theorem to find his kinetic energy and speed at the low point of his swing.

(*a*) The forces acting on Tarzan are a gravitational force $\vec{F}_g$ (his weight) exerted by the earth and the tension force $\vec{T}$ exerted by the vine on which he is swinging.

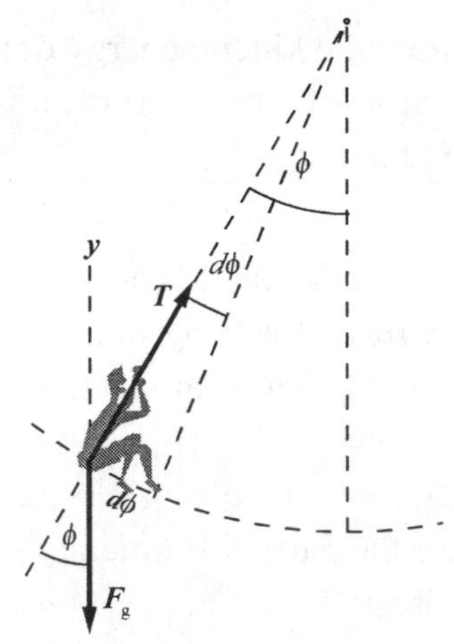

(*b*) The work dW_{total} done by the gravitational force $\vec{F}_g$ as Tarzan swings through an arc of length *ds* is:

$$dW_{total} = \vec{F}_g \cdot d\vec{s} = F_g ds\cos(90° - \phi) = F_g ds\sin\phi$$

Note that, because $d\phi$ is decreasing as Tarzan swings toward his equilibrium position:

$$ds = -\ell d\phi$$

Substituting for *ds* and F_g yields:

$$dW_{total} = \boxed{-mg\ell\sin\phi\, d\phi}$$

(*c*) Apply the work-kinetic energy theorem to Tarzan to obtain:

$$W_{total} = \Delta K = K_f - K_i$$

or, because $K_i = 0$,

$$W_{total} = K_f$$

Substituting for W_{total} yields:

$$K_f = -\int_{50°}^{0°} mg\ell\sin\phi\, d\phi = -mg\ell\left[-\cos\phi\right]_{50°}^{0°}$$

Substitute numerical values and evaluate the integral to find K_f:

$$\begin{aligned}K_f &= -(100\text{ kg})(9.81\text{ m/s}^2)(7.0\text{ m})[-\cos\phi]_{50°}^{0°}\\ &= (100\text{ kg})(9.81\text{ m/s}^2)(7.0\text{ m})(1-\cos 50°)\\ &= 2.453\text{ kJ}\\ &= \boxed{2.5\text{ kJ}}\end{aligned}$$

Express K_f as a function of v_f:

$$K_f = \tfrac{1}{2}mv_f^2 \Rightarrow v_f = \sqrt{\frac{2K_f}{m}}$$

Substitute numerical values and evaluate v_f:

$$v_f = \sqrt{\frac{2(2.453\text{ kJ})}{100\text{ kg}}} = \boxed{7.0\text{ m/s}}$$

Scalar (Dot) Products

43 •• **[SSM]** (*a*) Given two nonzero vectors $\vec{A}$ and $\vec{B}$, show that if $|\vec{A}+\vec{B}| = |\vec{A}-\vec{B}|$, then $\vec{A}\perp\vec{B}$. (*b*) Given a vector $\vec{A} = 4\hat{i} - 3\hat{j}$, find a vector in the *xy* plane that is perpendicular to $\vec{A}$ and has a magnitude of 10. Is this the only vector that satisfies the specified requirements? Explain.

Picture the Problem We can use the definitions of the magnitude of a vector and the dot product to show that if $|\vec{A}+\vec{B}| = |\vec{A}-\vec{B}|$, then $\vec{A}\perp\vec{B}$.

(*a*) Express $|\vec{A}+\vec{B}|^2$:

$$|\vec{A}+\vec{B}|^2 = (\vec{A}+\vec{B})^2$$

Express $|\vec{A}-\vec{B}|$:

$$|\vec{A}-\vec{B}|^2 = (\vec{A}-\vec{B})^2$$

Equate these expressions to obtain:

$$(\vec{A}+\vec{B})^2 = (\vec{A}-\vec{B})^2$$

Expand both sides of the equation to obtain:

$$A^2 + 2\vec{A}\cdot\vec{B} + B^2 = A^2 - 2\vec{A}\cdot\vec{B} + B^2$$

Simplify to obtain:

$$4\vec{A}\cdot\vec{B} = 0 \text{ or } \vec{A}\cdot\vec{B} = 0$$

From the definition of the dot product we have:

$$\vec{A}\cdot\vec{B} = AB\cos\theta$$

where θ is the angle between $\vec{A}$ and $\vec{B}$.

Because neither $\vec{A}$ nor $\vec{B}$ is the zero vector: $\cos\theta = 0 \Rightarrow \theta = 90°$ and $\vec{A} \perp \vec{B}$.

(*b*) Let the required vector be $\vec{B}$. The condition that $\vec{A}$ and $\vec{B}$ be perpendicular is that:

$$\vec{A}\cdot\vec{B} = 0 \qquad (1)$$

Express $\vec{B}$ as:

$$\vec{B} = B_x\hat{i} + B_y\hat{j} \qquad (2)$$

Substituting for $\vec{A}$ and $\vec{B}$ in equation (1) yields:

$$\left(4\hat{i} - 3\hat{j}\right)\cdot\left(B_x\hat{i} + B_y\hat{j}\right) = 0$$

or

$$4B_x - 3B_y = 0 \qquad (3)$$

Because the magnitude of $\vec{B}$ is 10:

$$B_x^2 + B_y^2 = 100 \qquad (4)$$

Solving equation (3) for B_x and substituting in equation (4) yields:

$$\left(\tfrac{3}{4}B_y\right)^2 + B_y^2 = 100$$

Solve for B_y to obtain:

$$B_y = \pm 8$$

Substituting for B_y in equation (3) and solving for B_x yields:

$$B_x = \pm 6$$

Substitute for B_x and B_y in equation (2) to obtain:

$$\vec{B} = \boxed{\pm 6\hat{i} \pm 8\hat{j}}$$

No. Because of the plus-and-minus signs in our expression for $\vec{B}$, the required vector is not the only vector that satisfies the specified requirements.

47 ••• **[SSM]** When a particle moves in a circle that is centered at the origin and the magnitude of its position vector $\vec{r}$ is constant. (*a*) Differentiate $\vec{r}\cdot\vec{r} = r^2 =$ constant with respect to time to show that $\vec{v}\cdot\vec{r} = 0$, and therefore $\vec{v} \perp \vec{r}$. (*b*) Differentiate $\vec{v}\cdot\vec{r} = 0$ with respect to time and show that $\vec{a}\cdot\vec{r} + v^2 = 0$, and therefore $a_r = -v^2/r$. (*c*) Differentiate $\vec{v}\cdot\vec{v} = v^2$ with respect to time to show that $\vec{a}\cdot\vec{v} = dv/dt$, and therefore $a_t = dv/dt$.

Picture the Problem The rules for the differentiation of vectors are the same as those for the differentiation of scalars and scalar multiplication is commutative.

(*a*) Differentiate $\vec{r}\cdot\vec{r}=r^2=$ constant:

$$\frac{d}{dt}(\vec{r}\cdot\vec{r})=\vec{r}\cdot\frac{d\vec{r}}{dt}+\frac{d\vec{r}}{dt}\cdot\vec{r}=2\vec{v}\cdot\vec{r}$$

$$=\frac{d}{dt}(\text{constant})=0$$

Because $\vec{v}\cdot\vec{r}=0$:

$$\boxed{\vec{v}\perp\vec{r}}$$

(*b*) Differentiate $\vec{v}\cdot\vec{v}=v^2=$ constant with respect to time:

$$\frac{d}{dt}(\vec{v}\cdot\vec{v})=\vec{v}\cdot\frac{d\vec{v}}{dt}+\frac{d\vec{v}}{dt}\cdot\vec{v}=2\vec{a}\cdot\vec{v}$$

$$=\frac{d}{dt}(\text{constant})=0$$

Because $\vec{a}\cdot\vec{v}=0$:

$$\boxed{\vec{a}\perp\vec{v}}$$

The results of (*a*) and (*b*) tell us that $\vec{a}$ is perpendicular to $\vec{v}$ and parallel (or antiparallel to $\vec{r}$.

(*c*) Differentiate $\vec{v}\cdot\vec{r}=0$ with respect to time:

$$\frac{d}{dt}(\vec{v}\cdot\vec{r})=\vec{v}\cdot\frac{d\vec{r}}{dt}+\vec{r}\cdot\frac{d\vec{v}}{dt}$$

$$=v^2+\vec{r}\cdot\vec{a}=\frac{d}{dt}(0)=0$$

Because $v^2+\vec{r}\cdot\vec{a}=0$:

$$\boxed{\vec{r}\cdot\vec{a}=-v^2} \qquad (1)$$

Express a_r in terms of θ, where θ is the angle between $\vec{r}$ and $\vec{a}$:

$$a_r=a\cos\theta$$

Express $\vec{r}\cdot\vec{a}$:

$$\vec{r}\cdot\vec{a}=ra\cos\theta=ra_r$$

Substitute in equation (1) to obtain:

$$ra_r=-v^2$$

Solving for a_r yields:

$$a_r=\boxed{-\frac{v^2}{r}}$$

Work and Power

51 • **[SSM]** You are in charge of installing a small food-service elevator (called a *dumbwaiter* in the food industry) in a campus cafeteria. The elevator is connected by a pulley system to a motor, as shown in Figure 6-34. The motor raises and lowers the dumbwaiter. The mass of the dumbwaiter is 35 kg. In operation, it moves at a speed of 0.35 m/s upward, without accelerating (except for a brief initial period, which we can neglect, just after the motor is turned on). Electric motors typically have an efficiency of 78%. If you purchase a motor with an efficiency of 78%, what minimum power rating should the motor have? Assume that the pulleys are frictionless.

Picture the Problem Choose a coordinate system in which upward is the positive y direction. We can find P_{in} from the given information that $\boldsymbol{P}_{\text{out}} = 0.78\boldsymbol{P}_{\text{in}}$. We can express P_{out} as the product of the tension in the cable T and the constant speed v of the dumbwaiter. We can apply Newton's 2nd law to the dumbwaiter to express T in terms of its mass m and the gravitational field g.

Express the relationship between the motor's input and output power:

$$\boldsymbol{P}_{\text{out}} = 0.78\boldsymbol{P}_{\text{in}} \Rightarrow \boldsymbol{P}_{\text{in}} = 1.282\boldsymbol{P}_{\text{out}} \quad (1)$$

Express the power required to move the dumbwaiter at a constant speed v:

$$P_{\text{out}} = Tv$$

Apply $\sum F_y = ma_y$ to the dumbwaiter:

$$T - mg = ma_y$$

or, because $a_y = 0$,

$$T = mg$$

Substitute for T in equation (1) to obtain:

$$P_{\text{in}} = 1.282Tv = 1.282mgv$$

Substitute numerical values and evaluate P_{in}:

$$P_{\text{in}} = (1.282)(35\,\text{kg})(9.81\,\text{m/s}^2)(0.35\,\text{m/s}) = \boxed{0.15\,\text{kW}}$$

General Problems

63 •• **[SSM]** A single horizontal force in the $+x$ direction acts on a cart of mass m. The cart starts from rest at $x = 0$, and the speed of the cart increases with x as $v = Cx$, where C is a constant. (*a*) Find the force acting on the cart as a function of x. (*b*) Find the work done by the force in moving the cart from $x = 0$ to $x = x_1$.

Picture the Problem We can use the definition of work to obtain an expression for the position-dependent force acting on the cart. The work done on the cart can be calculated from its change in kinetic energy.

(*a*) Express the force acting on the cart in terms of the work done on it:

$$F(x) = \frac{dW}{dx} \quad (1)$$

Apply the work-kinetic energy theorem to the cart to obtain:

$W = \Delta K = K - K_i$

or, because the cart starts from rest,

$W = K$

Substitute for W in equation (1) to obtain:

$$F(x) = \frac{dK}{dx} = \frac{d}{dx}\left(\tfrac{1}{2}mv^2\right) = \frac{d}{dx}\left[\tfrac{1}{2}m(Cx)^2\right] = \boxed{mC^2x}$$

(*b*) The work done by this force changes the kinetic energy of the cart:

$$W = \Delta K = \tfrac{1}{2}mv_1^2 - \tfrac{1}{2}mv_0^2 = \tfrac{1}{2}mv_1^2 - 0 = \tfrac{1}{2}m(Cx_1)^2 = \boxed{\tfrac{1}{2}mC^2x_1^2}$$

71 ••• **[SSM]** A force acting on a particle in the xy plane at coordinates (x, y) is given by $\vec{F} = (F_0/r)(y\hat{i} - x\hat{j})$, where F_0 is a positive constant and r is the distance of the particle from the origin. (*a*) Show that the magnitude of this force is F_0 and that its direction is perpendicular to $\boldsymbol{r} = x\hat{i} + y\hat{j}$. (*b*) Find the work done by this force on a particle that moves once around a circle of radius 5.0 m that is centered at the origin.

Picture the Problem (*a*) We can use the definition of the magnitude of vector to show that the magnitude of $\vec{F}$ is F_0 and the definition of the scalar product to show that its direction is perpendicular to $\vec{r}$. (*b*) The work done as the particle moves in a circular path can be found from its definition.

(*a*) Express the magnitude of $\vec{F}$:

$$\left|\vec{F}\right| = \sqrt{F_x^2 + F_y^2} = \sqrt{\left(\frac{F_0}{r}y\right)^2 + \left(-\frac{F_0}{r}x\right)^2} = \frac{F_0}{r}\sqrt{x^2 + y^2}$$

Because $r = \sqrt{x^2 + y^2}$:

$$\left|\vec{\mathbf{F}}\right| = \frac{F_0}{r}\sqrt{x^2 + y^2} = \frac{F_0}{r}r = \boxed{F_0}$$

Form the scalar product of $\vec{F}$ and $\vec{r}$:

$$\vec{F}\cdot\vec{r} = \left(\frac{F_0}{r}\right)\left(y\hat{i} - x\hat{j}\right)\cdot\left(x\hat{i} + y\hat{j}\right) = \left(\frac{F_0}{r}\right)(xy - xy) = 0$$

Because $\vec{F}\cdot\vec{r}=0$: $\boxed{\vec{F}\perp\vec{r}}$

(*b*) The work done in an angular displacement from θ_1 to θ_2 is given by:

$$W_{1\,\text{rev}}=\int_{\theta_1}^{\theta_2}\vec{F}\cdot d\vec{s}$$
$$=\int_{\theta_1}^{\theta_2}\left(\frac{F_0}{r}\right)\left(y\,\hat{i}-x\,\hat{j}\right)\cdot d\vec{s}$$
$$=\frac{F_0}{r}\int_{\theta_1}^{\theta_2}\left(r\sin\theta\,\hat{i}-r\cos\theta\,\hat{j}\right)\cdot d\vec{s}$$
$$=F_0\int_{\theta_1}^{\theta_2}\left(\sin\theta\,\hat{i}-\cos\theta\,\hat{j}\right)\cdot d\vec{s}$$

Express the radial vector $\vec{r}$ in terms of its magnitude r and the angle θ it makes with the positive x axis:

$$\vec{r}=r\cos\theta\,\hat{i}+r\sin\theta\,\hat{j}$$

The differential of $\vec{r}$ is tangent to the circle (you can easily convince yourself of this by forming the dot product of $\vec{r}$ and $d\vec{r}$) and so we can use it as $d\vec{s}$ in our expression for the work done by $\vec{F}$ in one revolution:

$$d\vec{r}=d\vec{s}=-r\sin\theta\,d\theta\,\hat{i}+r\cos\theta\,d\theta\,\hat{j}$$
$$=\left(-r\sin\theta\,\hat{i}+r\cos\theta\,\hat{j}\right)d\theta$$

Substitute for $d\vec{s}$, simplifying, and integrate to obtain:

$$W_{1\,\text{rev,ccw}}=F_0\int_0^{2\pi}\left(\sin\theta\,\hat{i}-\cos\theta\,\hat{j}\right)\cdot\left(-r\sin\theta\,\hat{i}+r\cos\theta\,\hat{j}\right)d\theta$$
$$=-rF_0\int_0^{2\pi}\left(\sin^2\theta+\cos^2\theta\right)d\theta=-rF_0\int_0^{2\pi}d\theta=-rF_0\theta\Big]_0^{2\pi}=-2\pi rF_0$$

Substituting the numerical value of r yields:

$$W_{1\text{rev, ccw}}=-2\pi(5.0\text{ m})F_0$$
$$=\boxed{(-10\pi\text{ m})F_0}$$

If the rotation is clockwise, the integral becomes:

$$W_{1\,\text{rev,cw}}=-rF_0\theta\Big]_{2\pi}^{0}=2\pi rF_0$$

Substituting the numerical value of r yields:

$$W_{1\text{rev, cw}}=2\pi(5.0\text{ m})F_0=\boxed{(10\pi\text{ m})F_0}$$

Chapter 7
Conservation of Energy

Conceptual Problems

1 • **[SSM]** Two cylinders of unequal mass are connected by a massless cord that passes over a frictionless peg (Figure 7-34). After the system is released from rest, which of the following statements are true? (U is the gravitational potential energy and K is the kinetic energy of the system.) (*a*) $\Delta U < 0$ and $\Delta K > 0$, (*b*) $\Delta U = 0$ and $\Delta K > 0$, (*c*) $\Delta U < 0$ and $\Delta K = 0$, (*d*) $\Delta U = 0$ and $\Delta K = 0$, (*e*) $\Delta U > 0$ and $\Delta K < 0$.

Determine the Concept Because the peg is frictionless, mechanical energy is conserved as this system evolves from one state to another. The system moves and so we know that $\Delta K > 0$. Because $\Delta K + \Delta U = constant$, $\Delta U < 0$. $\boxed{(a)}$ is correct.

Estimation and Approximation

15 •• **[SSM]** Assume that your maximum metabolic rate (the maximum rate at which your body uses its chemical energy) is 1500 W (about 2.7 hp). Assuming a 40 percent efficiency for the conversion of chemical energy into mechanical energy, estimate the following: (*a*) the shortest time you could run up four flights of stairs if each flight is 3.5 m high, (*b*) the shortest time you could climb the Empire State Building (102 stories high) using your Part (*a*) result. Comment on the feasibility of you actually achieving Part (*b*) result.

Picture the Problem The rate at which you expend energy, that is do work, is defined as *power* and is the ratio of the work done to the time required to do the work.

(*a*) Relate the rate at which you can expend energy to the work done in running up the four flights of stairs:

$$\varepsilon P = \frac{\Delta W}{\Delta t} \Rightarrow \Delta t = \frac{\Delta W}{\varepsilon P}$$

where e is the efficiency for the conversion of chemical energy into mechanical energy.

The work you do in climbing the stairs increases your gravitational potential energy:

$$\Delta W = mgh$$

Substitute for ΔW to obtain:

$$\Delta t = \frac{mgh}{\varepsilon P} \qquad (1)$$

Assuming that your mass is 70 kg, substitute numerical values in equation (1) and evaluate Δt:

$$\Delta t = \frac{(70\,\text{kg})(9.81\,\text{m/s}^2)(4\times 3.5\,\text{m})}{(0.40)(1500\,\text{W})} \approx \boxed{16\,\text{s}}$$

(*b*) Substituting numerical values in equation (1) yields:

$$\Delta t = \frac{(70\,\text{kg})(9.81\,\text{m/s}^2)(102\times 3.5\,\text{m})}{(0.40)(1500\,\text{W})} = 409\,\text{s} \approx \boxed{6.8\,\text{min}}$$

The time of about 6.8 min is clearly not reasonable. The fallacy is that you cannot do work at the given rate of 250 W for more than very short intervals of time.

17 •• **[SSM]** The chemical energy released by burning a gallon of gasoline is approximately 1.3×10^5 kJ. Estimate the total energy used by all of the cars in the United States during the course of one year. What fraction does this represent of the total energy use by the United States in one year (currently about 5×10^{20} J)?

Picture the Problem There are about 3×10^8 people in the United States. On the assumption that the average family has 4 people in it and that they own two cars, we have a total of 1.5×10^8 automobiles on the road (excluding those used for industry). We'll assume that each car uses about 15 gal of fuel per week.

Calculate, based on the assumptions identified above, the total annual consumption of energy derived from gasoline:

$$\left(1.5\times 10^8\,\text{auto}\right)\left(15\frac{\text{gal}}{\text{auto}\cdot\text{week}}\right)\left(52\frac{\text{weeks}}{y}\right)\left(1.3\times 10^5\,\frac{\text{kJ}}{\text{gal}}\right) = \boxed{1.5\times 10^{19}\,\text{J/y}}$$

Express this rate of energy use as a fraction of the total annual energy use by the United States:

$$\frac{1.5\times 10^{19}\,\text{J/y}}{5\times 10^{20}\,\text{J/y}} \approx \boxed{3\%}$$

Force, Potential Energy, and Equilibrium

25 •• **[SSM]** The force F_x is associated with the potential-energy function $U = C/x$, where C is a positive constant. (*a*) Find the force F_x as a function of x. (*b*) Is this force directed toward the origin or away from it in the region $x > 0$? Repeat the question for the region $x < 0$. (*c*) Does the potential energy U increase or decrease as x increases in the region $x > 0$? (*d*) Answer Parts (*b*) and (*c*) where C is a negative constant.

Picture the Problem F_x is defined to be the negative of the derivative of the potential-energy function with respect to x, that is $F_x = -dU/dx$. Consequently, given U as a function of x, we can find F_x by differentiating U with respect to x.

(*a*) Evaluate $F_x = -\frac{dU}{dx}$:

$$F_x = -\frac{d}{dx}\left(\frac{C}{x}\right) = \boxed{\frac{C}{x^2}}$$

(*b*) Because $C > 0$, if $x > 0$, F_x is positive and $\vec{F}$ points away from the origin. If $x < 0$, F_x is still positive and $\vec{F}$ points toward the origin.

(*c*) Because U is inversely proportional to x and $C > 0$, $U(x)$ decreases with increasing x.

(*d*) When $C < 0$, if $x > 0$, F_x is negative and $\vec{F}$ points toward the origin. If $x < 0$, F_x is negative and $\vec{F}$ points away from the origin.

Because U is inversely proportional to x and $C < 0$, $U(x)$ becomes less negative as x increases and $U(x)$ increases with increasing x.

29 •• **[SSM]** The potential energy of an object constrained to the x axis is given by $U(x) = 8x^2 - x^4$, where U is in joules and x is in meters. (*a*) Determine the force F_x associated with this potential energy function. (*b*) Assuming no other forces act on the object, at what positions is this object in equilibrium? (*c*) Which of these equilibrium positions are stable and which are unstable?

Picture the Problem F_x is defined to be the negative of the derivative of the potential-energy function with respect to x, that is $F_x = -dU/dx$. Consequently, given U as a function of x, we can find F_x by differentiating U with respect to x. To determine whether the object is in stable or unstable equilibrium at a given point, we'll evaluate d^2U/dx^2 at the point of interest.

(*a*) Evaluate the negative of the derivative of U with respect to x:

$$F_x = -\frac{dU}{dx} = -\frac{d}{dx}\left(8x^2 - x^4\right) = 4x^3 - 16x = \boxed{4x(x+2)(x-2)}$$

(*b*) The object is in equilibrium wherever $F_{net} = F_x = 0$:

$4x(x+2)(x-2) = 0 \Rightarrow$ the equilibrium points are $\boxed{x = -2\,\text{m}, 0, \text{and } 2\,\text{m}.}$

(*c*) To decide whether the equilibrium at a particular point is stable or unstable, evaluate the 2nd derivative of the potential energy function at the point of interest:

$$\frac{d^2U}{dx^2} = \frac{d}{dx}\left(16x - 4x^3\right) = 16 - 12x^2$$

Evaluating d^2U/dx^2 at $x = -2$ m, 0 and $x = 2$ m yields the following results:

x, m	d^2U/dx^2	Equilibrium
−2	−32	Unstable
0	16	Stable
2	−32	Unstable

Remarks: You could also decide whether the equilibrium positions are stable or unstable by plotting *F*(*x*) and examining the curve at the equilibrium positions.

33 •• **[SSM]** A straight rod of negligible mass is mounted on a frictionless pivot, as shown in Figure 7-38. Blocks have masses m_1 and m_2 are attached to the rod at distances ℓ_1 and ℓ_2. (*a*) Write an expression for the gravitational potential energy of the blocks-Earth system as a function of the angle θ made by the rod and the horizontal. (*b*) For what angle θ is this potential energy a minimum? Is the statement "systems tend to move toward a configuration of minimum potential energy" consistent with your result? (*c*) Show that if $m_1\ell_1 = m_2\ell_2$, the potential energy is the same for all values of θ. (When this holds, the system will balance at any angle θ. This result is known as *Archimedes' law of the lever*.)

Picture the Problem The gravitational potential energy of this system of two objects is the sum of their individual potential energies and is dependent on an arbitrary choice of where, or under what condition(s), the gravitational potential energy is zero. The best choice is one that simplifies the mathematical details of the expression for *U*. In this problem let's choose $U = 0$ where $\theta = 0$.

(*a*) Express *U* for the 2-object system as the sum of their gravitational potential energies; noting that because the object whose mass is m_2 is above the position we have chosen for $U = 0$, its potential energy is positive while that of the object whose mass is m_1 is negative:

$$\begin{aligned} U(\theta) &= U_1 + U_2 \\ &= m_2 g\ell_2 \sin\theta - m_1 g\ell_1 \sin\theta \\ &= \boxed{(m_2\ell_2 - m_1\ell_1) g \sin\theta} \end{aligned}$$

(*b*) Differentiate *U* with respect to θ and set this derivative equal to zero to identify extreme values:

$$\frac{dU}{d\theta} = (m_2\ell_2 - m_1\ell_1)g\cos\theta = 0$$

from which we can conclude that $\cos\theta = 0$ and $\theta = \cos^{-1}0$.

To be physically meaningful, $-\pi/2 \le \theta \le \pi/2$. Hence:

$$\theta = \pm\pi/2$$

Express the 2nd derivative of *U* with respect to θ and evaluate this derivative at $\theta = \pm\pi/2$:

$$\frac{d^2U}{d\theta^2} = -(m_2\ell_2 - m_1\ell_1)g\sin\theta$$

If we assume, in the expression for *U* that we derived in (*a*), that $m_2\ell_2 - m_1\ell_1 > 0$, then $U(\theta)$ is a sine function and, in the interval of interest,

$$-\pi/2 \le \theta \le \pi/2,$$

takes on its minimum value when $\theta = -\pi/2$:

$$\left.\frac{d^2U}{d\theta^2}\right|_{-\pi/2} > 0 \text{ and}$$

$$\boxed{U \text{ is a minimum at } \theta = -\pi/2}$$

$$\left.\frac{d^2U}{d\theta^2}\right|_{\pi/2} < 0 \text{ and}$$

$$\boxed{U \text{ is a maximum at } \theta = \pi/2}$$

(c) If $m2\ell2 = m1\ell1$, then:

$$m_1\ell_1 - m_2\ell_2 = 0$$

and

$$\boxed{U = 0 \text{ independent of } \theta.}$$

Remarks: An alternative approach to establishing that *U* is a maximum at $\theta = \pi/2$ is to plot its graph and note that, in the interval of interest, *U* is concave downward with its maximum value at $\theta = \pi/2$. Similarly, it can be shown that *U* is a minimum at $\theta = -\pi/2$ (Part (*b*)).

The Conservation of Mechanical Energy

41 • **[SSM]** A 16-kg child on a 6.0-m-long playground swing moves with a speed of 3.4 m/s when the swing seat passes through its lowest point. What is the angle that the swing makes with the vertical when the swing is at its highest point? Assume that the effects due to air resistance are negligible, and assume that the child is not pumping the swing.

Picture the Problem Let the system consist of the earth and the child. Then $W_{ext} = 0$. Choose $U_g = 0$ at the child's lowest point as shown in the diagram to the right. Then the child's initial energy is entirely kinetic and its energy when it is at its highest point is entirely gravitational potential. We can determine h from conservation of mechanical energy and then use trigonometry to determine θ.

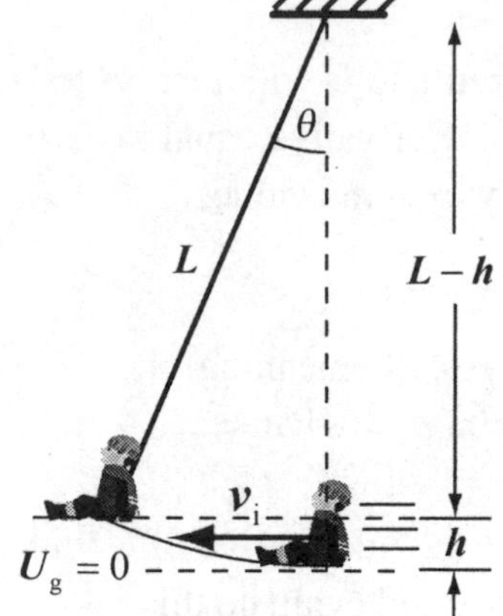

Using the diagram, relate θ to h and L:

$$\theta = \cos^{-1}\left(\frac{L-h}{L}\right) = \cos^{-1}\left(1-\frac{h}{L}\right) \quad (1)$$

Apply conservation of mechanical energy to the system to obtain:

$$W_{ext} = \Delta K + \Delta U = 0$$

or, because $K_f = U_{g,i} = 0$,

$$-K_i + U_{g,f} = 0$$

Substituting for K_i and $U_{g,f}$ yields:

$$-\tfrac{1}{2}mv_i^2 + mgh = 0 \Rightarrow h = \frac{v_i^2}{2g}$$

Substitute for h in equation (1) to obtain:

$$\theta = \cos^{-1}\left(1-\frac{v_i^2}{2gL}\right)$$

Substitute numerical values and evaluate θ:

$$\theta = \cos^{-1}\left(1-\frac{(3.4\,\text{m/s})^2}{2(9.81\,\text{m/s}^2)(6.0\,\text{m})}\right) = \boxed{26^\circ}$$

45 •• **[SSM]** A ball at the end of a string moves in a vertical circle with constant mechanical energy E. What is the difference between the tension at the bottom of the circle and the tension at the top?

Picture the Problem The diagram represents the ball traveling in a circular path with constant energy. U_g has been chosen to be zero at the lowest point on the circle and the superimposed free-body diagrams show the forces acting on the ball at the top (T) and bottom (B) of the circular path. We'll apply Newton's 2nd law to the ball at the top and bottom of its path to obtain a relationship between T_T and T_B and conservation of mechanical energy to relate the speeds of the ball at these two locations.

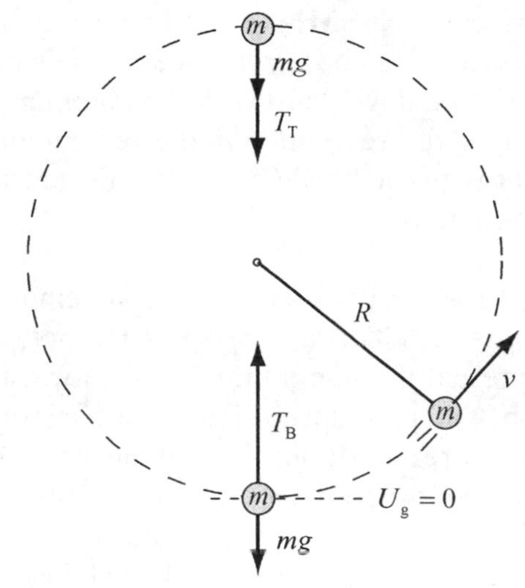

Apply $\sum F_{radial} = ma_{radial}$ to the ball at the bottom of the circle and solve for T_B:

$$T_B - mg = m\frac{v_B^2}{R}$$

and

$$T_B = mg + m\frac{v_B^2}{R} \qquad (1)$$

Apply $\sum F_{radial} = ma_{radial}$ to the ball at the top of the circle and solve for T_T:

$$T_T + mg = m\frac{v_T^2}{R}$$

and

$$T_T = -mg + m\frac{v_T^2}{R} \qquad (2)$$

Subtract equation (2) from equation (1) to obtain:

$$T_B - T_T = mg + m\frac{v_B^2}{R} - \left(-mg + m\frac{v_T^2}{R}\right) = m\frac{v_B^2}{R} - m\frac{v_T^2}{R} + 2mg \qquad (3)$$

Using conservation of mechanical energy, relate the energy of the ball at the bottom of its path to its mechanical energy at the top of the circle:

$$\tfrac{1}{2}mv_B^2 = \tfrac{1}{2}mv_T^2 + mg(2R)$$

or

$$m\frac{v_B^2}{R} - m\frac{v_T^2}{R} = 4mg$$

Substituting in equation (3) yields:

$$T_B - T_T = \boxed{6mg}$$

55 •• **[SSM]** A pendulum consists of a string of length L and a bob of mass m. The bob is rotated until the string is horizontal. The bob is then projected downward with the minimum initial speed needed to enable the bob to make a full revolution in the vertical plane. (*a*) What is the maximum kinetic energy of the bob? (*b*) What is the tension in the string when the kinetic energy is maximum?

Picture the Problem Let the system consist of the earth and pendulum bob. Then $W_{ext} = 0$. Choose $U_g = 0$ at the bottom of the circle and let points 1, 2 and 3 represent the bob's initial point, lowest point and highest point, respectively. The bob will gain speed and kinetic energy until it reaches point 2 and slow down until it reaches point 3; so it has its maximum kinetic energy when it is at point 2. We can use Newton's 2nd law at points 2 and 3 in conjunction with conservation of mechanical energy to find the maximum kinetic energy of the bob and the tension in the string when the bob has its maximum kinetic energy.

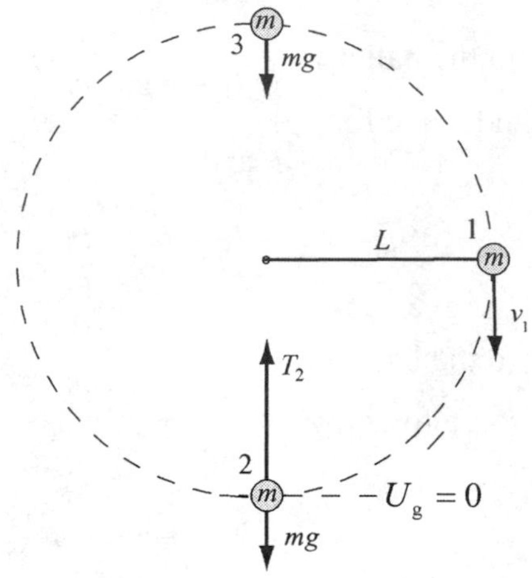

(*a*) Apply $\sum F_{radial} = ma_{radial}$ to the bob at the top of the circle and solve for v_3^2:

$$mg = m\frac{v_3^2}{L} \Rightarrow v_3^2 = gL$$

Apply conservation of mechanical energy to the system to express the relationship between K_2, K_3 and U_3:

$$K_3 - K_2 + U_3 - U_2 = 0$$

or, because $U_2 = 0$,

$$K_3 - K_2 + U_3 = 0$$

Solving for K_2 yields:

$$K_2 = K_{max} = K_3 + U_3$$

Substituting for K_3 and U_3 yields:

$$K_{max} = \tfrac{1}{2}mv_3^2 + mg(2L)$$

Substitute for v_3^2 and simplify to obtain:

$$K_{max} = \tfrac{1}{2}m(gL) + 2mgL = \boxed{\tfrac{5}{2}mgL}$$

(*b*) Apply $\sum F_{\text{radial}} = ma_c$ to the bob at the bottom of the circle and solve for T_2:

$$F_{\text{net}} = T_2 - mg = m\frac{v_2^2}{L}$$

and

$$T_2 = mg + m\frac{v_2^2}{L} \qquad (1)$$

Use conservation of mechanical energy to relate the energies of the bob at points 2 and 3 and solve for K_2:

$$K_3 - K_2 + U_3 - U_2 = 0 \text{ where } U_2 = 0$$

and

$$K_2 = K_3 + U_3 = \tfrac{1}{2}mv_3^2 + mg(2L)$$

Substitute for v_3^2 and K_2 to obtain:

$$\tfrac{1}{2}mv_2^2 = \tfrac{1}{2}m(gL) + mg(2L) \Rightarrow v_2^2 = 5gL$$

Substitute for v_2^2 in equation (1) and simplify to obtain:

$$T_2 = mg + m\frac{5gL}{L} = \boxed{6mg}$$

59 ••• **[SSM]** A pendulum is suspended from the ceiling and attached to a spring fixed to the floor directly below the pendulum support (Figure 7-48). The mass of the pendulum bob is *m*, the length of the pendulum is *L*, and the force constant is *k*. The unstressed length of the spring is *L*/2 and the distance between the floor and ceiling is 1.5*L*. The pendulum is pulled aside so that it makes an angle θ with the vertical and is then released from rest. Obtain an expression for the speed of the pendulum bob as the bob passes through a point directly below the pendulum support.

Picture the Problem Choose $U_g = 0$ at point 2, the lowest point of the bob's trajectory and let the system consist of the earth, ceiling, spring, and pendulum bob. Given this choice, there are no external forces doing work to change the energy of the system. The bob's initial energy is partially gravitational potential and partially potential energy stored in the stretched spring. As the bob swings down to point 2 this energy is transformed into kinetic energy. By equating these energies, we can derive an expression for the speed of the bob at point 2.

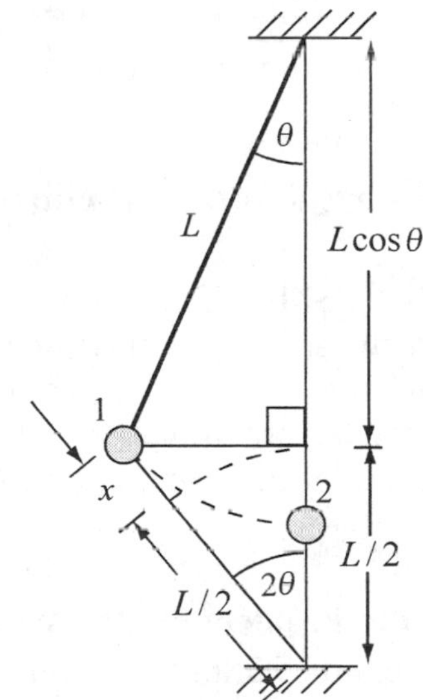

Apply conservation of mechanical energy to the system as the pendulum bob swings from point 1 to point 2:

$$W_{ext} = \Delta K + \Delta U_g + \Delta U_s = 0$$

or, because $K_1 = U_{g,2} = U_{s,2} = 0$,

$$K_2 - U_{g,1} - U_{s,1} = 0$$

Substituting for K_2, $U_{g,1}$, and $U_{s,2}$ yields:

$$\tfrac{1}{2}mv_2^2 - mgL(1-\cos\theta) - \tfrac{1}{2}kx^2 = 0 \quad (1)$$

Apply the Pythagorean theorem to the lower triangle in the diagram to obtain:

$$\left(x + \tfrac{1}{2}L\right)^2 = L^2\left[\sin^2\theta + \left(\tfrac{3}{2} - \cos\theta\right)^2\right] = L^2\left[\sin^2\theta + \tfrac{9}{4} - 3\cos\theta + \cos^2\theta\right]$$
$$= L^2\left(\tfrac{13}{4} - 3\cos\theta\right)$$

Take the square root of both sides of the equation to obtain:

$$x + \tfrac{1}{2}L = L\sqrt{\left(\tfrac{13}{4} - 3\cos\theta\right)}$$

Solving for x yields:

$$x = L\left[\sqrt{\left(\tfrac{13}{4} - 3\cos\theta\right)} - \tfrac{1}{2}\right]$$

Substitute for x in equation (1) to obtain:

$$\tfrac{1}{2}mv_2^2 = \tfrac{1}{2}kL^2\left[\sqrt{\left(\tfrac{13}{4} - 3\cos\theta\right)} - \tfrac{1}{2}\right]^2 + mgL(1-\cos\theta)$$

Solving for v_2 yields:

$$v_2 = \boxed{L\sqrt{2\frac{g}{L}(1-\cos\theta) + \frac{k}{m}\left(\sqrt{\tfrac{13}{4} - 3\cos\theta} - \tfrac{1}{2}\right)^2}}$$

Total Energy and Non-conservative Forces

65 •• **[SSM]** The 2.0-kg block in Figure 7-49 slides down a frictionless curved ramp, starting from rest at a height of 3.0 m. The block then slides 9.0 m on a rough horizontal surface before coming to rest. (*a*) What is the speed of the block at the bottom of the ramp? (*b*) What is the energy dissipated by friction? (*c*) What is the coefficient of kinetic friction between the block and the horizontal surface?

Picture the Problem Let the system include the block, the ramp and horizontal surface, and the earth. Given this choice, there are no external forces acting that will change the energy of the system. Because the curved ramp is frictionless, mechanical energy is conserved as the block slides down it. We can calculate its speed at the bottom of the ramp by using conservation of energy. The potential energy of the block at the top of the ramp or, equivalently, its kinetic energy at the bottom of the ramp is converted into thermal energy during its slide along the horizontal surface.

(*a*) Let the numeral 1 designate the initial position of the block and the numeral 2 its position at the foot of the ramp. Choose $U_g = 0$ at point 2 and use conservation of energy to relate the block's potential energy at the top of the ramp to its kinetic energy at the bottom:

$$W_{ext} = \Delta E_{mech} + \Delta E_{therm}$$

or, because $W_{ext} = K_i = U_f = \Delta E_{therm} = 0$,

$$0 = \tfrac{1}{2}mv_2^2 - mg\Delta h = 0 \Rightarrow v_2 = \sqrt{2g\Delta h}$$

Substitute numerical values and evaluate v_2:

$$v_2 = \sqrt{2(9.81\,\text{m/s}^2)(3.0\,\text{m})} = 7.67\,\text{m/s}$$
$$= \boxed{7.7\,\text{m/s}}$$

(*b*) The energy dissipated by friction is responsible for changing the thermal energy of the system:

$$W_f + \Delta K + \Delta U = \Delta E_{therm} + \Delta K + \Delta U$$
$$= 0$$

Because ΔK = 0 for the slide:

$$W_f = -\Delta U = -(U_2 - U_1) = U_1$$

Substituting for U1 yields:

$$W_f = mg\Delta h$$

Substitute numerical values and evaluate U1:

$$W_f - (2.0\,\text{kg})(9.81\,\text{m/s}^2)(3.0\,\text{m}) = 58.9\,\text{J}$$
$$= \boxed{59\,\text{J}}$$

(c) The energy dissipated by friction is given by:

$$\Delta E_{therm} = f\Delta s = \mu_k mg\Delta x$$

Solving for μk yields:

$$\mu_k = \frac{\Delta E_{therm}}{mg\Delta x}$$

Substitute numerical values and evaluate μk:

$$\mu_k = \frac{58.9\,\text{J}}{(2.0\,\text{kg})(9.81\,\text{m/s}^2)(9.0\,\text{m})}$$
$$= \boxed{0.33}$$

69 •• **[SSM]** The initial speed of a 2.4-kg box traveling up a plane inclined 37° to the horizontal is 3.8 m/s. The coefficient of kinetic friction between the box and the plane is 0.30. (*a*) How far along the incline does the box travel before coming to a stop? (*b*) What is its speed when it has traveled half the distance found in Part (*a*)?

Picture the Problem The box will slow down and stop due to the dissipation of thermal energy. Let the system be the earth, the box, and the inclined plane and apply the work-energy theorem with friction. With this choice of the system, there are no external forces doing work to change the energy of the system. The pictorial representation shows the forces acting on the box when it is moving up the incline.

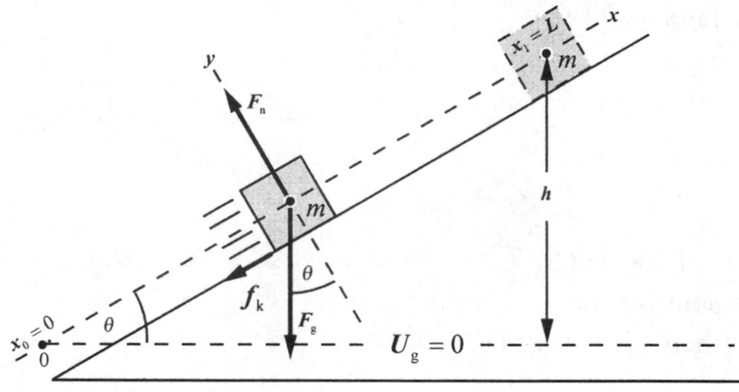

(*a*) Apply the work-energy theorem with friction to the system:

$$W_{ext} = \Delta E_{mech} + \Delta E_{therm} = \Delta K + \Delta U + \Delta E_{therm}$$

Substitute for ΔK, ΔU, and ΔE_{therm} to obtain:

$$0 = \tfrac{1}{2}mv_1^2 - \tfrac{1}{2}mv_0^2 + mgh + \mu_k F_n L \quad (1)$$

Referring to the free-body diagram, relate the normal force to the weight of the box and the angle of the incline:

$$F_n = mg\cos\theta$$

Relate h to the distance L along the incline:

$$h = L\sin\theta$$

Substitute in equation (1) to obtain:

$$\mu_k mgL\cos\theta + \tfrac{1}{2}mv_1^2 - \tfrac{1}{2}mv_0^2 + mgL\sin\theta = 0 \quad (2)$$

Solving equation (2) for L yields:

$$L=\frac{v_0^2}{2g(\mu_k\cos\theta+\sin\theta)}$$

Substitute numerical values and evaluate L:

$$L=\frac{(3.8\,\text{m/s})^2}{2(9.81\,\text{m/s}^2)[(0.30)\cos37°+\sin37°]}=0.8747\,\text{m}=\boxed{0.87\,\text{m}}$$

(*b*) Let $v_{\frac{1}{2}L}$ represent the box's speed when it is halfway up the incline. Then equation (2) becomes:

$$\mu_k mg\left(\tfrac{1}{2}L\right)\cos\theta+\tfrac{1}{2}mv_{\frac{1}{2}L}^2-\tfrac{1}{2}mv_0^2+mg\left(\tfrac{1}{2}L\right)\sin\theta=0$$

Solving for $v_{\frac{1}{2}L}$ yields :

$$v_{\frac{1}{2}L}=\sqrt{v_0^2-gL(\sin\theta+\mu_k\cos\theta)}$$

Substitute numerical values and evaluate $v_{\frac{1}{2}L}$:

$$v_f=\sqrt{(3.8\,\text{m/s})^2-(9.81\,\text{m/s}^2)(0.8747\,\text{m})[\sin37°+(0.30)\cos37°]}=\boxed{2.7\,\text{m/s}}$$

Mass and Energy

75 • **[SSM]** You are designing the fuel requirements for a small fusion electric-generating plant. Assume 33% conversion to electric energy. For the deuterium–tritium (D–T) fusion reaction in Example 7-18, calculate the number of reactions per second that are necessary to generate 1.00 kW of electric power.

Picture the Problem The number of reactions per second is given by the ratio of the power generated to the energy released per reaction. The number of reactions that must take place to produce a given amount of energy is the ratio of the energy per second (power) to the energy released per second.

In Example 7-18 it is shown that the energy per reaction is 17.59 MeV. Convert this energy to joules:

$$17.59\,\text{MeV}=(17.59\,\text{MeV})(1.602\times10^{-19}\,\text{J/eV})=28.18\times10^{-13}\,\text{J}$$

Assuming 33% conversion to electric energy, the number of reactions per second is:

$$\frac{1000\,\text{J/s}}{(0.33)(28.18\times10^{-13}\ \text{J/reaction})}\approx\boxed{1.1\times10^{15}\ \text{reactions/s}}$$

General Problems

85 • **[SSM]** You are in charge of "solar-energizing" your grandfather's farm. At the farm's location, an average of 1.0 kW/m^2 reaches the surface during the daylight hours on a clear day. If this could be converted at 25% efficiency to electric energy, how large a collection area would you need to run a 4.0-hp irrigation water pump during the daylight hours?

Picture the Problem The solar constant is the average energy per unit area and per unit time reaching the upper atmosphere. This physical quantity can be thought of as the power per unit area and is known as *intensity*.

Letting I_{surface} represent the intensity of the solar radiation at the surface of the earth, express I_{surface} as a function of power and the area on which this energy is incident:

$$\varepsilon I_{\text{surface}}=\frac{P}{A}\Rightarrow A=\frac{P}{\varepsilon I_{\text{surface}}}$$

where ε is the efficiency of conversion to electric energy.

Substitute numerical values and evaluate A:

$$A=\frac{4.0\ \text{hp}\times\dfrac{746\ \text{W}}{\text{hp}}}{(0.25)(1.0\,\text{kW/m}^2)}=\boxed{12\ \text{m}^2}$$

93 •• **[SSM]** In a volcanic eruption, a 2-kg piece of porous volcanic rock is thrown straight upward with an initial speed of 40 m/s. It travels upward a distance of 50 m before it begins to fall back to Earth. (*a*) What is the initial kinetic energy of the rock? (*b*) What is the increase in thermal energy due to air resistance during ascent? (*c*) If the increase in thermal energy due to air resistance on the way down is 70% of that on the way up, what is the speed of the rock when it returns to its initial position?

Picture the Problem Let the system consist of the earth, rock and air. Given this choice, there are no external forces to do work on the system and $W_{\text{ext}}=0$. Choose $U_g=0$ to be where the rock begins its upward motion. The initial kinetic energy of the rock is partially transformed into potential energy and partially dissipated by air resistance as the rock ascends. During its descent, its potential energy is partially transformed into kinetic energy and partially dissipated by air resistance.

(*a*) The initial kinetic energy of the rock is given by:

$$K_i = \tfrac{1}{2}mv_i^2$$

Substitute numerical values and evaluate K_i:

$$K_i = \tfrac{1}{2}(2.0\,\text{kg})(40\,\text{m/s})^2 = \boxed{1.6\,\text{kJ}}$$

(*b*) Apply the work-energy theorem with friction to relate the energies of the system as the rock ascends:

$$\Delta K + \Delta U + \Delta E_{\text{therm}} = 0$$

or, because $K_f = 0$,

$$-K_i + \Delta U + \Delta E_{\text{therm}} = 0$$

Solving for ΔE_{therm} yields:

$$\Delta E_{\text{therm}} = K_i - \Delta U$$

Substitute numerical values and evaluate ΔE_{therm}:

$$\Delta E_{\text{therm}} = 1.6\,\text{kJ} - (2.0\,\text{kg})(9.81\,\text{m/s}^2)(50\,\text{m}) = 0.619\,\text{kJ} = \boxed{0.6\,\text{kJ}}$$

(*c*) Apply the work-energy theorem with friction to relate the energies of the system as the rock descends:

$$\Delta K + \Delta U + 0.70\Delta E_{\text{therm}} = 0$$

Because $K_i = U_f = 0$:

$$K_f - U_i + 0.70\Delta E_{\text{therm}} = 0$$

Substitute for the energies to obtain:

$$\tfrac{1}{2}mv_f^2 - mgh + 0.70\Delta E_{\text{therm}} = 0$$

Solve for v_f to obtain:

$$v_f = \sqrt{2gh - \frac{1.40\Delta E_{\text{therm}}}{m}}$$

Substitute numerical values and evaluate v_f:

$$v_f = \sqrt{2(9.81\,\text{m/s}^2)(50\,\text{m}) - \frac{1.40(0.619\,\text{kJ})}{2.0\,\text{kg}}} = \boxed{23\,\text{m/s}}$$

95 •• **[SSM]** A block of mass *m* is suspended from a wall bracket by a spring and is free to move vertically (Figure 7-54). The +*y* direction is downward and the origin is at the position of the block when the spring is unstressed. (*a*) Show that the potential energy as a function of position may be expressed as $U = \tfrac{1}{2}ky^2 - mgy$, (*b*) Using a **spreadsheet** program or graphing calculator, make a graph of *U* as a function of *y* with $k = 2$ N/m and $mg = 1$ N. (*c*) Explain how this graph shows that there is a position of stable equilibrium for a positive value of y. Using the Part (*a*) expression for U, determine (symbolically) the value of y when the block is at its equilibrium position. (*d*) From the expression for U, find the net force acting on m at any position y. (*e*) The block is released from rest with the spring unstressed; if there is no friction, what is the maximum value of y that will be reached by the mass? Indicate y_{max} on your graph/spreadsheet.

Picture the Problem Given the potential energy function as a function of y, we can find the net force acting on a given system from $F = -dU/dy$. The maximum extension of the spring; that is, the lowest position of the mass on its end, can be found by applying the work-energy theorem. The equilibrium position of the system can be found by applying the work-energy theorem with friction … as can the amount of thermal energy produced as the system oscillates to its equilibrium position. In Part (*c*), setting dU/dy equal to zero and solving the resulting equation for y will yield the value of y when the block is in its equilibrium position

(*a*) The potential energy of the oscillator is the sum of the gravitational potential energy of block and the energy stored in the stretched spring:

$$U = U_g + U_s$$

Letting the zero of gravitational potential energy be at the oscillator's equilibrium position yields:

$$U = \boxed{\tfrac{1}{2}ky^2 - mgy}$$

where y is the distance the spring is stretched.

(*b*) A graph of U as a function of y follows. Because k and m are not specified, k has been set equal to 2 and mg to 1.

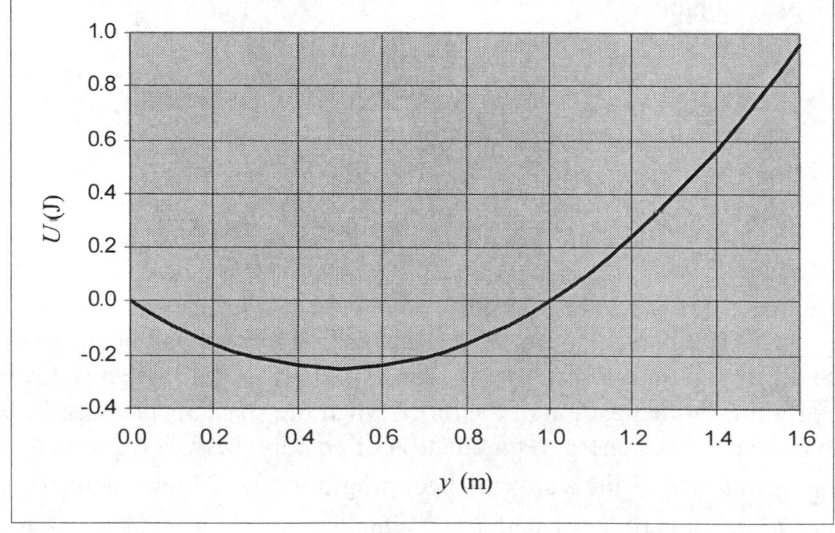

(*c*) The fact that U is a minimum near $y = 0.5$ m tells us that this is a position of stable equilibrium.

Differentiate U with respect to y to obtain:

$$\frac{dU}{dy} = \frac{d}{dy}\left(\tfrac{1}{2}ky^2 - mgy\right) = ky - mg$$

Setting this expression equal to zero for extrema yields:

$$ky - mg = 0 \Rightarrow y = \boxed{\frac{mg}{k}}$$

(*d*) Evaluate the negative of the derivative of U with respect to y:

$$F = -\frac{dU}{dy} = -\frac{d}{dy}\left(\tfrac{1}{2}ky^2 - mgy\right)$$

$$= \boxed{-ky + mg}$$

(*e*) Apply conservation of energy to the movement of the mass from $y = 0$ to $y = y_{max}$:

$$\Delta K + \Delta U + \Delta E_{therm} = 0$$

Because $\Delta K = 0$ (the object starts from rest and is momentarily at rest at $y = y_{max}$) and (no friction), it follows that:

$$\Delta U = U(y_{max}) - U(0) = 0$$

Because $U(0) = 0$:

$$U(y_{max}) = 0 \Rightarrow \tfrac{1}{2}ky_{max}^2 - mgy_{max} = 0$$

Solve for y_{max} to obtain:

$$y_{max} = \boxed{\frac{2mg}{k}}$$

On the graph, y_{max} is at (1.0, 0.0).

99 ••• **[SSM]** To measure the combined force of friction (rolling friction plus air drag) on a moving car, an automotive engineering team you are on turns off the engine and allows the car to coast down hills of known steepness. The team collects the following data: (1) On a 2.87° hill, the car can coast at a steady 20 m/s. (2) On a 5.74° hill, the steady coasting speed is 30 m/s. The total mass of the car is 1000 kg. (*a*) What is the magnitude of the combined force of friction at 20 m/s (F_{20}) and at 30 m/s (F_{30})? (*b*) How much power must the engine deliver to drive the car on a level road at steady speeds of 20 m/s (P_{20}) and 30 m/s (P_{30})? (*c*) The maximum power the engine can deliver is 40 kW. What is the angle of the steepest incline up which the car can maintain a steady 20 m/s? (*d*) Assume that the engine delivers the same total useful work from each liter of gas, no matter what the speed. At 20 m/s on a level road, the car gets 12.7 km/L. How many kilometers per liter does it get if it goes 30 m/s instead?

Picture the Problem We can use Newton's 2nd law to determine the force of friction as a function of the angle of the hill for a given constant speed. The power output of the engine is given by $P = \vec{F}_\text{f} \cdot \vec{v}$.

FBD for (*a*):

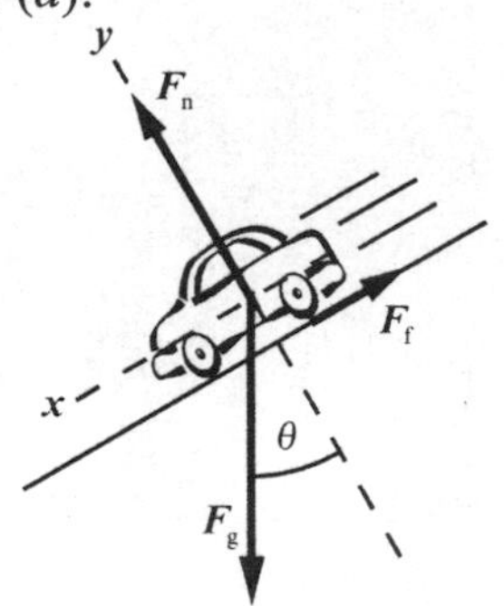

FBD for (*c*):

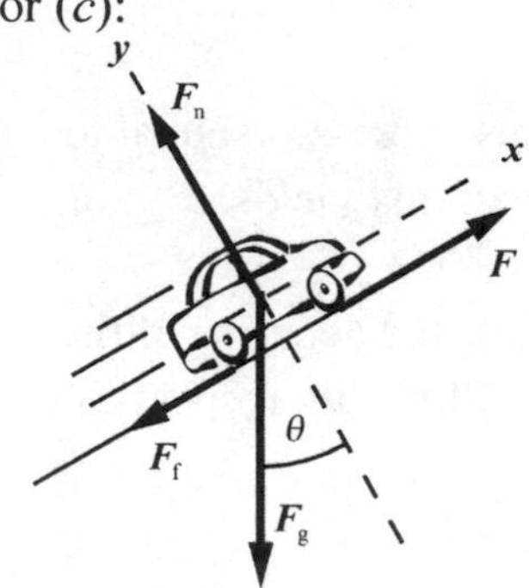

(*a*) Apply $\sum F_x = ma_x$ to the car:

$$\boldsymbol{mg}\sin\theta - \boldsymbol{F} = 0 \Rightarrow \boldsymbol{F} = \boldsymbol{mg}\sin\theta$$

Evaluate F for the two speeds:

$$F_{20} = (1000\,\text{kg})(9.81\,\text{m/s}^2)\sin(2.87°) = \boxed{491\,\text{N}}$$

and

$$F_{30} = (1000\,\text{kg})(9.81\,\text{m/s}^2)\sin(5.74°) = \boxed{981\,\text{N}}$$

(*b*) The power an engine must deliver on a level road in order to overcome friction loss is given by:

$$\boldsymbol{P} = \boldsymbol{F}_\text{f}\boldsymbol{v}$$

Evaluate this expression for $v = 20$ m/s and 30 m/s:

$$P_{20} = (491\,\text{N})(20\,\text{m/s}) = \boxed{9.8\,\text{kW}}$$

and

$$P_{30} = (981\,\text{N})(30\,\text{m/s}) = \boxed{29\,\text{kW}}$$

(*c*) Apply $\sum F_x = ma_x$ to the car:

$$\sum \boldsymbol{F}_x = \boldsymbol{F} - \boldsymbol{mg}\sin\theta - \boldsymbol{F}_\text{f} = 0$$

Solving for F yields:

$$\boldsymbol{F} = \boldsymbol{mg}\sin\theta + \boldsymbol{F}_\text{f}$$

Relate F to the power output of the engine and the speed of the car:

$$\boldsymbol{F} = \frac{\boldsymbol{P}}{\boldsymbol{v}}$$

Equate these expressions for F to obtain:

$$\frac{P}{v} = mg\sin\theta + F_f$$

Solving for θ yields:

$$\theta = \sin^{-1}\left[\frac{\frac{P}{v} - F_f}{mg}\right]$$

Substitute numerical values and evaluate θ for $F_f = F_{20}$:

$$\theta = \sin^{-1}\left[\frac{\frac{40\,\text{kW}}{20\,\text{m/s}} - 491\,\text{N}}{(1000\,\text{kg})(9.81\,\text{m/s}^2)}\right]$$

$$= \boxed{8.8°}$$

(*d*) Express the equivalence of the work done by the engine in driving the car at the two speeds:

$$W_{\text{engine}} = F_{20}(\Delta s)_{20} = F_{30}(\Delta s)_{30}$$

Let ΔV represent the volume of fuel consumed by the engine driving the car on a level road and divide both sides of the work equation by ΔV to obtain:

$$F_{20}\frac{(\Delta s)_{20}}{\Delta V} = F_{30}\frac{(\Delta s)_{30}}{\Delta V}$$

Solve for $\frac{(\Delta s)_{30}}{\Delta V}$:

$$\frac{(\Delta s)_{30}}{\Delta V} = \frac{F_{20}}{F_{30}}\frac{(\Delta s)_{20}}{\Delta V}$$

Substitute numerical values and evaluate $\frac{(\Delta s)_{30}}{\Delta V}$:

$$\frac{(\Delta s)_{30}}{\Delta V} = \frac{491\,\text{N}}{981\,\text{N}}(12.7\,\text{km/L})$$

$$= \boxed{6.36\,\text{km/L}}$$

105 •• **[SSM]** A high school teacher once suggested measuring the magnitude of free-fall acceleration by the following method: Hang a mass on a very fine thread (length L) to make a pendulum with the mass a height H above the floor when at its lowest point P. Pull the pendulum back so that the thread makes an angle θ_0 with the vertical. Just above point P, place a razor blade that is positioned to cut through the thread as the mass swings through point P. Once the thread is cut, the mass is projected horizontally, and hits the floor a horizontal distance D from point P. The idea was that the measurement of D as a function of θ_0 should somehow determine g. Apart from some obvious experimental difficulties, the experiment had one fatal flaw: D does not depend on g! Show that this is true, and that D depends only on the angle θ_0.

Picture the Problem The pictorial representation shows the bob swinging through an angle θ before the thread is cut and the ball is launched horizontally. Let its speed at position 1 be v. We can use conservation of mechanical energy to relate v to the change in the potential energy of the bob as it swings through the angle θ. We can find its flight time Δt from a constant-acceleration equation and then express D as the product of v and Δt.

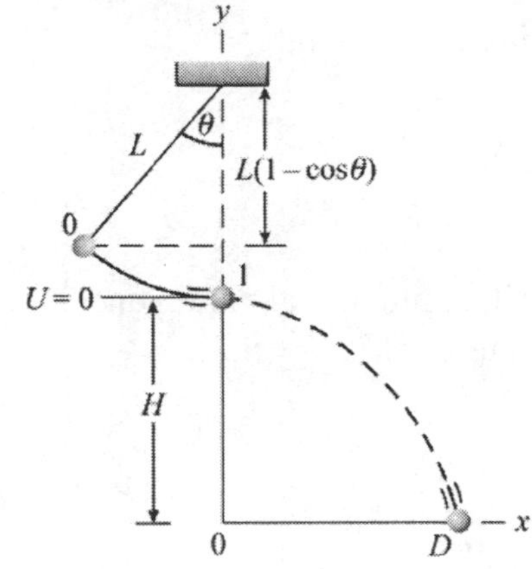

Relate the distance D traveled horizontally by the bob to its launch speed v and time of flight Δt:

$$D = v\Delta t \qquad (1)$$

Use conservation of mechanical energy to relate its launch speed v to the length of the pendulum L and the angle θ:

$$K_1 - K_0 + U_1 - U_0 = 0$$

or, because $U_1 = K_0 = 0$,

$$K_1 - U_0 = 0$$

Substitute for K_1 and U_0 to obtain:

$$\tfrac{1}{2}mv^2 - mgL(1-\cos\theta) = 0$$

Solving for v yields:

$$v = \sqrt{2gL(1-\cos\theta)}$$

In the absence of air resistance, the horizontal and vertical motions of the bob are independent of each other and we can use a constant-acceleration equation to express the time of flight (the time to fall a distance H):

$$\Delta y = v_{0y}\Delta t + \tfrac{1}{2}a_y(\Delta t)^2$$

or, because $\Delta y = -H$, $a_y = -g$, and $v_{0y} = 0$,

$$-H = -\tfrac{1}{2}g(\Delta t)^2 \Rightarrow \Delta t = \sqrt{2H/g}$$

Substitute in equation (1) and simplify to obtain:

$$D = \sqrt{2gL(1-\cos\theta)}\sqrt{\frac{2H}{g}} = \boxed{2\sqrt{HL(1-\cos\theta)}}$$

which shows that, while D depends on θ, it is independent of g.

Chapter 8
Conservation of Linear Momentum

Conceptual Problems

7 •• **[SSM]** Much early research in rocket motion was done by Robert Goddard, physics professor at Clark College in Worcester, Massachusetts. A quotation from a 1920 editorial in the *New York Times* illustrates the public opinion of his work: "That Professor Goddard with his 'chair' at Clark College and the countenance of the Smithsonian Institution does not know the relation between action and reaction, and the need to have something better than a vacuum against which to react—to say that would be absurd. Of course, he only seems to lack the knowledge ladled out daily in high schools." The belief that a rocket needs something to push against was a prevalent misconception before rockets in space were commonplace. Explain why that belief is wrong.

Determine the Concept In a way, the rocket does need something to push upon. It pushes the exhaust in one direction, and the exhaust pushes it in the opposite direction. However, the rocket does not push against the air.

15 •• **[SSM]** In Problem 14 for the performer falling from a height of 25 m, estimate the ratio of the collision time with the safety net to the collision time with the concrete. *Hint: Use the procedure outlined in Step 4 of the Problem-Solving Strategy located in Section 8-3.*

Determine the Concept The stopping time for the performer is the ratio of the distance traveled during stopping to the average speed during stopping.

Letting d_{net} be the distance the net gives on impact, $d_{concrete}$ the distance the concrete gives, and $v_{av,\text{ with net}}$ and $v_{av,\text{without net}}$ the average speeds during stopping, express the ratio of the impact times:

$$r = \frac{\Delta t_{net}}{\Delta t_{concrete}} = \frac{\dfrac{d_{net}}{v_{av,\text{ with net}}}}{\dfrac{d_{concrete}}{v_{av,\text{ without net}}}} \quad (1)$$

Assuming constant acceleration, the average speed of the performer during stopping is given by:

$$v_{av} = \frac{v_f + v}{2}$$

or, because $v_f = 0$ in both cases,

$$v_{av} = \tfrac{1}{2}v$$

where v is the impact speed.

Substituting in equation (1) and simplifying yields:

$$r = \frac{\dfrac{d_{\text{net}}}{\frac{1}{2}v}}{\dfrac{d_{\text{concrete}}}{\frac{1}{2}v}} = \frac{d_{\text{net}}}{d_{\text{concrete}}}$$

Assuming that the net gives about 1 m and concrete about 0.1 mm yields:

$$r = \frac{1\text{ m}}{0.1\text{ mm}} \approx \boxed{10^4}$$

Conceptual Problems from Optional Sections

29 ••• **[SSM]** To show that even really intelligent people can make mistakes, consider the following problem which was asked of a freshman class at Caltech on an exam (paraphrased): *A sailboat is sitting in the water on a windless day. In order to make the boat move, a misguided sailor sets up a fan in the back of the boat to blow into the sails to make the boat move forward. Explain why the boat won't move.* The idea was that the net force of the wind pushing the sail forward would be counteracted by the force pushing the fan back (Newton's third law). However, as one of the students pointed out to his professor, the sailboat *could* in fact move forward. Why is that?

Determine the Concept Think of the sail facing the fan (like the sail on a square rigger might), and think of the stream of air molecules hitting the sail. Imagine that they bounce off the sail elastically–their net change in momentum is then roughly twice the change in momentum that they experienced going through the fan. Thus the change in momentum of the air is backward, so to conserve momentum of the air-fan-boat system the change in momentum of the fan-boat system will be forward.

Conservation of Linear Momentum

33 • **[SSM]** Tyrone, an 85-kg teenager, runs off the end of a horizontal pier and lands on a free-floating 150-kg raft that was initially at rest. After he lands on the raft, the raft, with him on it, moves away from the pier at 2.0 m/s. What was Tyrone's speed as he ran off the end of the pier?

Picture the Problem Let the system include the raft, the earth, and Tyrone and apply conservation of linear momentum to find Tyrone's speed when he ran off the end of the pier.

Apply conservation of linear momentum to the system consisting of the raft and Tyrone to obtain:

$$\Delta\vec{p}_{\text{system}} = \Delta\vec{p}_{\text{Tyrone}} + \Delta\vec{p}_{\text{raft}} = 0$$

or, because the motion is one-dimensional,

$$p_{\text{f,Tyrone}} - p_{\text{i, Tyrone}} + p_{\text{f,raft}} - p_{\text{i,raft}} = 0$$

Because the raft is initially at rest: $p_{\text{f,Tyrone}} - p_{\text{i, Tyrone}} + p_{\text{f,raft}} = 0$

Use the definition of linear momentum to obtain:

$$m_{\text{Tyrone}} v_{\text{f,Tyrone}} - m_{\text{Tyrone}} v_{\text{i,Tyrone}} + m_{\text{raft}} v_{\text{f,raft}} = 0$$

Solve for $v_{\text{i,Tyrone}}$ to obtain:

$$v_{\text{i,Tyrone}} = \frac{m_{\text{raft}}}{m_{\text{Tyrone}}} v_{\text{f,raft}} + v_{\text{f,Tyrone}}$$

Letting v represent the common final speed of the raft and Tyrone yields:

$$v_{\text{i,Tyrone}} = \left(1 + \frac{m_{\text{raft}}}{m_{\text{Tyrone}}}\right) v$$

Substitute numerical values and evaluate $v_{\text{i,Tyrone}}$:

$$v_{\text{i, Tyrone}} = \left(1 + \frac{150\ \text{kg}}{85\ \text{kg}}\right)(2.0\ \text{m/s}) = \boxed{5.5\ \text{m/s}}$$

Kinetic Energy of a System of Particles

41 •• **[SSM]** A 3.0-kg block is traveling to the right (the $+x$ direction) at 5.0 m/s, and a second 3.0-kg block is traveling to the left at 2.0 m/s. (*a*) Find the total kinetic energy of the two blocks. (*b*) Find the velocity of the center of mass of the two-block system. (*c*) Find the velocity of each block relative to the center of mass. (*d*) Find the kinetic energy of the blocks relative to the center of mass. (*e*) Show that your answer for Part (*a*) is greater than your answer for Part (*d*) by an amount equal to the kinetic energy associated with the motion of the center of mass.

Picture the Problem Choose a coordinate system in which the positive x direction is to the right. Use the expression for the total momentum of a system to find the velocity of the center of mass and the definition of relative velocity to express the sum of the kinetic energies relative to the center of mass.

(*a*) The total kinetic energy is the sum of the kinetic energies of the blocks: $K = K_1 + K_2 = \tfrac{1}{2} m_1 v_1^2 + \tfrac{1}{2} m_2 v_2^2$

Substitute numerical values and evaluate K:

$$K = \tfrac{1}{2}(3.0\ \text{kg})(5.0\ \text{m/s})^2 + \tfrac{1}{2}(3.0\ \text{kg})(2.0\ \text{m/s})^2 = 43.5\ \text{J} = \boxed{44\ \text{J}}$$

(*b*) Relate the velocity of the center of mass of the system to its total momentum:

$$M\vec{v}_{cm} = m_1\vec{v}_1 + m_2\vec{v}_2$$

Solving for $\vec{v}_{cm}$ yields:

$$\vec{v}_{cm} = \frac{m_1\vec{v}_1 + m_2\vec{v}_2}{m_1 + m_2}$$

Substitute numerical values and evaluate $\vec{v}_{cm}$:

$$\vec{v}_{cm} = \frac{(3.0\,\text{kg})(5.0\,\text{m/s})\hat{i} + (3.0\,\text{kg})(2.0\,\text{m/s})\hat{i}}{3.0\,\text{kg} + 3.0\,\text{kg}} = \boxed{(1.5\,\text{m/s})\hat{i}}$$

(*c*) The velocity of an object relative to the center of mass is given by:

$$\vec{v}_{rel} = \vec{v} - \vec{v}_{cm}$$

Substitute numerical values to obtain:

$$\vec{v}_{1,rel} = (5.0\,\text{m/s})\hat{i} - (1.5\,\text{m/s})\hat{i} = \boxed{(3.5\,\text{m/s})\hat{i}}$$

$$\vec{v}_{2,rel} = (-2.0\,\text{m/s})\hat{i} - (1.5\,\text{m/s})\hat{i} = \boxed{(-3.5\,\text{m/s})\hat{i}}$$

(*d*) Express the sum of the kinetic energies relative to the center of mass:

$$K_{rel} = K_{1,rel} + K_{2,rel} = \tfrac{1}{2}m_1v_{1,rel}^2 + \tfrac{1}{2}m_2v_{2,rel}^2$$

Substitute numerical values and evaluate K_{rel}:

$$K_{rel} = \tfrac{1}{2}(3.0\,\text{kg})(3.5\,\text{m/s})^2 + \tfrac{1}{2}(3.0\,\text{kg})(-3.5\,\text{m/s})^2 = \boxed{37\,\text{J}}$$

(*e*) K_{cm} is given by:

$$K_{cm} = \tfrac{1}{2}m_{tot}v_{cm}^2$$

Substitute numerical values and evaluate K_{cm} :

$$K_{cm} = \tfrac{1}{2}(6.0\,\text{kg})(1.5\,\text{m/s})^2 = 6.75\,\text{J} = 43.5\,\text{J} - 36.75\,\text{J} = \boxed{K - K_{rel}}$$

Impulse and Average Force

43 • **[SSM]** You kick a soccer ball whose mass is 0.43 kg. The ball leaves your foot with an initial speed of 25 m/s. (*a*) What is the magnitude of the impulse associated with the force of your foot on the ball? (*b*) If your foot is in contact with the ball for 8.0 ms, what is the magnitude of the average force exerted by your foot on the ball?

Picture the Problem The impulse imparted to the ball by the kicker equals the *change* in the ball's momentum. The impulse is also the product of the average force exerted on the ball by the kicker and the time during which the average force acts.

(*a*) Relate the magnitude of the impulse delivered to the ball to its change in momentum:

$$I = |\Delta\vec{p}| = p_f - p_i$$

or, because $v_i = 0$,

$$I = mv_f$$

Substitute numerical values and evaluate I:

$$I = (0.43\,\text{kg})(25\,\text{m/s}) = 10.8\,\text{N}\cdot\text{s} = \boxed{11\,\text{N}\cdot\text{s}}$$

(*b*) The impulse delivered to the ball as a function of the average force acting on it is given by:

$$I = F_{av}\Delta t \Rightarrow F_{av} = \frac{I}{\Delta t}$$

Substitute numerical values and evaluate F_{av}:

$$F_{av} = \frac{10.8\,\text{N}\cdot\text{s}}{0.0080\,\text{s}} = \boxed{1.3\,\text{kN}}$$

Collisions in One Dimension

53 • **[SSM]** A 2000-kg car traveling to the right at 30 m/s is chasing a second car of the same mass that is traveling in the same direction at 10 m/s. (*a*) If the two cars collide and stick together, what is their speed just after the collision? (*b*) What fraction of the initial kinetic energy of the cars is lost during this collision? Where does it go?

Picture the Problem We can apply conservation of linear momentum to this perfectly inelastic collision to find the after-collision speed of the two cars. The ratio of the transformed kinetic energy to kinetic energy before the collision is the fraction of kinetic energy lost in the collision.

(*a*) Letting V be the velocity of the two cars after their collision, apply conservation of linear momentum to their perfectly inelastic collision:

$$p_{\text{initial}} = p_{\text{final}}$$

or

$$mv_1 + mv_2 = (m+m)V \Rightarrow V = \frac{v_1 + v_2}{2}$$

Substitute numerical values and evaluate V:

$$V = \frac{30\,\text{m/s} + 10\,\text{m/s}}{2} = \boxed{20\,\text{m/s}}$$

(*b*) The ratio of the kinetic energy that is lost to the kinetic energy of the two cars before the collision is:

$$\frac{\Delta K}{K_{\text{initial}}} = \frac{K_{\text{final}} - K_{\text{initial}}}{K_{\text{initial}}} = \frac{K_{\text{final}}}{K_{\text{initial}}} - 1$$

Substitute for the kinetic energies and simplify to obtain:

$$\frac{\Delta K}{K_{\text{initial}}} = \frac{\frac{1}{2}(2m)V^2}{\frac{1}{2}mv_1^2 + \frac{1}{2}mv_2^2} - 1 = \frac{2V^2}{v_1^2 + v_2^2} - 1$$

Substitute numerical values and evaluate $\Delta K/K_{\text{initial}}$:

$$\frac{\Delta K}{K_{\text{initial}}} = \frac{2(20\,\text{m/s})^2}{(30\,\text{m/s})^2 + (10\,\text{m/s})^2} - 1 = -0.20$$

20% of the initial kinetic energy is transformed into heat, sound, and the deformation of the materials from which the car is constructed.

63 •• **[SSM]** A16-g bullet is fired into the bob of a 1.5-kg ballistic pendulum (Figure 8-18). When the bob is at its maximum height, the strings make an angle of 60° with the vertical. The pendulum strings are 2.3 m long. Find the speed of the bullet prior to impact.

Picture the Problem The following pictorial representation shows the bullet about to imbed itself in the bob of the ballistic pendulum and then, later, when the bob plus bullet have risen to their maximum height. We can use conservation of momentum during the collision to relate the speed of the bullet to the initial speed of the bob plus bullet (V). The initial kinetic energy of the bob plus bullet is transformed into gravitational potential energy when they reach their maximum height. Hence we apply conservation of mechanical energy to relate V to the angle through which the bullet plus bob swings and then solve the momentum and energy equations simultaneously for the speed of the bullet.

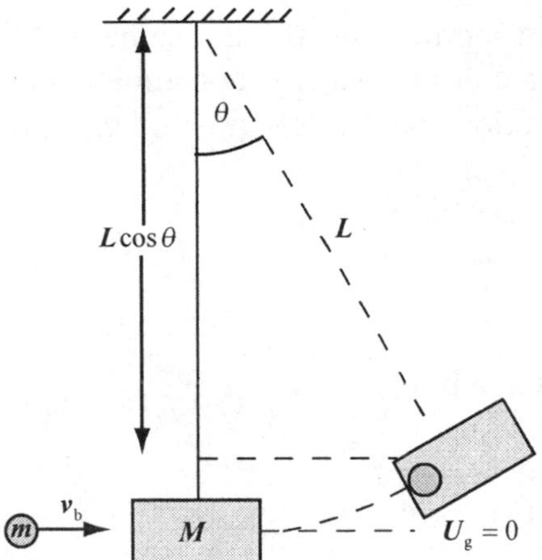

Use conservation of momentum to relate the speed of the bullet just before impact to the initial speed of the bob-bullet:

$$mv_b = (m+M)V \Rightarrow v_b = \left(1+\frac{M}{m}\right)V \quad (1)$$

Use conservation of energy to relate the initial kinetic energy of the bob-bullet to their final potential energy:

$$\Delta K + \Delta U = 0$$

or, because $K_f = U_i = 0$,

$$-K_i + U_f = 0$$

Substitute for K_i and U_f to obtain:

$$-\tfrac{1}{2}(m+M)V^2 + (m+M)gL(1-\cos\theta) = 0$$

Solving for V yields:

$$V = \sqrt{2gL(1-\cos\theta)}$$

Substitute for V in equation (1) to obtain:

$$v_b = \left(1+\frac{M}{m}\right)\sqrt{2gL(1-\cos\theta)}$$

Substitute numerical values and evaluate v_b:

$$v_b = \left(1+\frac{1.5\,\text{kg}}{0.016\,\text{kg}}\right)\sqrt{2(9.81\,\text{m/s}^2)(2.3\,\text{m})(1-\cos 60°)} = \boxed{0.45\,\text{km/s}}$$

69 •• **[SSM]** A 0.425-kg ball with a speed of 1.30 m/s rolls across a level surface toward an open 0.327-kg box that is resting on its side. The ball enters the box, and the box (with the ball inside slides across the surface a distance of $x = 0.520$ m. What is the coefficient of kinetic friction between the box and the table?

Picture the Problem The collision of the ball with the box is perfectly inelastic and we can find the speed of the box-and-ball immediately after their collision by applying conservation of momentum. If we assume that the kinetic friction force is constant, we can use a constant-acceleration equation to find the acceleration of the box and ball combination and the definition of μ_k to find its value.

Using its definition, express the coefficient of kinetic friction of the table:

$$\mu_k = \frac{f_k}{F_n} = \frac{(M+m)|a|}{(M+m)g} = \frac{|a|}{g} \quad (1)$$

Use conservation of momentum to relate the speed of the ball just before the collision to the speed of the ball+box immediately after the collision:

$$MV = (m+M)v \Rightarrow v = \frac{MV}{m+M} \quad (2)$$

Use a constant-acceleration equation to relate the sliding distance of the ball+box to its initial and final velocities and its acceleration:

$$v_f^2 = v_i^2 + 2a\Delta x$$

or, because $v_f = 0$ and $v_i = v$,

$$0 = v^2 + 2a\Delta x \Rightarrow a = -\frac{v^2}{2\Delta x}$$

Substitute for a in equation (1) to obtain:

$$\mu_k = \frac{v^2}{2g\Delta x}$$

Use equation (2) to eliminate v:

$$\mu_k = \frac{1}{2g\Delta x}\left(\frac{MV}{m+M}\right)^2 = \frac{1}{2g\Delta x}\left(\frac{V}{\frac{m}{M}+1}\right)^2$$

Substitute numerical values and evaluate μ_k:

$$\mu_k = \frac{1}{2(9.81\,\text{m/s}^2)(0.520\,\text{m})}\left(\frac{1.30\,\text{m/s}}{\frac{0.327\,\text{kg}}{0.425\,\text{kg}}+1}\right)^2 = \boxed{0.0529}$$

71 •• **[SSM]** Scientists estimate that the meteorite responsible for the creation of Barringer Meteorite Crater in Arizona weighed roughly 2.72×10^5 tonne (1 tonne = 1000 kg) and was traveling at a speed of 17.9 km/s. Take Earth's orbital speed to be about 30.0 km/s. (*a*) What should the direction of impact be if Earth's orbital speed is to be changed by the maximum possible amount? (*b*) Assuming the condition of collision in Part (*a*), estimate the maximum percentage change in Earth's orbital speed as a result of this collision.

(*c*) What mass asteroid, having a speed equal to Earth's orbital speed, would be necessary to change Earth's orbital speed by 1.00%?

Picture the Problem Let the system include Earth and the asteroid. Choose a coordinate system in which the direction of Earth's orbital speed is the $+x$ direction. We can apply conservation of linear momentum to the perfectly inelastic collision of Earth and the asteroid to find the percentage change in Earth's orbital speed as well as the mass of an asteroid that would change Earth's orbital speed by 1.00%. Note that the following solution neglects the increase in Earth's orbital speed due to the gravitational pull of the asteroid during descent.

(*a*) For maximum slowing of Earth, the collision would have to have taken place with the meteorite impacting Earth along a line exactly opposite Earth's orbital velocity vector. In this case, we have a head-on inelastic collision.

(*b*) Express the percentage change in Earth's orbital speed as a result of the collision:

$$\left|\frac{\Delta v}{v_{\text{Earth}}}\right| = \left|\frac{v_{\text{Earth}} - v_{\text{f}}}{v_{\text{Earth}}}\right| = \left|1 - \frac{v_{\text{f}}}{v_{\text{Earth}}}\right| \quad (1)$$

where v_{f} is Earth's orbital speed after the collision.

Apply the conservation of linear momentum to the system to obtain:

$$\Delta\vec{p} = \vec{p}_{\text{f}} - \vec{p}_{\text{i}} = 0$$

or, because the asteroid and the earth are moving horizontally,

$$p_{\text{f},x} - p_{\text{i},x} = 0$$

Because the collision is perfectly inelastic:

$$(m_{\text{Earth}} + m_{\text{asteroid}})v_{\text{f}} - (m_{\text{Earth}}v_{\text{Earth}} - m_{\text{asteroid}}v_{\text{asteroid}}) = 0$$

Solving for v_{f} yields:

$$v_{\text{f}} = \frac{m_{\text{Earth}}v_{\text{Earth}} - m_{\text{asteroid}}v_{\text{asteroid}}}{m_{\text{Earth}} + m_{\text{asteroid}}} = \frac{m_{\text{Earth}}v_{\text{Earth}}}{m_{\text{Earth}} + m_{\text{asteroid}}} - \frac{m_{\text{asteroid}}v_{\text{asteroid}}}{m_{\text{Earth}} + m_{\text{asteroid}}}$$

$$= \frac{v_{\text{Earth}}}{1 + \dfrac{m_{\text{asteroid}}}{m_{\text{Earth}}}} - \frac{\dfrac{m_{\text{asteroid}}}{m_{\text{Earth}}}v_{\text{asteroid}}}{1 + \dfrac{m_{\text{asteroid}}}{m_{\text{Earth}}}}$$

Because $m_{\text{asteroid}} << m_{\text{Earth}}$:

$$v_f \approx v_{\text{Earth}} - \frac{\frac{m_{\text{asteroid}}}{m_{\text{Earth}}} v_{\text{asteroid}}}{1+\frac{m_{\text{asteroid}}}{m_{\text{Earth}}}} = v_{\text{Earth}} - \frac{m_{\text{asteroid}}}{m_{\text{Earth}}} v_{\text{asteroid}} \left(1+\frac{m_{\text{asteroid}}}{m_{\text{Earth}}}\right)^{-1}$$

Expanding $\left(1+\frac{m_{\text{asteroid}}}{m_{\text{Earth}}}\right)^{-1}$ binomially yields:

$$\left(1+\frac{m_{\text{asteroid}}}{m_{\text{Earth}}}\right)^{-1} = 1 - \frac{m_{\text{asteroid}}}{m_{\text{Earth}}} + \text{higher order terms}$$

Substitute for $\left(1+\frac{m_{\text{asteroid}}}{m_{\text{Earth}}}\right)^{-1}$ in the expression for v_f to obtain:

$$v_f \approx v_{\text{Earth}} - \frac{m_{\text{asteroid}}}{m_{\text{Earth}}} v_{\text{asteroid}} \left(1-\frac{m_{\text{asteroid}}}{m_{\text{Earth}}}\right)$$

$$\approx v_{\text{Earth}} - \frac{m_{\text{asteroid}}}{m_{\text{Earth}}} v_{\text{asteroid}}$$

Substitute for v_f in equation (1) to obtain:

$$\left|\frac{\Delta v}{v_{\text{Earth}}}\right| = \left|1 - \frac{v_{\text{Earth}} - \frac{m_{\text{asteroid}}}{m_{\text{Earth}}} v_{\text{asteroid}}}{v_{\text{Earth}}}\right| = \left|\frac{\frac{m_{\text{asteroid}}}{m_{\text{Earth}}} v_{\text{asteroid}}}{v_{\text{Earth}}}\right|$$

Using data found in the appendices of your text or given in the problem statement, substitute numerical values and evaluate $\left|\frac{\Delta v}{v_{\text{Earth}}}\right|$:

$$\left|\frac{\Delta v}{v_{\text{Earth}}}\right| = \left|\frac{\frac{\left(2.72\times10^{5}\ \text{tonne}\times\frac{10^{3}\ \text{kg}}{1\ \text{tonne}}\right)(17.9\ \text{km/s})}{5.98\times10^{24}\ \text{kg}}}{30.0\ \text{km/s}}\right| = \boxed{2.71\times10^{-15}\%}$$

(*c*) If the asteroid is to change Earth's orbital speed by 1%:

$$\frac{\frac{m_{\text{asteroid}}}{m_{\text{Earth}}} v_{\text{asteroid}}}{v_{\text{Earth}}} = \frac{1}{100}$$

Solve for $m_{asteroid}$ to obtain:

$$m_{asteroid} = \frac{v_{Earth} m_{Earth}}{100 v_{asteroid}}$$

Substitute numerical values and evaluate $m_{asteroid}$:

$$m_{asteroid} = \frac{(30.0\text{ km/s})(5.98\times10^{24}\text{ kg})}{100(17.9\text{ km/s})} = \boxed{1.00\times10^{23}\text{ kg}}$$

Remarks: The mass of this asteroid is approximately that of the moon!

Explosions and Radioactive Decay

73 •• **[SSM]** The beryllium isotope ^{8}Be is unstable and decays into two α particles ($m_\alpha = 6.64 \times 10^{-27}$ kg) and releases 1.5×10^{-14} J of energy. Determine the velocities of the two α particles that arise from the decay of a ^{8}Be nucleus at rest, assuming that all the energy appears as kinetic energy of the particles.

Picture the Problem This nuclear reaction is $^{4}\text{Be} \rightarrow 2\alpha + 1.5 \times 10^{-14}$ J. In order to conserve momentum, the alpha particles will have move in opposite directions with the same velocities. We'll use conservation of energy to find their speeds.

Letting E represent the energy released in the reaction, express conservation of energy for this process:

$$2K_\alpha = 2\left(\tfrac{1}{2} m_\alpha v_\alpha^2\right) = E \Rightarrow v_\alpha = \sqrt{\frac{E}{m_\alpha}}$$

Substitute numerical values and evaluate v_α:

$$v_\alpha = \sqrt{\frac{1.5\times10^{-14}\text{ J}}{6.64\times10^{-27}\text{ kg}}} = \boxed{1.5\times10^{6}\text{ m/s}}$$

Coefficient of Restitution

83 •• **[SSM]** To make puck handling easy, hockey pucks are kept frozen until they are used in the game. (*a*) Explain why room temperature pucks would be more difficult to handle on the end of a stick than a frozen puck. (*Hint: Hockey pucks are made of rubber.*) (*b*) A room-temperature puck rebounds 15 cm when dropped onto a wooden surface from 100 cm. If a frozen puck has only half the coefficient of restitution of a room-temperature one, predict how high the frozen puck would rebound under the same conditions.

Picture the Problem The coefficient of restitution is defined as the ratio of the velocity of recession to the velocity of approach. These velocities can be determined from the heights from which the ball was dropped and the height to which it rebounded by using conservation of mechanical energy.

(a) At room-temperature rubber will bounce more when it hits a stick than it will at freezing temperatures.

(b) The mechanical energy of the rebounding puck is constant:

$$\Delta \boldsymbol{K} + \Delta \boldsymbol{U} = 0$$

or, because $K_f = U_i = 0$,

$$-\boldsymbol{K}_i + \boldsymbol{U}_f = 0$$

If the puck's speed of recession is v_{rec} and it rebounds to a height h, then:

$$-\tfrac{1}{2}\boldsymbol{m}\boldsymbol{v}_{rec}^2 + \boldsymbol{mgh} = 0 \Rightarrow \boldsymbol{h} = \frac{\boldsymbol{v}_{rec}^2}{2\boldsymbol{g}}$$

The coefficient of restitution is the ratio of the speeds of approach and recession:

$$e = \frac{v_{rec}}{v_{app}} \Rightarrow \boldsymbol{v}_{rec} = \boldsymbol{e}\boldsymbol{v}_{app} \quad (1)$$

Substitute for v_{rec} to obtain:

$$\boldsymbol{h} = \frac{\boldsymbol{e}^2\boldsymbol{v}_{app}^2}{2\boldsymbol{g}} \quad (2)$$

Letting $U_g = 0$ at the surface of from which the puck rebounds, the mechanical energy of the puck-Earth system is:

$$\Delta K + \Delta U = 0$$

Because $K_i = U_f = 0$,

$$\boldsymbol{K}_f - \boldsymbol{U}_i = 0$$

Substituting for K_f and U_i yields:

$$\tfrac{1}{2}mv_{app}^2 - mgh_{app} = 0 \Rightarrow v_{app} = \sqrt{2gh_{app}}$$

In like manner, show that:

$$v_{rec} = \sqrt{2gh_{rec}}$$

Substitute for v_{rec} and v_{app} in equation (1) and simplify to obtain:

$$e = \frac{\sqrt{2gh_{rec}}}{\sqrt{2gh_{app}}} = \sqrt{\frac{h_{rec}}{h_{app}}}$$

Substitute numerical values and evaluate $e_{room\ temp}$:

$$e_{room\ temp} = \sqrt{\frac{15\,\text{cm}}{100\,\text{cm}}} = 0.387$$

For the falling puck, v_{app} is given by:

$$v_{app} = \sqrt{2gH}$$

where H is the height from which the puck was dropped.

Substituting for v_{app} in equation (2) and simplifying yields:

$$h = \frac{2e^2 gH}{2g} = e^2 H \qquad (3)$$

For the room-temperature puck:

$$e_{frozen} = \tfrac{1}{2} e_{room\ temp}$$

Substituting for e in equation (3) yields:

$$h = \tfrac{1}{4} e^2_{room\ temp} H$$

Substitute numerical values and evaluate h:

$$h = \tfrac{1}{4}(0.387)^2(100\ \text{cm}) = \boxed{3.8\ \text{cm}}$$

Remarks: The puck that rebounds only 3.8 cm is a much "deader" and, therefore, much better puck.

Collisions in More Than One Dimension

87 •• **[SSM]** A puck of mass 5.0 kg moving at 2.0 m/s approaches an identical puck that is stationary on frictionless ice. After the collision, the first puck leaves with a speed v_1 at 30° to the original line of motion; the second puck leaves with speed v_2 at 60°, as in Figure 8-50. (*a*) Calculate v_1 and v_2. (*b*) Was the collision elastic?

Picture the Problem Let the direction of motion of the puck that is moving before the collision be the $+x$ direction. Applying conservation of momentum to the collision in both the x and y directions will lead us to two equations in the unknowns v_1 and v_2 that we can solve simultaneously. We can decide whether the collision was elastic by either calculating the system's kinetic energy before and after the collision or by determining whether the angle between the final velocities is 90°.

(*a*) Use conservation of linear momentum in the x direction to obtain:

$$p_{xi} = p_{xf}$$

or

$$mv = mv_1 \cos 30° + mv_2 \cos 60°$$

Simplify further to obtain:

$$v = v_1 \cos 30° + v_2 \cos 60° \qquad (1)$$

Use conservation of momentum in the y direction to obtain a second equation relating the velocities of the collision participants before and after the collision:

$$p_{yi} = p_{yf}$$

or

$$0 = mv_1 \sin 30° - mv_2 \sin 60°$$

Simplifying further yields:

$$0 = v_1 \sin 30° - v_2 \sin 60° \qquad (2)$$

Solve equations (1) and (2) simultaneously to obtain:

$$v_1 = \boxed{1.7\,\text{m/s}} \text{ and } v_2 = \boxed{1.0\,\text{m/s}}$$

(*b*) Because the angle between $\vec{v}_1$ and $\vec{v}_2$ is 90°, the collision was elastic.

*Center-of-Mass Reference Frame

93 •• **[SSM]** Repeat Problem 92 with the second block having a mass of 5.0 kg and moving to the right at 3.0 m/s.

Picture the Problem Let the numerals 3 and 5 denote the blocks whose masses are 3.0 kg and 5.0 kg respectively. We can use $\sum_i m_i\vec{v}_i = M\vec{v}_{cm}$ to find the velocity of the center-of-mass of the system and simply follow the directions in the problem step by step.

(*a*) Express the total momentum of this two-particle system in terms of the velocity of its center of mass:

$$\vec{P} = \sum_i m_i\vec{v}_i = m_3\vec{v}_3 + m_5\vec{v}_5 = M\vec{v}_{cm} = (m_3 + m_5)\vec{v}_{cm}$$

Solve for $\vec{v}_{cm}$:

$$\vec{v}_{cm} = \frac{m_3\vec{v}_3 + m_5\vec{v}_5}{m_3 + m_5}$$

Substitute numerical values and evaluate $\vec{v}_{cm}$:

$$\vec{v}_{cm} = \frac{(3.0\,\text{kg})(-5.0\,\text{m/s})\hat{i} + (5.0\,\text{kg})(3.0\,\text{m/s})\hat{i}}{3.0\,\text{kg} + 5.0\,\text{kg}} = \boxed{0}$$

(*b*) Find the velocity of the 3.0-kg block in the center of mass reference frame:

$$\vec{u}_3 = \vec{v}_3 - \vec{v}_{cm} = (-5.0\,\text{m/s})\,\hat{i} - 0 = \boxed{(-5.0\,\text{m/s})\,\hat{i}}$$

Find the velocity of the 5.0-kg block in the center of mass reference frame:

$$\vec{u}_5 = \vec{v}_5 - \vec{v}_{cm} = (3.0\,\text{m/s})\hat{i} - 0 = \boxed{(3.0\,\text{m/s})\hat{i}}$$

(*c*) Express the after-collision velocities of both blocks in the center of mass reference frame:

$$\vec{u}_3' = \boxed{(5.0\,\text{m/s})\hat{i}}$$

and

$$u_5' = \boxed{0.75\,\text{m/s}}$$

(*d*) Transform the after-collision velocity of the 3.0-kg block from the center of mass reference frame to the original reference frame:

$$\vec{v}_3' = \vec{u}_3' + \vec{v}_{\rm cm} = (5.0\,{\rm m/s})\hat{\boldsymbol{i}} + 0 = \boxed{(5.0\,{\rm m/s})\hat{\boldsymbol{i}}}$$

Transform the after-collision velocity of the 5.0-kg block from the center of mass reference frame to the original reference frame:

$$\vec{v}_5' = \vec{u}_5' + \vec{v}_{\rm cm} = (-3.0\,{\rm m/s})\hat{\boldsymbol{i}} + 0 = \boxed{(-3.0\,{\rm m/s})\hat{\boldsymbol{i}}}$$

(*e*) Express $K_{\rm i}$ in the original frame of reference:

$$K_{\rm i} = \tfrac{1}{2}m_3v_3^2 + \tfrac{1}{2}m_5v_5^2$$

Substitute numerical values and evaluate $K_{\rm i}$:

$$K_{\rm i} = \tfrac{1}{2}\left[(3.0\,{\rm kg})(5.0\,{\rm m/s})^2 + (5.0\,{\rm kg})(3.0\,{\rm m/s})^2\right] = \boxed{60\,{\rm J}}$$

Express $K_{\rm f}$ in the original frame of reference:

$$K_{\rm f} = \tfrac{1}{2}m_3v_3'^2 + \tfrac{1}{2}m_5v_5'^2$$

Substitute numerical values and evaluate $K_{\rm f}$:

$$K_{\rm f} = \tfrac{1}{2}\left[(3.0\,{\rm kg})(5.0\,{\rm m/s})^2 + (5.0\,{\rm kg})(3.0\,{\rm m/s})^2\right] = \boxed{60\,{\rm J}}$$

*Systems With Continuously Varying Mass: Rocket Propulsion

97 ••• **[SSM]** The initial *thrust-to-weight ratio* τ_0 of a rocket is $\tau_0 = F_{\rm th}/(m_0 g)$, where $F_{\rm th}$ is the rocket's thrust and m_0 the initial mass of the rocket, including the propellant. (*a*) For a rocket launched straight up from the earth's surface, show that $\tau_0 = 1 + (a_0/g)$, where a_0 is the initial acceleration of the rocket. For manned rocket flight, τ_0 cannot be made much larger than 4 for the comfort and safety of the astronauts. (The astronauts will feel that their weight as the rocket lifts off is equal to τ_0 times their normal weight.) (*b*) Show that the final velocity of a rocket launched from the earth's surface, in terms of τ_0 and $I_{\rm sp}$ (see Problem 96) can be written as

$$v_{\rm f} = gI_{\rm sp}\left[\ln\left(\frac{m_0}{m_{\rm f}}\right) - \frac{1}{\tau_0}\left(1 - \frac{m_{\rm f}}{m_0}\right)\right]$$

where $m_{\rm f}$ is the mass of the rocket (not including the spent propellant). (*c*) Using a **spreadsheet** program or **graphing calculator**, graph $v_{\rm f}$ as a function of the mass ratio $m_0/m_{\rm f}$ for $I_{\rm sp}$ = 250 s and τ_0 = 2 for values of the mass ratio from 2 to 10.

(Note that the mass ratio cannot be less than 1.) (*d*) To lift a rocket into orbit, a final velocity after burnout of $v_f = 7.0$ km/s is needed. Calculate the mass ratio required of a single stage rocket to do this, using the values of specific impulse and thrust ratio given in Part (*b*). For engineering reasons, it is difficult to make a rocket with a mass ratio much greater than 10. Can you see why multistage rockets are usually used to put payloads into orbit around the earth?

Picture the Problem We can use the rocket equation and the definition of rocket thrust to show that $\tau_0 = 1 + a_0/g$. In Part (*b*) we can express the burn time t_b in terms of the initial and final masses of the rocket and the rate at which the fuel burns, and then use this equation to express the rocket's final velocity in terms of I_{sp}, τ_0, and the mass ratio m_0/m_f. In Part (*d*) we'll need to use trial-and-error methods or a graphing calculator to solve the transcendental equation giving v_f as a function of m_0/m_f.

(*a*) Express the rocket equation:

$$-mg + Ru_{ex} = ma$$

From the definition of rocket thrust we have:

$$F_{th} = Ru_{ex}$$

Substitute for Ru_{ex} to obtain:

$$-mg + F_{th} = ma$$

Solve for F_{th} at takeoff:

$$F_{th} = m_0 g + m_0 a_0$$

Divide both sides of this equation by $m_0 g$ to obtain:

$$\frac{F_{th}}{m_0 g} = 1 + \frac{a_0}{g}$$

Because $\tau_0 = F_{th}/(m_0 g)$:

$$\tau_0 = \boxed{1 + \frac{a_0}{g}}$$

(*b*) Use Equation 8-39 to express the final speed of a rocket that starts from rest with mass m_0:

$$v_f = u_{ex} \ln\left(\frac{m_0}{m_f}\right) - gt_b, \quad (1)$$

where t_b is the burn time.

Express the burn time in terms of the burn rate R (assumed constant):

$$t_b = \frac{m_0 - m_f}{R} = \frac{m_0}{R}\left(1 - \frac{m_f}{m_0}\right)$$

Multiply t_b by one in the form $\frac{gF_{th}}{gF_{th}}$ and simplify to obtain:

$$t_b = \frac{gF_{th}}{gF_{th}}\frac{m_0}{R}\left(1-\frac{m_f}{m_0}\right) = \frac{gm_0}{F_{th}}\frac{F_{th}}{gR}\left(1-\frac{m_f}{m_0}\right) = \frac{I_{sp}}{\tau_0}\left(1-\frac{m_f}{m_0}\right)$$

Substitute in equation (1):

$$v_f = u_{ex}\ln\left(\frac{m_0}{m_f}\right) - \frac{gI_{sp}}{\tau_0}\left(1-\frac{m_f}{m_0}\right)$$

From Problem 96 we have:

$u_{ex} = gI_{sp}$,
where u_{ex} is the exhaust velocity of the propellant.

Substitute for u_{ex} and factor to obtain:

$$v_f = gI_{sp}\ln\left(\frac{m_0}{m_f}\right) - \frac{gI_{sp}}{\tau_0}\left(1-\frac{m_f}{m_0}\right) = \boxed{gI_{sp}\left[\ln\left(\frac{m_0}{m_f}\right) - \frac{1}{\tau_0}\left(1-\frac{m_f}{m_0}\right)\right]}$$

(*c*) A spreadsheet program to calculate the final velocity of the rocket as a function of the mass ratio m_0/m_f is shown below. The constants used in the velocity function and the formulas used to calculate the final velocity are as follows:

Cell	Content/Formula	Algebraic Form
B1	250	I_{sp}
B2	9.81	g
B3	2	τ_0
D9	D8 + 0.25	m_0/m_f
E8	\$B\$2*\$B\$1*(LOG(D8) – (1/\$B\$3)*(1/D8))	$gI_{sp}\left[\ln\left(\frac{m_0}{m_f}\right)-\frac{1}{\tau_0}\left(1-\frac{m_f}{m_0}\right)\right]$

	A	B	C	D	E
1	I_{sp} =	250	s		
2	g =	9.81	m/s^2		
3	τ_0=	2			
4					
5					
6					
7				mass ratio	v_f

8				2.00	1.252E+02
9				2.25	3.187E+02
10				2.50	4.854E+02
11				2.75	6.316E+02
12				3.00	7.614E+02
36				9.00	2.204E+03
37				9.25	2.237E+03
38				9.50	2.269E+03
39				9.75	2.300E+03
40				10.00	2.330E+03
41				725.00	7.013E+03

A graph of final velocity as a function of mass ratio follows.

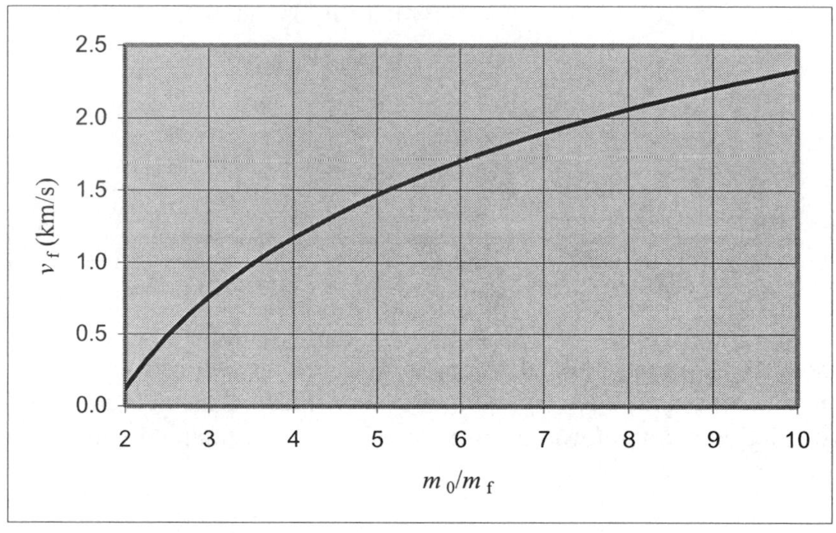

(*d*) Substitute the data given in part (*c*) in the equation derived in Part (*b*) to obtain:

$$7.00\,\text{km/s} = \left(9.81\,\text{m/s}^2\right)\left(250\,\text{s}\right)\left(\ln\left(\frac{m_0}{m_f}\right) - \frac{1}{2}\left(1 - \frac{m_f}{m_0}\right)\right)$$

or

$$2.854 = \ln x - 0.5 + \frac{0.5}{x} \quad \text{where } x = m_0/m_f.$$

Use trial-and-error methods or a graphing calculator to solve this transcendental equation for the root greater than 1: $x \approx \boxed{28}$, a value considerably larger than the practical limit of 10 for single-stage rockets.

General Problems

99 • **[SSM]** A 250-g model-train car traveling at 0.50 m/s links up with a 400-g car that is initially at rest. What is the speed of the cars immediately after they link up? Find the pre- and post-collision kinetic energies of the two-car system.

Picture the Problem Let the direction the 250-g car is moving before the collision be the $+x$ direction. Let the numeral 1 refer to the 250-kg car, the numeral 2 refer to the 400-kg car, and V represent the velocity of the linked cars. Let the system include Earth and the cars. We can use conservation of momentum to find their speed after they have linked together and the definition of kinetic energy to find their pre- and post-collision kinetic energies.

Use conservation of momentum to relate the speeds of the cars immediately before and immediately after their collision:

$$p_{ix} = p_{fx}$$

or

$$m_1 v_1 = (m_1 + m_2)V \Rightarrow V = \frac{m_1 v_1}{m_1 + m_2}$$

Substitute numerical values and evaluate V:

$$V = \frac{(0.250\,\text{kg})(0.50\,\text{m/s})}{0.250\,\text{kg} + 0.400\,\text{kg}} = 0.192\,\text{m/s} = \boxed{0.19\,\text{m/s}}$$

Find the pre-collision kinetic energy of the cars:

$$K_{\text{pre}} = \tfrac{1}{2}m_1 v_1^2 = \tfrac{1}{2}(0.250\,\text{kg})(0.50\,\text{m/s})^2 = \boxed{31\,\text{mJ}}$$

Find the post-collision kinetic energy of the coupled cars:

$$K_{\text{post}} = \tfrac{1}{2}(m_1 + m_2)V^2 = \tfrac{1}{2}(0.250\,\text{kg} + 0.400\,\text{kg})(0.192\,\text{m/s})^2 = \boxed{12\,\text{mJ}}$$

105 ••• **[SSM]** One popular, if dangerous, classroom demonstration involves holding a baseball an inch or so directly above a basketball, holding the basketball a few feet above a hard floor, and dropping the two balls simultaneously. The two balls will collide just after the basketball bounces from the floor; the baseball will then rocket off into the ceiling tiles with a hard "thud" while the basketball will stop in midair. (The author of this problem once broke a light doing this.) (*a*) Assuming that the collision of the basketball with the floor is elastic, what is the relation between the velocities of the balls just before they collide? (*b*) Assuming the collision between the two balls is elastic, use the result of Part (*a*) and the conservation of momentum and energy to show that, if the basketball is three times as heavy as the baseball, the final velocity of the basketball will be zero. (This is approximately the true mass ratio, which is why the demonstration is so dramatic.) (*c*) If the speed of the baseball is v just before the collision, what is its speed just after the collision?

Picture the Problem Let the numeral 1 refer to the basketball and the numeral 2 to the baseball. The left-hand side of the diagram shows the balls after the basketball's elastic collision with the floor and just before they collide. The right-hand side of the diagram shows the balls just after their collision. We can apply conservation of momentum and the definition of an elastic collision to obtain equations relating the initial and final velocities of the masses of the colliding objects that we can solve for v_{1f} and v_{2f}.

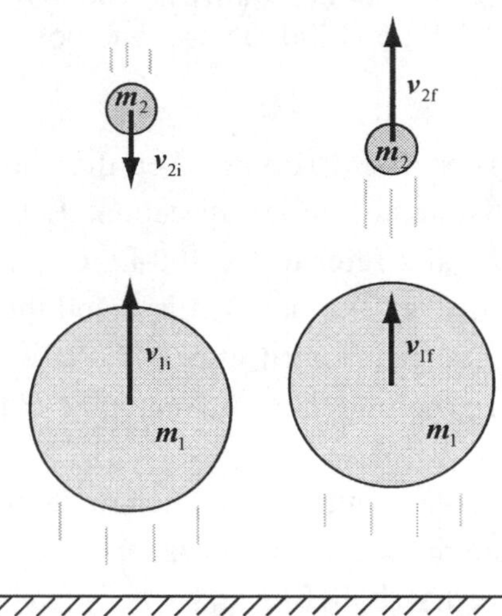

(*a*) Because both balls are in free-fall, and both are in the air for the same amount of time, they have the same velocity just before the basketball rebounds. After the basketball rebounds elastically, its velocity will have the same magnitude, but the opposite direction than just before it hit the ground. The velocity of the basketball will be equal in magnitude but opposite in direction to the velocity of the baseball.

(*b*) Apply conservation of linear momentum to the collision of the balls to obtain:

$$m_1v_{1f} + m_2v_{2f} = m_1v_{1i} + m_2v_{2i} \quad (1)$$

Relate the initial and final kinetic energies of the balls in their elastic collision:

$$\tfrac{1}{2}m_1v_{1f}^2 + \tfrac{1}{2}m_2v_{2f}^2 = \tfrac{1}{2}m_1v_{1i}^2 + \tfrac{1}{2}m_2v_{2i}^2$$

Rearrange this equation and factor to obtain:

$$m_2\left(v_{2f}^2 - v_{2i}^2\right) = m_1\left(v_{1i}^2 - v_{1f}^2\right)$$

or

$$m_2\left(v_{2f} - v_{2i}\right)\left(v_{2f} + v_{2i}\right) = m_1\left(v_{1i} - v_{1f}\right)\left(v_{1i} + v_{1f}\right) \quad (2)$$

Rearrange equation (1) to obtain:

$$m_2\left(v_{2f} - v_{2i}\right) = m_1\left(v_{1i} - v_{1f}\right) \quad (3)$$

Divide equation (2) by equation (3) to obtain:

$$v_{2f} + v_{2i} = v_{1i} + v_{1f}$$

Rearrange this equation to obtain equation (4):

$$v_{1f} - v_{2f} = v_{2i} - v_{1i} \quad (4)$$

Multiply equation (4) by m_2 and add it to equation (1) to obtain:

$$(m_1 + m_2)v_{1f} = (m_1 - m_2)v_{1i} + 2m_2 v_{2i}$$

Solve for v_{1f} to obtain:

$$v_{1f} = \frac{m_1 - m_2}{m_1 + m_2}v_{1i} + \frac{2m_2}{m_1 + m_2}v_{2i}$$

or, because $v_{2i} = -v_{1i}$,

$$v_{1f} = \frac{m_1 - m_2}{m_1 + m_2}v_{1i} - \frac{2m_2}{m_1 + m_2}v_{1i} = \frac{m_1 - 3m_2}{m_1 + m_2}v_{1i}$$

For $m_1 = 3m_2$ and $v_{1i} = v$:

$$v_{1f} = \frac{3m_2 - 3m_2}{3m_2 + m_2}v = \boxed{0}$$

(*c*) Multiply equation (4) by m_1 and subtract it from equation (1) to obtain:

$$(m_1 + m_2)v_{2f} = (m_2 - m_1)v_{2i} + 2m_1 v_{1i}$$

Solve for v_{2f} to obtain:

$$v_{2f} = \frac{2m_1}{m_1 + m_2}v_{1i} + \frac{m_2 - m_1}{m_1 + m_2}v_{2i}$$

or, because $v_{2i} = -v_{1i}$,

$$v_{2f} = \frac{2m_1}{m_1 + m_2}v_{1i} - \frac{m_2 - m_1}{m_1 + m_2}v_{1i} = \frac{3m_1 - m_2}{m_1 + m_2}v_{1i}$$

For $m_1 = 3m_2$ and $v_{1i} = v$:

$$v_{2f} = \frac{3(3m_2) - m_2}{3m_2 + m_2}v = \boxed{2v}$$

107 ••• [SSM] In the "slingshot effect," the transfer of energy in an elastic collision is used to boost the energy of a space probe so that it can escape from the solar system. All speeds are relative to an inertial frame in which the center of the sun remains at rest. Figure 8-55 shows a space probe moving at 10.4 km/s toward Saturn, which is moving at 9.6 km/s toward the probe. Because of the gravitational attraction between Saturn and the probe, the probe swings around Saturn and heads back in the opposite direction with speed v_f. (*a*) Assuming this collision to be a one-dimensional elastic collision with the mass of Saturn much greater than that of the probe, find v_f. (*b*) By what factor is the kinetic energy of the probe increased? Where does this energy come from?

Picture the Problem Let the direction the probe is moving after its elastic collision with Saturn be the positive direction. The probe gains kinetic energy at the expense of the kinetic energy of Saturn. We'll relate the velocity of approach relative to the center of mass to u_{rec} and then to v. Let the $+x$ direction be in the direction of the motion of Saturn.

(*a*) Relate the velocity of recession to the velocity of recession relative to the center of mass:

$$v = u_{rec} + v_{cm} \quad (1)$$

Find the velocity of approach:

$$u_{app} = -9.6\,\text{km/s} - 10.4\,\text{km/s} = -20.0\,\text{km/s}$$

Relate the relative velocity of approach to the relative velocity of recession for an elastic collision:

$$u_{rec} = -u_{app} = 20.0\,\text{km/s}$$

Because Saturn is so much more massive than the space probe:

$$v_{cm} = v_{Saturn} = 9.6\,\text{km/s}$$

Substitute numerical values in equation (1) and evaluate v:

$$v = 20\,\text{km/s} + 9.6\,\text{km/s} = \boxed{30\,\text{km/s}}$$

(*b*) Express the ratio of the final kinetic energy to the initial kinetic energy and simplify:

$$\frac{K_f}{K_i} = \frac{\frac{1}{2}Mv_{rec}^2}{\frac{1}{2}Mv_i^2} = \left(\frac{v_{rec}}{v_i}\right)^2$$

Substitute numerical values and evaluate K_f/K_i:

$$\frac{K_f}{K_i} = \left(\frac{29.6\,\text{km/s}}{10.4\,\text{km/s}}\right)^2 = \boxed{8.1}$$

The energy comes from an immeasurably small slowing of Saturn.

109 ••• **[SSM]** Your accident reconstruction team has been hired by the local police to analyze the following accident. A careless driver rear-ended a car that was halted at a stop sign. Just before impact, the driver slammed on his brakes, locking the wheels. The driver of the struck car had his foot solidly on the brake pedal, locking his brakes. The mass of the struck car was 900 kg, and that of the initially moving vehicle was 1200 kg. On collision, the bumpers of the two cars meshed. Police determine from the skid marks that after the collision the two cars moved 0.76 m together. Tests revealed that the coefficient of kinetic friction between the tires and pavement was 0.92. The driver of the moving car claims that he was traveling at less than 15 km/h as he approached the intersection. Is he telling the truth?

Picture the Problem Let the direction the moving car was traveling before the collision be the $+x$ direction. Let the numeral 1 denote this car and the numeral 2 the car that is stopped at the stop sign and the system include both cars and Earth. We can use conservation of momentum to relate the speed of the initially-moving car to the speed of the meshed cars immediately after their perfectly inelastic collision and conservation of energy to find the initial speed of the meshed cars.

Using conservation of momentum, relate the before-collision velocity to the after-collision velocity of the meshed cars:

$$p_i = p_f$$

or

$$m_1 v_1 = (m_1 + m_2)V$$

Solving for v_1 and simplifying yields:

$$v_1 = \frac{m_1 + m_2}{m_1} V = \left(1 + \frac{m_2}{m_1}\right) V \quad (1)$$

Using conservation of energy, relate the initial kinetic energy of the meshed cars to the work done by friction in bringing them to a stop:

$$\Delta K + \Delta E_{\text{thermal}} = 0$$

or, because $K_f = 0$ and $\Delta E_{\text{thermal}} = f\Delta s$,

$$-K_i + f_k \Delta s = 0$$

Substitute for K_i and, using $f_k = \mu_k F_n = \mu_k Mg$, eliminate f_k to obtain:

$$-\tfrac{1}{2} MV^2 + \mu_k Mg\Delta x = 0$$

Solving for V yields:

$$V = \sqrt{2\mu_k g\Delta x}$$

Substitute for V in equation (1) to obtain:

$$v_1 = \left(1 + \frac{m_2}{m_1}\right)\sqrt{2\mu_k g\Delta x}$$

Substitute numerical values and evaluate v_1:

$$v_1 = \left(1 + \frac{900\,\text{kg}}{1200\,\text{kg}}\right)\sqrt{2(0.92)(9.81\,\text{m/s}^2)(0.76\,\text{m})} = 6.48\,\text{m/s} = 23\,\text{km/h}$$

The driver was not telling the truth. He was traveling at 23 km/h.

111 ••• **[SSM]** A 1.00-kg block and a second block of mass M are both initially at rest on a frictionless inclined plane (Figure 8-56) Mass M rests against a spring that has a force constant of 11.0 kN/m. The distance along the plane between the two blocks is 4.00 m. The 1.00-kg block is released, making an elastic collision with the unknown block. The 1.00-kg block then rebounds a distance of 2.56 m back up the inclined plane. The block of mass M comes momentarily comes to rest 4.00 cm from its initial position. Find M.

Picture the Problem Choose the zero of gravitational potential energy at the location of the spring's maximum compression. Let the system include the spring, the blocks, and Earth. Then the net external force is zero as is work done against friction. We can use conservation of energy to relate the energy transformations taking place during the evolution of this system.

Apply conservation of energy to the system: $\Delta K + \Delta U_g + \Delta U_s = 0$

Because $\Delta K = 0$: $\Delta U_g + \Delta U_s = 0$

Express the change in the gravitational potential energy: $\Delta U_g = -mg\Delta h - Mgx\sin\theta$

Express the change in the potential energy of the spring: $\Delta U_s = \frac{1}{2}kx^2$

Substitute to obtain: $-mg\Delta h - Mgx\sin\theta + \frac{1}{2}kx^2 = 0$

Solving for M and simplifying yields: $M = \dfrac{\frac{1}{2}kx^2 - mg\Delta h}{gx\sin 30°} = \dfrac{kx}{g} - \dfrac{2m\Delta h}{x}$

Relate Δh to the initial and rebound positions of the block whose mass is m:

$$\Delta h = (4.00\,\text{m} - 2.56\,\text{m})\sin 30° = 0.72\,\text{m}$$

Substitute numerical values and evaluate M:

$$M = \frac{(11.0\times10^3\,\text{N/m})(0.0400\,\text{m})}{9.81\,\text{m/s}^2} - \frac{2(1.00\,\text{kg})(0.72\,\text{m})}{0.0400\,\text{m}} = \boxed{8.9\,\text{kg}}$$

115 ••• **[SSM]** Two astronauts at rest face each other in space. One, with mass m_1, throws a ball of mass m_b to the other, whose mass is m_2. She catches the ball and throws it back to the first astronaut. Following each throw the ball has a speed of v relative to the thrower. After each has made one throw and one catch, (*a*) How fast are the astronauts moving? (*b*) How much has the two-astronaut system's kinetic energy changed and where did this energy come from?

Picture the Problem Let the direction that astronaut 1 first throws the ball be the positive direction and let v_b be the initial speed of the ball in the laboratory frame. Note that each collision is perfectly inelastic. We can apply conservation of momentum and the definition of the speed of the ball relative to the thrower to each of the perfectly inelastic collisions to express the final speeds of each astronaut after one throw and one catch.

(*a*) Use conservation of linear momentum to relate the speeds of astronaut 1 and the ball after the first throw:

$$m_1v_1 + m_bv_b = 0 \quad (1)$$

Relate the speed of the ball in the laboratory frame to its speed relative to astronaut 1:

$$v = v_b - v_1 \quad (2)$$

Eliminate v_b between equations (1) and (2) and solve for v_1:

$$v_1 = -\frac{m_b}{m_1 + m_b}v \quad (3)$$

Substitute equation (3) in equation (2) and solve for v_b:

$$v_b = \frac{m_1}{m_1 + m_b}v \quad (4)$$

Apply conservation of linear momentum to express the speed of astronaut 2 and the ball after the first catch:

$$0 = m_bv_b = (m_2 + m_b)v_2 \quad (5)$$

Solving for v_2 yields:

$$v_2 = \frac{m_b}{m_2 + m_b}v_b \quad (6)$$

Express v_2 in terms of v by substituting equation (4) in equation (6):

$$v_2 = \frac{m_b}{m_2 + m_b}\frac{m_1}{m_1 + m_b}v = \left[\frac{m_bm_1}{(m_2 + m_b)(m_1 + m_b)}\right]v \quad (7)$$

Use conservation of momentum to express the speed of astronaut 2 and the ball after she throws the ball:

$$(m_2 + m_b)v_2 = m_bv_{bf} + m_2v_{2f} \quad (8)$$

Relate the speed of the ball in the laboratory frame to its speed relative to astronaut 2:

$$v = v_{2f} - v_{bf} \quad (9)$$

Eliminate v_{bf} between equations (8) and (9) and solve for v_{2f}:

$$v_{2f} = \boxed{\left(\frac{m_b}{m_2 + m_b}\right)\left(1 + \frac{m_1}{m_1 + m_b}\right)v} \quad (10)$$

Substitute equation (10) in equation (9) and solve for v_{bf}:

$$v_{bf} = \left[\frac{m_b}{m_2 + m_b} - 1\right]\left[1 + \frac{m_1}{m_1 + m_b}\right]v \quad (11)$$

Apply conservation of momentum to express the speed of astronaut 1 and the ball after she catches the ball:

$$(m_1 + m_b)v_{1f} = m_b v_{bf} + m_1 v_1 \quad (12)$$

Using equations (3) and (11), eliminate v_{bf} and v_1 in equation (12) and solve for v_{1f}:

$$v_{1f} = \boxed{-\frac{m_2 m_b(2m_1 + m_b)}{(m_1 + m_b)^2(m_2 + m_b)}v}$$

(*b*) The change in the kinetic energy of the system is:

$$\Delta K = K_f - K_i$$

or, because $K_i = 0$,

$$\Delta K = K_f = K_{1f} + K_{2f} = \tfrac{1}{2}m_1 v_{1f}^2 + \tfrac{1}{2}m_2 v_{2f}^2$$

Substitute for v_{1f} and v_{2f} to obtain:

$$\Delta K = \tfrac{1}{2}m_1\left(-\frac{m_2 m_b(2m_1 + m_b)}{(m_1 + m_b)^2(m_2 + m_b)}\right)^2 v^2 + \tfrac{1}{2}m_2\left(\frac{m_b}{m_2 + m_b}\right)^2\left(1 + \frac{m_1}{m_1 + m_b}\right)^2 v^2$$

Simplify to obtain:

$$\Delta K = \boxed{\frac{1}{2}\frac{m_2 m_b^2(2m_1 + m_b)^2}{(m_2 + m_b)^2(m_1 + m_b)^2}\left(1 + \frac{m_1 m_2}{(m_1 + m_b)^2}\right)v^2}$$

This additional energy came from chemical energy in the astronaut's bodies.

Chapter 9
Rotation

Conceptual Problems

7 • **[SSM]** During a baseball game, the pitcher has a blazing fastball. You have not been able to swing the bat in time to hit the ball. You are now just trying to make the bat contact the ball, hit the ball foul, and avoid a strikeout. So, you decide to take your coach's advice and grip the bat high rather than at the very end. This change should increase bat speed; thus you will be able to swing the bat quicker and increase your chances of hitting the ball. Explain how this theory works in terms of the moment of inertia, angular acceleration, and torque of the bat.

Determine the Concept The closer the rotation axis to the center of mass, the smaller the moment of inertia of the bat. By choking up, you are rotating the bat about an axis closer to the center of mass, thus reducing the bat's moment of inertia. The smaller the moment of inertia the larger the angular acceleration (a quicker bat).

11 • **[SSM]** The motor of a merry-go-round exerts a constant torque on it. As it speeds up from rest, the power output of the motor (*a*) is constant, (*b*) increases linearly with the angular speed of the merry-go-round, (*c*) is zero. (*d*) None of the above.

Determine the Concept The power delivered by the constant torque is the product of the torque and the angular speed of the merry-go-round. Because the constant torque causes the merry-go-round to accelerate, the power output increases linearly with the angular speed of the merry-go-round. $\boxed{(b)}$ is correct.

21 •• **[SSM]** A spool is free to rotate about a fixed axis (see Figure 9-42*a*), and a string wrapped around the axle causes the spool to rotate in a counterclockwise direction. However, if the spool is set on a horizontal tabletop (Figure 9-42*b*), the spool instead (given sufficient frictional force between the table and the spool) rotates in a clockwise direction, and rolls to the right. By considering torque about the appropriate axes, show that these conclusions are consistent with Newton's second law for rotations.

Determine the Concept First, visualize the situation. The string pulling to the right exerts a torque on the spool with a moment arm equal in length to the radius of the inner portion of the spool. When the spool is freely rotating about that axis, then the torque due to the pulling string causes a counter clockwise rotation. Second, in the situation in which the spool is resting on the horizontal tabletop, one should (for ease of understanding) consider torques not about the central axle of the spool, but about the point of contact with the tabletop. In this situation, there is only one force that can produce a torque – the applied force. The motion of the spool can then be understood in terms of the force applied by the string and the moment arm equal to the difference between the outer radius and the inner radius. This torque will cause a clockwise rotation about the point of contact between spool and table – and thus the spool rolls to the right (whereas we might have thought the spool would rotate in a counter-clockwise sense, and thus move left).

Angular Velocity, Angular Speed and Angular Acceleration

29 • **[SSM]** A wheel released from rest is rotating with constant angular acceleration of 2.6 rad/s^2. At 6.0 s after its release: (*a*) What is its angular speed? (*b*) Through what angle has the wheel turned? (*c*) How many revolutions has it completed? (*d*) What is the linear speed, and what is the magnitude of the linear acceleration, of a point 0.30 m from the axis of rotation?

Picture the Problem Because the angular acceleration is constant, we can find the various physical quantities called for in this problem by using constant-acceleration equations.

(*a*) Using a constant-acceleration equation, relate the angular speed of the wheel to its angular acceleration:

$$\omega = \omega_0 + \alpha\Delta t$$

or, when $\omega_0 = 0$,

$$\omega = \alpha\Delta t$$

Evaluate ω when $\Delta t = 6.0$ s:

$$\omega = \left(2.6\,\text{rad/s}^2\right)(6.0\,\text{s}) = 15.6\,\text{rad/s}$$

$$= \boxed{16\,\text{rad/s}}$$

(*b*) Using another constant-acceleration equation, relate the angular displacement to the wheel's angular acceleration and the time it has been accelerating:

$$\Delta\theta = \omega_0\Delta t + \tfrac{1}{2}\alpha(\Delta t)^2$$

or, when $\omega_0 = 0$,

$$\Delta\theta = \tfrac{1}{2}\alpha(\Delta t)^2$$

Evaluate $\Delta\theta$ when $\Delta t = 6.0$ s:

$$\Delta\theta(6\,\text{s}) = \tfrac{1}{2}\left(2.6\,\text{rad/s}^2\right)(6.0\,\text{s})^2 = 46.8\,\text{rad}$$

$$= \boxed{47\,\text{rad}}$$

(*c*) Convert $\Delta\theta(6.0\,\text{s})$ from radians to revolutions:

$$\Delta\theta(6.0\,\text{s}) = 46.8\,\text{rad} \times \frac{1\,\text{rev}}{2\pi\,\text{rad}} = \boxed{7.4\,\text{rev}}$$

(*d*) Relate the angular speed of the particle to its tangential speed and evaluate the latter when $\Delta t = 6.0$ s:

$$v = r\omega = (0.30\,\text{m})(15.6\,\text{rad/s}) = \boxed{4.7\,\text{m/s}}$$

Relate the resultant acceleration of the point to its tangential and centripetal accelerations when $\Delta t = 6.0$ s:

$$a = \sqrt{a_t^2 + a_c^2} = \sqrt{(r\alpha)^2 + (r\omega^2)^2} = r\sqrt{\alpha^2 + \omega^4}$$

Substitute numerical values and evaluate *a*:

$$a = (0.30\,\text{m})\sqrt{(2.6\,\text{rad/s}^2)^2 + (15.6\,\text{rad/s})^4} = \boxed{73\,\text{m/s}^2}$$

37 •• **[SSM]** The tape in a standard VHS videotape cassette has a total length of 246 m, which is enough for the tape to play for 2.0 h (Figure 9-44). As the tape starts, the full reel has a 45-mm outer radius and a 12-mm inner radius. At some point during the play, both reels have the same angular speed. Calculate this angular speed in radians per second and in revolutions per minute. (*Hint: Between the two reels the tape moves at constant speed.*)

Picture the Problem The two tapes will have the same tangential and angular velocities when the two reels are the same size, i.e., have the same area. We can calculate the tangential speed of the tape from its length and running time and relate the angular speed to the constant tangential speed and the radius of the reels when they are turning with the same angular speed.

Relate the angular speed of the tape to its tangential speed:

$$\omega = \frac{v}{r} \qquad (1)$$

Letting R_f represent the outer radius of the reel when the reels have the same area, express the condition that they have the same speed:

$$\pi R_f^2 - \pi r^2 = \tfrac{1}{2}(\pi R^2 - \pi r^2)$$

Solving for R_f yields:

$$R_f = \sqrt{\frac{R^2 + r^2}{2}}$$

Substitute numerical values and evaluate R_f:

$$R_f = \sqrt{\frac{(45\,\text{mm})^2 + (12\,\text{mm})^2}{2}} = 32.9\,\text{mm}$$

Find the tangential speed of the tape from its length and running time:

$$v = \frac{L}{\Delta t} = \frac{246\,\text{m} \times \dfrac{100\,\text{cm}}{\text{m}}}{2.0\,\text{h} \times \dfrac{3600\,\text{s}}{\text{h}}} = 3.42\,\text{cm/s}$$

Substitute in equation (1) and evaluate ω:

$$\omega = \frac{3.42\,\text{cm/s}}{32.9\,\text{mm} \times \dfrac{1\,\text{cm}}{10\,\text{mm}}} = 1.04\,\text{rad/s}$$

$$= \boxed{1.0\,\text{rad/s}}$$

Convert 1.04 rad/s to rev/min:

$$1.04\,\text{rad/s} = 1.04\frac{\text{rad}}{\text{s}} \times \frac{1\,\text{rev}}{2\pi\,\text{rad}} \times \frac{60\,\text{s}}{\text{min}}$$

$$= \boxed{9.9\,\text{rev/min}}$$

Calculating the Moment of Inertia

41 • **[SSM]** Four particles, one at each of the four corners of a square with 2.0-m long edges, are connected by massless rods (Figure 9-45). The masses of the particles are $m_1 = m_3 = 3.0$ kg and $m_2 = m_4 = 4.0$ kg. Find the moment of inertia of the system about the z axis.

Picture the Problem The moment of inertia of a system of particles with respect to a given axis is the sum of the products of the mass of each particle and the square of its distance from the given axis.

Use the definition of the moment of inertia of a system of four particles to obtain:

$$I = \sum_i m_i r_i^2$$
$$= m_1 r_1^2 + m_2 r_2^2 + m_3 r_3^2 + m_4 r_4^2$$

Substitute numerical values and evaluate $I_{z\text{ axis}}$:

$$I_{z\text{ axis}} = (3.0\,\text{kg})(2.0\,\text{m})^2 + (4.0\,\text{kg})\left(2\sqrt{2}\,\text{m}\right)^2 + (4.0\,\text{kg})(2.0\,\text{m})^2 + (3.0\,\text{kg})(0)^2$$
$$= \boxed{60\,\text{kg}\cdot\text{m}^2}$$

53 ••• **[SSM]** Use integration to show that the moment of inertia of a thin spherical shell of radius R and mass m about an axis through its center is $2mR^2/3$.

Picture the Problem We can derive the given expression for the moment of inertia of a spherical shell by following the procedure outlined in the problem statement.

Find the moment of inertia of a sphere, with respect to an axis through a diameter, in Table 9-1:

$$I = \tfrac{2}{5} mR^2$$

Express the mass of the sphere as a function of its density and radius:

$$m = \tfrac{4}{3}\pi \rho R^3$$

Substitute for m to obtain:

$$I = \tfrac{8}{15}\pi \rho R^5$$

Express the differential of this expression:

$$dI = \tfrac{8}{3}\pi \rho R^4 dR \qquad (1)$$

Express the increase in mass dm as the radius of the sphere increases by dR:

$$dm = 4\pi \rho R^2 dR \qquad (2)$$

Eliminate dR between equations (1) and (2) to obtain:

$$dI = \tfrac{2}{3} R^2 dm$$

Integrate over the mass of the spherical shell to obtain:

$$I_{\text{spherical shell}} = \boxed{\tfrac{2}{3} mR^2}$$

Torque, Moment of Inertia, and Newton's Second Law for Rotation

59 • **[SSM]** A 2.5-kg 11-cm-radius cylinder, initially at rest, is free to rotate about the axis of the cylinder. A rope of negligible mass is wrapped around it and pulled with a force of 17 N. Assuming that the rope does not slip, find (*a*) the torque exerted on the cylinder by the rope, (*b*) the angular acceleration of the cylinder, and (*c*) the angular speed of the cylinder after 0.50 s.

Picture the Problem We can find the torque exerted by the 17-N force from the definition of torque. The angular acceleration resulting from this torque is related to the torque through Newton's 2nd law in rotational form. Once we know the angular acceleration, we can find the angular speed of the cylinder as a function of time.

(*a*) The torque exerted by the rope is:

$$\tau = F\ell = (17\,\text{N})(0.11\,\text{m}) = 1.87\,\text{N}\cdot\text{m} = \boxed{1.9\,\text{N}\cdot\text{m}}$$

(*b*) Use Newton's 2nd law in rotational form to relate the acceleration resulting from this torque to the torque:

$$\alpha = \frac{\tau}{I}$$

Express the moment of inertia of the cylinder with respect to its axis of rotation:

$$I = \tfrac{1}{2}MR^2$$

Substitute for I and simplify to obtain:

$$\alpha = \frac{2\tau}{MR^2}$$

Substitute numerical values and evaluate α:

$$\alpha = \frac{2(1.87\,\text{N}\cdot\text{m})}{(2.5\,\text{kg})(0.11\,\text{m})^2} = 124\,\text{rad/s}^2$$

$$= \boxed{1.2\times10^2\,\text{rad/s}^2}$$

(*c*) Using a constant-acceleration equation, express the angular speed of the cylinder as a function of time:

$$\omega = \omega_0 + \alpha t$$

or, because $\omega_0 = 0$,

$$\omega = \alpha t$$

Substitute numerical values and evaluate ω (5.0 s):

$$\omega(5.0\,\text{s}) = (124\,\text{rad/s}^2)(5.0\,\text{s})$$

$$= \boxed{6.2\times10^2\,\text{rad/s}}$$

Energy Methods Including Rotational Kinetic Energy

65 • [SSM] A 1.4-kg 15-cm-diameter solid sphere is rotating about its diameter at 70 rev/min. (*a*) What is its kinetic energy? (*b*) If an additional 5.0 mJ of energy are added to the kinetic energy, what is the new angular speed of the sphere?

Picture the Problem We can find the kinetic energy of this rotating ball from its angular speed and its moment of inertia. In Part (*b*) we can use the work-kinetic energy theorem to find the angular speed of the sphere when additional kinetic energy has been added to the sphere.

(*a*) The initial rotational kinetic energy of the ball is:

$$K_\text{i} = \tfrac{1}{2}I\omega_\text{i}^2$$

Express the moment of inertia of the ball with respect to its diameter:

$$I = \tfrac{2}{5}MR^2$$

Substitute for I to obtain: $K_i = \frac{1}{5}MR^2\omega_i^2$

Substitute numerical values and evaluate K:

$$K_i = \tfrac{1}{5}(1.4\,\text{kg})(0.075\,\text{m})^2\left(70\frac{\text{rev}}{\text{min}}\times\frac{2\pi\,\text{rad}}{\text{rev}}\times\frac{1\,\text{min}}{60\,\text{s}}\right)^2 = 84.6\,\text{mJ} = \boxed{85\,\text{mJ}}$$

(*b*) Apply the work-kinetic energy theorem to the sphere to obtain:

$$W = \Delta K = K_f - K_i$$

or

$$W = \tfrac{1}{2}I\omega_f^2 - K_i \Rightarrow \omega_f = \sqrt{\frac{2(W+K_i)}{I}}$$

Substitute for I and simplify to obtain:

$$\omega_f = \sqrt{\frac{2(W+K_i)}{\frac{2}{5}MR^2}} = \sqrt{\frac{5(W+K_i)}{MR^2}}$$

Substitute numerical values and evaluate ω_f:

$$\omega_f = \sqrt{\frac{5(84.6\,\text{mJ}+5.0\,\text{mJ})}{(1.4\,\text{kg})(7.5\,\text{cm})^2}}$$
$$= 7.542\frac{\text{rad}}{\text{s}}\times\frac{1\,\text{rev}}{2\pi\,\text{rad}}\times\frac{60\,\text{s}}{\text{min}}$$
$$= \boxed{72\,\text{rev/min}}$$

67 •• **[SSM]** A 2000-kg block is lifted at a constant speed of 8.0 cm/s by a steel cable that passes over a massless pulley to a motor-driven winch (Figure 9-53). The radius of the winch drum is 30 cm. (*a*) What is the tension in the cable? (*b*) What torque does the cable exert on the winch drum? (*c*) What is the angular speed of the winch drum? (*d*) What power must be developed by the motor to drive the winch drum?

Picture the Problem Because the load is not being accelerated, the tension in the cable equals the weight of the load. The role of the massless pulley is to change the direction the force (tension) in the cable acts.

(*a*) Because the block is lifted at constant speed:

$$T = mg = (2000\,\text{kg})(9.81\,\text{m/s}^2)$$
$$= \boxed{19.6\,\text{kN}}$$

(*b*) Apply the definition of torque at the winch drum:

$$\tau = Tr = (19.6\,\text{kN})(0.30\,\text{m})$$
$$= \boxed{5.9\,\text{kN}\cdot\text{m}}$$

(*c*) Relate the angular speed of the winch drum to the rate at which the load is being lifted (the tangential speed of the cable on the drum):

$$\omega = \frac{v}{r} = \frac{0.080\,\text{m/s}}{0.30\,\text{m}} = \boxed{0.27\,\text{rad/s}}$$

(*d*) The power developed by the motor in terms is the product of the tension in the cable and the speed with which the load is being lifted:

$$P = Tv = (19.6\,\text{kN})(0.080\,\text{m/s}) = \boxed{1.6\,\text{kW}}$$

Pulleys, Yo-Yos, and Hanging Things

71 •• **[SSM]** The system shown in Figure 9-55consists of a 4.0-kg block resting on a frictionless horizontal ledge. This block is attached to a string that passes over a pulley, and the other end of the string is attached to a hanging 2.0-kg block. The pulley is a uniform disk of radius 8.0 cm and mass 0.60 kg. Find the acceleration of each block and the tension in the string.

Picture the Problem The diagrams show the forces acting on each of the masses and the pulley. We can apply Newton's 2nd law to the two blocks and the pulley to obtain three equations in the unknowns T_1, T_2, and a.

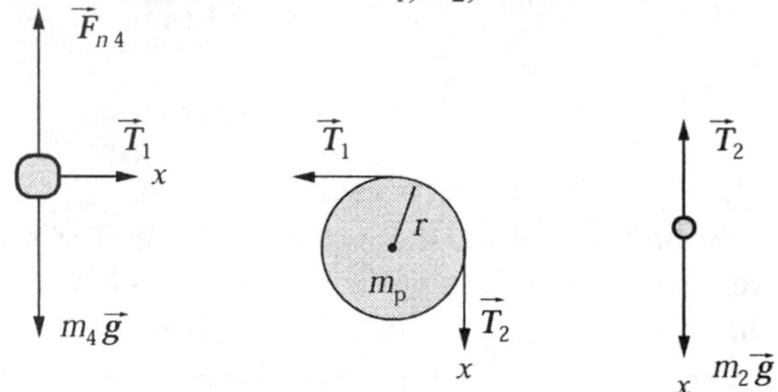

Apply Newton's 2nd law to the two blocks and the pulley:

$$\sum F_x = T_1 = m_4 a, \qquad (1)$$

$$\sum \tau_p = (T_2 - T_1)r = I_p\alpha, \qquad (2)$$

and

$$\sum F_x = m_2 g - T_2 = m_2 a \qquad (3)$$

Substitute for I_p and α in equation (2) to obtain:

$$T_2 - T_1 = \tfrac{1}{2}M_p a \qquad (4)$$

Eliminate T_1 and T_2 between equations (1), (3) and (4) and solve for a:

$$a = \frac{m_2 g}{m_2 + m_4 + \frac{1}{2}M_p}$$

Substitute numerical values and evaluate a:

$$a = \frac{(2.0\,\text{kg})(9.81\,\text{m/s}^2)}{2.0\,\text{kg} + 4.0\,\text{kg} + \frac{1}{2}(0.60\,\text{kg})}$$
$$= 3.11\,\text{m/s}^2 = \boxed{3.1\,\text{m/s}^2}$$

Using equation (1), evaluate T_1:

$$T_1 = (4.0\,\text{kg})(3.11\,\text{m/s}^2) = \boxed{12\,\text{N}}$$

Solve equation (3) for T_2:

$$T_2 = m_2(g - a)$$

Substitute numerical values and evaluate T_2:

$$T_2 = (2.0\,\text{kg})(9.81\,\text{m/s}^2 - 3.11\,\text{m/s}^2)$$
$$= \boxed{13\,\text{N}}$$

Remarks: Note that the only effect of the pulley is to change the direction of the force in the string.

79 •• **[SSM]** Two objects are attached to ropes that are attached to two wheels on a common axle, as shown in Figure 9-60. The two wheels are attached together so that they form a single rigid object. The moment of inertia of the rigid object is 40 kg·m^2. The radii of the wheels are R_1 = 1.2 m and R_2 = 0.40 m. (*a*) If m_1= 24 kg, find m_2 such that there is no angular acceleration of the wheels. (*b*) If 12 kg is placed on top of m_1, find the angular acceleration of the wheels and the tensions in the ropes.

Picture the Problem The following diagram shows the forces acting on both objects and the pulley for the conditions of Part (*b*). By applying Newton's 2nd law of motion, we can obtain a system of three equations in the unknowns T_1, T_2, and α that we can solve simultaneously.

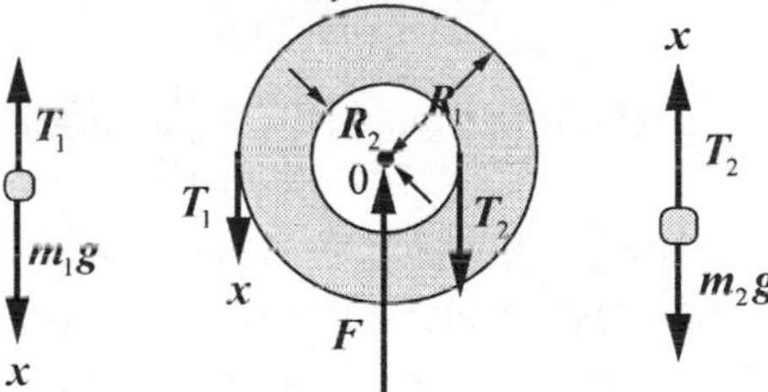

(*a*) When the system does not accelerate, $\boldsymbol{T_1 = m_1 g}$ and $\boldsymbol{T_2 = m_2 g}$. Under these conditions:

$$\sum \tau_0 = m_1 g R_1 - m_2 g R_2 = 0$$

Solving for m_2 yields:

$$m_2 = m_1 \frac{R_1}{R_2}$$

Substitute numerical values and evaluate m_2:

$$m_2 = (24\,\text{kg})\frac{1.2\,\text{m}}{0.40\,\text{m}} = \boxed{72\,\text{kg}}$$

(*b*) Apply Newton's 2nd law to the objects and the pulley:

$$\sum F_x = m_1 g - T_1 = m_1 a, \qquad (1)$$
$$\sum \tau_0 = T_1 R_1 - T_2 R_2 = I_0 \alpha, \qquad (2)$$
and
$$\sum F_x = T_2 - m_2 g = m_2 a \qquad (3)$$

Eliminate a in favor of α in equations (1) and (3) and solve for T_1 and T_2:

$$T_1 = m_1(g - R_1\alpha) \qquad (4)$$
and
$$T_2 = m_2(g + R_2\alpha) \qquad (5)$$

Substitute for T_1 and T_2 in equation (2) and solve for α to obtain:

$$\alpha = \frac{(m_1 R_1 - m_2 R_2)g}{m_1 R_1^2 + m_2 R_2^2 + I_0}$$

Substitute numerical values and evaluate α:

$$\alpha = \frac{[(36\,\text{kg})(1.2\,\text{m}) - (72\,\text{kg})(0.40\,\text{m})](9.81\,\text{m/s}^2)}{(36\,\text{kg})(1.2\,\text{m})^2 + (72\,\text{kg})(0.40\,\text{m})^2 + 40\,\text{kg}\cdot\text{m}^2} = 1.37\,\text{rad/s}^2 = \boxed{1.4\,\text{rad/s}^2}$$

Substitute numerical values in equation (4) to find T_1:

$$T_1 = (36\,\text{kg})[9.81\,\text{m/s}^2 - (1.2\,\text{m})(1.37\,\text{rad/s}^2)] = \boxed{0.29\,\text{kN}}$$

Substitute numerical values in equation (5) to find T_2:

$$T_2 = (72\,\text{kg})[9.81\,\text{m/s}^2 + (0.40\,\text{m})(1.37\,\text{rad/s}^2)] = \boxed{0.75\,\text{kN}}$$

81 •• **[SSM]** A uniform cylinder of mass m_1 and radius R is pivoted on frictionless bearings. A massless string wrapped around the cylinder is connected to a block of mass m_2 that is on a frictionless incline of angle θ, as shown in Figure 9-62. The system is released from rest with the block a vertical distance h above the bottom of the incline. (*a*) What is the acceleration of the block? (*b*) What is the tension in the string? (*c*) What is the speed of the block as it reaches the bottom of the incline? (*d*) Evaluate your answers for the special case where $\theta = 90°$ and $m_1 = 0$. Are your answers what you would expect for this special case? Explain.

Picture the Problem Let the zero of gravitational potential energy be at the bottom of the incline. By applying Newton's 2nd law to the cylinder and the block we can obtain simultaneous equations in a, T, and α from which we can express a and T. By applying the conservation of energy, we can derive an expression for the speed of the block when it reaches the bottom of the incline.

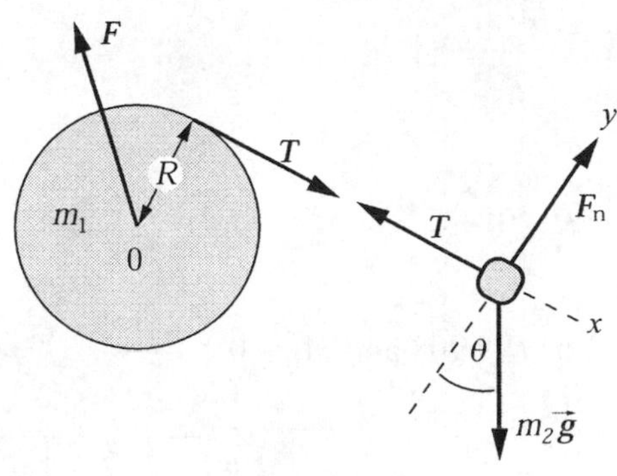

(*a*) Apply Newton's 2nd law to the cylinder and the block:

$$\sum \tau_0 = TR = I_0\alpha \qquad (1)$$

and

$$\sum F_x = m_2 g \sin\theta - T = m_2 a \qquad (2)$$

Substitute for α and I_0 in equation (1), solve for T, and substitute in equation (2) and solve for a to obtain:

$$a = \boxed{\dfrac{g\sin\theta}{1+\dfrac{m_1}{2m_2}}}$$

(*b*) Substituting for a in equation (2) and solve for T yields:

$$T = \boxed{\dfrac{\frac{1}{2}m_1 g\sin\theta}{1+\dfrac{m_1}{2m_2}}}$$

(*c*) Noting that the block is released from rest, express the total energy of the system when the block is at height h:

$$E = U + K = \boxed{m_2 gh}$$

(*d*) Use the fact that this system is conservative to express the total energy at the bottom of the incline:

$$E_{\text{bottom}} = \boxed{m_2 gh}$$

(*e*) Express the total energy of the system when the block is at the bottom of the incline in terms of its kinetic energies:

$$E_{\text{bottom}} = K_{\text{tran}} + K_{\text{rot}} = \tfrac{1}{2}m_2 v^2 + \tfrac{1}{2}I_0\omega^2$$

Substitute for ω and I_0 to obtain:

$$\tfrac{1}{2}m_2 v^2 + \tfrac{1}{2}\left(\tfrac{1}{2}m_1 r^2\right)\frac{v^2}{r^2} = m_2 gh$$

Solving for v yields:

$$v = \boxed{\sqrt{\frac{2gh}{1+\frac{m_1}{2m_2}}}}$$

For $\theta = 0$:

$$\boldsymbol{a} = \boxed{0} \text{ and } \boldsymbol{T} = \boxed{0}$$

(*f*) For $\theta = 90°$ and $m_1 = 0$:

$$\boldsymbol{a} = \boxed{g},\ T = \boxed{0},\ \text{and } v = \boxed{\sqrt{2gh}}$$

Objects Rotating and Rolling Without Slipping

85 •• **[SSM]** In 1993 a giant 400-kg yo-yo with a radius of 1.5 m was dropped from a crane at a height of 57 m. One end of the string was tied to the top of the crane, so the yo-yo unwound as it descended. Assuming that the axle of the yo-yo had a radius of 0.10 m, estimate its linear speed at the end of the fall.

Picture the Problem The forces acting on the yo-yo are shown in the figure. We can use a constant-acceleration equation to relate the velocity of descent at the end of the fall to the yo-yo's acceleration and Newton's 2nd law in both translational and rotational form to find the yo-yo's acceleration.

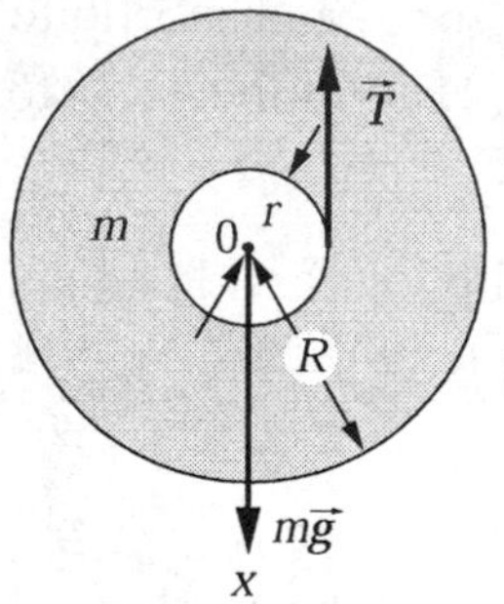

Using a constant-acceleration equation, relate the yo-yo's final speed to its acceleration and fall distance:

$$v^2 = v_0^2 + 2a\Delta h$$

or, because $v_0 = 0$,

$$v = \sqrt{2a\Delta h} \qquad (1)$$

Use Newton's 2nd law to relate the forces that act on the yo-yo to its acceleration:

$$\sum F_x = mg - T = ma \qquad (2)$$

and

$$\sum \tau_0 = Tr = I_0\alpha \qquad (3)$$

Use $a = r\alpha$ to eliminate α in equation (3)

$$Tr = I_0 \frac{a}{r} \qquad (4)$$

Eliminate T between equations (2) and (4) to obtain:

$$mg - \frac{I_0}{r^2}a = ma \qquad (5)$$

Substitute $\frac{1}{2}mR^2$ for I_0 in equation (5):

$$mg - \frac{\frac{1}{2}mR^2}{r^2}a = ma \Rightarrow a = \frac{g}{1+\frac{R^2}{2r^2}}$$

Substitute numerical values and evaluate a:

$$a = \frac{9.81\,\text{m/s}^2}{1+\frac{(1.5\,\text{m})^2}{2(0.10\,\text{m})^2}} = 0.0864\,\text{m/s}^2$$

Substitute in equation (1) and evaluate v:

$$v = \sqrt{2(0.0864\,\text{m/s}^2)(57\,\text{m})} = \boxed{3.1\,\text{m/s}}$$

93 •• **[SSM]** A uniform thin cylindrical shell and a solid cylinder roll horizontally without slipping. The speed of the cylindrical shell is v. The cylinder and the hollow cylinder encounter an incline that they climb without slipping. If the maximum height they reach is the same, find the initial speed v' of the solid cylinder.

Picture the Problem Let the subscripts u and h refer to the uniform and thin-walled spheres, respectively. Because the cylinders climb to the same height, their kinetic energies at the bottom of the incline must be equal.

Express the total kinetic energy of the thin-walled cylinder at the bottom of the inclined plane:

$$K_\text{h} = K_\text{trans} + K_\text{rot} = \tfrac{1}{2}m_\text{h}v^2 + \tfrac{1}{2}I_\text{h}\omega^2$$
$$= \tfrac{1}{2}m_\text{h}v^2 + \tfrac{1}{2}\left(m_\text{h}r^2\right)\frac{v^2}{r^2} = m_\text{h}v^2$$

Express the total kinetic energy of the solid cylinder at the bottom of the inclined plane:

$$\boldsymbol{K}_\text{u} = \boldsymbol{K}_\text{trans} + \boldsymbol{K}_\text{rot} = \tfrac{1}{2}\boldsymbol{m}_\text{u}\boldsymbol{v}'^2 + \tfrac{1}{2}\boldsymbol{I}_\text{u}\boldsymbol{\omega}'^2$$
$$= \tfrac{1}{2}\boldsymbol{m}_\text{u}\boldsymbol{v}'^2 + \tfrac{1}{2}\left(\tfrac{1}{2}\boldsymbol{m}_\text{u}\boldsymbol{r}^2\right)\frac{\boldsymbol{v}'^2}{\boldsymbol{r}^2} = \tfrac{3}{4}\boldsymbol{m}_\text{u}\boldsymbol{v}'^2$$

Because the cylinders climb to the same height:

$$\tfrac{3}{4}m_\text{u}v'^2 = m_\text{u}gh$$

and

$$m_\text{h}v^2 = m_\text{h}gh$$

Divide the first of these equations by the second:

$$\frac{\frac{3}{4}m_\text{u}v'^2}{m_\text{h}v^2} = \frac{m_\text{u}gh}{m_\text{h}gh}$$

Simplify to obtain:

$$\frac{3v'^2}{4v^2} = 1 \Rightarrow v' = \boxed{\sqrt{\frac{4}{3}}v}$$

99 ••• **[SSM]** Two large gears that are being designed as part of a large machine and are shown in Figure 9-66; each is free to rotate about a fixed axis through its center. The radius and moment of inertia of the smaller gear are 0.50 m and 1.0 kg·m^2, respectively, and the radius and moment of inertia of the larger gear are 1.0 m and 16 kg·m^2, respectively. The lever attached to the smaller gear is 1.0 m long and has a negligible mass. (*a*) If a worker will typically apply a force of 2.0 N to the end of the lever, as shown, what will be the angular accelerations of gears the two gears? (*b*) Another part of the machine (not shown) will apply a force tangentially to the outer edge of the larger gear to temporarily keep the gear system from rotating. What should the magnitude and direction of this force (clockwise or counterclockwise) be?

Picture the Problem The forces responsible for the rotation of the gears are shown in the diagram to the right. The forces acting through the centers of mass of the two gears have been omitted because they produce no torque. We can apply Newton's 2nd law in rotational form to obtain the equations of motion of the gears and the not slipping condition to relate their angular accelerations.

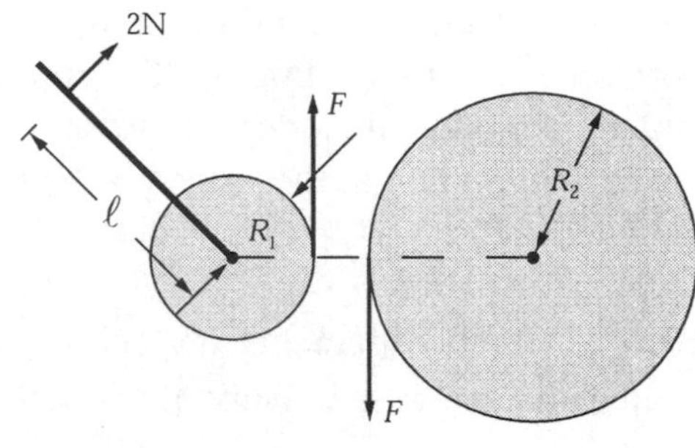

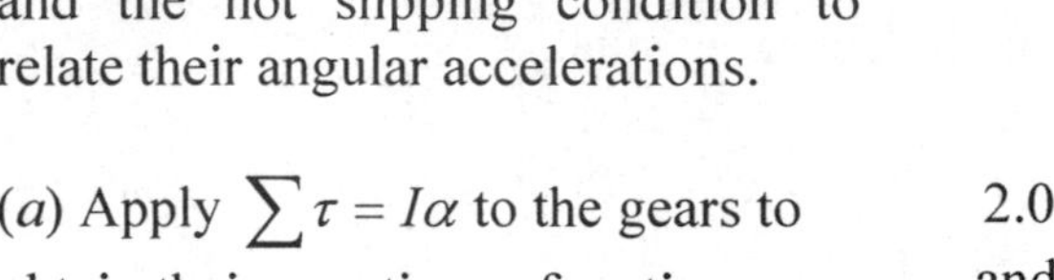

(*a*) Apply $\sum \tau = I\alpha$ to the gears to obtain their equations of motion:

$$2.0\ \text{N}\cdot\text{m} - \boldsymbol{F}\boldsymbol{R}_1 = \boldsymbol{I}_1\boldsymbol{\alpha}_1 \qquad (1)$$

and

$$FR_2 = I_2\alpha_2 \qquad (2)$$

where F is the force keeping the gears from slipping with respect to each other.

Because the gears do not slip relative to each other, the tangential accelerations of the points where they are in contact must be the same:

$$R_1\alpha_1 = R_2\alpha_2$$

or

$$\alpha_2 = \frac{R_1}{R_2}\alpha_1 = \tfrac{1}{2}\alpha_1 \qquad (3)$$

Divide equation (1) by R_1 to obtain:

$$\frac{2.0\ \text{N}\cdot\text{m}}{R_1} - F = \frac{I_1}{R_1}\alpha_1$$

Divide equation (2) by R_2 to obtain:

$$F = \frac{I_2}{R_2}\alpha_2$$

Adding these equations yields:

$$\frac{2.0\ \text{N}\cdot\text{m}}{R_1} = \frac{I_1}{R_1}\alpha_1 + \frac{I_2}{R_2}\alpha_2$$

Use equation (3) to eliminate α_2:

$$\frac{2.0\ \text{N}\cdot\text{m}}{R_1}=\frac{I_1}{R_1}\alpha_1+\frac{I_2}{2R_2}\alpha_1$$

Solving for α_1 yields:

$$\alpha_1=\frac{\frac{2.0\ \text{N}\cdot\text{m}}{R_1}}{I_1+\frac{R_1}{2R_2}I_2}$$

Substitute numerical values and evaluate α_1:

$$\alpha_1=\frac{2.0\ \text{N}\cdot\text{m}}{1.0\ \text{kg}\cdot\text{m}^2+\frac{0.50\ \text{m}}{2(1.0\ \text{m})}\left(16\ \text{kg}\cdot\text{m}^2\right)}$$
$$=0.400\ \text{rad/s}^2=\boxed{0.40\ \text{rad/s}^2}$$

Use equation (3) to evaluate α_2:

$$\alpha_2=\tfrac{1}{2}\left(0.400\ \text{rad/s}^2\right)=\boxed{0.20\ \text{rad/s}^2}$$

(*b*) To counterbalance the 2.0-N·m torque, a counter torque of 2.0 N·m must be applied to the first gear:

$$2.0\ \text{N}\cdot\text{m}-FR_1=0\Rightarrow F=\frac{2.0\ \text{N}\cdot\text{m}}{R_1}$$

Substitute numerical values and evaluate F:

$$F=\frac{2.0\ \text{N}\cdot\text{m}}{0.50\ \text{m}}=\boxed{4.0\ \text{N, clockwise}}$$

Rolling With Slipping

105 •• **[SSM]** A 0.16-kg billiard ball whose radius is 3.0 cm is given a sharp blow by a cue stick. The applied force is horizontal and the line of action of the force passes through the center of the ball. The speed of the ball just after the blow is 4.0 m/s, and the coefficient of kinetic friction between the ball and the billiard table is 0.60. (*a*) How long does the ball slide before it begins to roll without slipping? (*b*) How far does it slide? (*c*) What is its speed once it begins rolling without slipping?

Picture the Problem Because the impulse is applied through the center of mass, ω_0 = 0. We can use the results of Example 9-16 to find the rolling time without slipping, the distance traveled to rolling without slipping, and the velocity of the ball once it begins to roll without slipping.

(*a*) From Example 9-16 we have:

$$t_1=\frac{2}{7}\frac{v_0}{\mu_k g}$$

Substitute numerical values and evaluate t_1:

$$t_1=\frac{2}{7}\frac{4.0\ \text{m/s}}{(0.60)\left(9.81\ \text{m/s}^2\right)}=\boxed{0.19\ \text{s}}$$

(*b*) From Example 9-16 we have:

$$s_1 = \frac{12}{49}\frac{v_0^2}{\mu_k g}$$

Substitute numerical values and evaluate s_1:

$$s_1 = \frac{12}{49}\frac{(4.0\,\text{m/s})^2}{(0.60)(9.81\,\text{m/s}^2)} = \boxed{0.67\,\text{m}}$$

(*c*) From Example 9-16 we have:

$$v_1 = \frac{5}{7}v_0$$

Substitute numerical values and evaluate v_1:

$$v_1 = \frac{5}{7}(4.0\,\text{m/s}) = \boxed{2.9\,\text{m/s}}$$

General Problems

113 •• **[SSM]** A uniform 120-kg disk with a radius equal to 1.4 m initially rotates with an angular speed of 1100 rev/min. A constant tangential force is applied at a radial distance of 0.60 m from the axis. (*a*) How much work must this force do to stop the wheel? (*b*) If the wheel is brought to rest in 2.5 min, what torque does the force produce? What is the magnitude of the force? (*c*) How many revolutions does the wheel make in these 2.5 min?

Picture the Problem To stop the wheel, the tangential force will have to do an amount of work equal to the initial rotational kinetic energy of the wheel. We can find the stopping torque and the force from the average power delivered by the force during the slowing of the wheel. The number of revolutions made by the wheel as it stops can be found from a constant-acceleration equation.

(*a*) Relate the work that must be done to stop the wheel to its kinetic energy:

$$W = \tfrac{1}{2}I\omega^2 = \tfrac{1}{2}\left(\tfrac{1}{2}mr^2\right)\omega^2 = \tfrac{1}{4}mr^2\omega^2$$

Substitute numerical values and evaluate W:

$$W = \tfrac{1}{4}(120\,\text{kg})(1.4\,\text{m})^2\left[1100\frac{\text{rev}}{\text{min}}\times\frac{2\pi\,\text{rad}}{\text{rev}}\times\frac{1\,\text{min}}{60\,\text{s}}\right]^2 = 780\,\text{kJ} = \boxed{7.8\times10^2\,\text{kJ}}$$

(*b*) Express the stopping torque in terms of the average power required:

$$P_{\text{av}} = \tau\omega_{\text{av}} \Rightarrow \tau = \frac{P_{\text{av}}}{\omega_{\text{av}}}$$

Substitute numerical values and evaluate τ:

$$\tau=\frac{\dfrac{780\,\text{kJ}}{(2.5\,\text{min})(60\,\text{s/min})}}{\dfrac{(1100\,\text{rev/min})(2\pi\,\text{rad/rev})(1\,\text{min}/60\,\text{s})}{2}}=90.3\,\text{N}\cdot\text{m}=\boxed{90\,\text{N}\cdot\text{m}}$$

Relate the stopping torque to the magnitude of the required force and evaluate F:

$$F=\frac{\tau}{R}=\frac{90.3\,\text{N}\cdot\text{m}}{0.60\,\text{m}}=\boxed{0.15\text{ kN}}$$

(*c*) Using a constant-acceleration equation, relate the angular displacement of the wheel to its average angular speed and the stopping time:

$$\Delta\theta=\omega_{\text{av}}\Delta t$$

Substitute numerical values and evaluate $\Delta\theta$:

$$\Delta\theta=\left(\frac{1100\,\text{rev/min}}{2}\right)(2.5\,\text{min})$$
$$=\boxed{1.4\times10^{3}\text{ rev}}$$

119 •• **[SSM]** You are participating in league bowling with your friends. Time after time, you notice that your bowling ball rolls back to you without slipping on the flat section of track. When the ball encounters the slope that brings it up to the ball return, it is moving at 3.70 m/s. The length of the sloped part of the track is 2.50 m. The radius of the bowling ball is 11.5 cm. (*a*) What is the angular speed of the ball before it encounters the slope? (*b*) If the speed with which the ball emerges at the top of the incline is 0.40 m/s, what is the angle (assumed constant), that the sloped section of the track makes with the horizontal? (*c*) What is the magnitude of the angular acceleration of the ball while it is on the slope?

Picture the Problem The pictorial representation shows the bowling ball slowing down as it rolls up the slope. Let the system include the ball, the incline, and the earth. Then $W_{\text{ext}}=0$ and we can use conservation of mechanical energy to find the angle of the sloped section of the track.

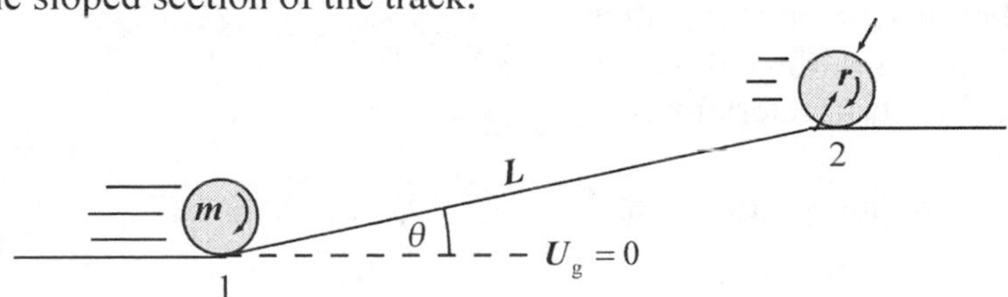

(*a*) Because the bowling ball rolls without slipping, its angular speed is directly proportional to its linear speed:

$$\omega = \frac{v}{r}$$

where r is the radius of the bowling ball.

Substitute numerical values and evaluate ω:

$$\omega = \frac{3.70\ \text{m/s}}{0.115\ \text{m}} = 32.17\ \text{rad/s}$$

$$= \boxed{32.2\ \text{rad/s}}$$

(*b*) Apply conservation of mechanical energy to the system as the bowling ball rolls up the incline:

$$W_{\text{ext}} = \Delta K + \Delta U$$

or, because $W_{\text{ext}} = 0$,

$$K_{t,2} - K_{t,1} + K_{r,2} - K_{r,1} + U_2 - U_1 = 0$$

Substituting for the kinetic and potential energies yields:

$$\tfrac{1}{2}mv_2^2 - \tfrac{1}{2}mv_1^2 + \tfrac{1}{2}I_{\text{ball}}\omega_2^2 - \tfrac{1}{2}I_{\text{ball}}\omega_1^2 + mgL\sin\theta = 0$$

Solving for θ yields:

$$\theta = \sin^{-1}\left[\frac{m(v_1^2 - v_2^2) + I_{\text{ball}}(\omega_1^2 - \omega_2^2)}{2mgL}\right]$$

Because $I_{\text{ball}} = \frac{2}{5}mr^2$:

$$\theta = \sin^{-1}\left[\frac{m(v_1^2 - v_2^2) + \frac{2}{5}mr^2(\omega_1^2 - \omega_2^2)}{2mgL}\right]$$

$$= \sin^{-1}\left[\frac{7(v_1^2 - v_2^2)}{10gL}\right]$$

Substitute numerical values and evaluate θ:

$$\theta = \sin^{-1}\left[\frac{7((3.70\ \text{m/s})^2 - (0.40\ \text{m/s})^2)}{10(9.81\ \text{m/s}^2)(2.50\ \text{m})}\right]$$

$$= \boxed{23°}$$

(*c*) The angular acceleration of the bowling ball is directly proportional to its translational acceleration:

$$\alpha = \frac{a}{r} \qquad (1)$$

Use a constant-acceleration equation to relate the speeds of the ball at points 1 and 2 to its acceleration:

$$v_2^2 = v_1^2 + 2aL \Rightarrow a = \frac{v_2^2 - v_1^2}{2L}$$

Substitute in equation (1) to obtain:

$$\alpha = \frac{v_2^2 - v_1^2}{2rL}$$

Substitute numerical values and evaluate $|\alpha|$:

$$|\alpha| = \left|\frac{(0.40\text{ m/s})^2 - (3.70\text{ m/s})^2}{2(0.115\text{ m})(2.50\text{ m})}\right|$$

$$= \boxed{24\text{ rad/s}^2}$$

121 •• **[SSM]** A popular classroom demonstration involves taking a meterstick and holding it horizontally at the 0.0-cm end with a number of pennies spaced evenly along its surface. If the hand is suddenly relaxed so that the meterstick pivots freely about the 0.0-cm mark under the influence of gravity, an interesting thing is seen during the first part of the stick's rotation: the pennies nearest the 0.0-cm mark remain on the meterstick, while those nearest the 100-cm mark are left behind by the falling meterstick. (This demonstration is often called the "faster than gravity" demonstration.) Suppose this demonstration is repeated without any pennies on the meterstick. (*a*) What would the initial acceleration of the 100.0-cm mark then be? (The initial acceleration is the acceleration just after the release.) (*b*) What point on the meterstick would then have an initial acceleration greater than *g*?

Picture the Problem The diagram shows the force the hand supporting the meterstick exerts at the pivot point and the force the earth exerts on the meterstick acting at the center of mass. We can relate the angular acceleration to the acceleration of the end of the meterstick using $a = L\alpha$ and use Newton's 2nd law in rotational form to relate α to the moment of inertia of the meterstick.

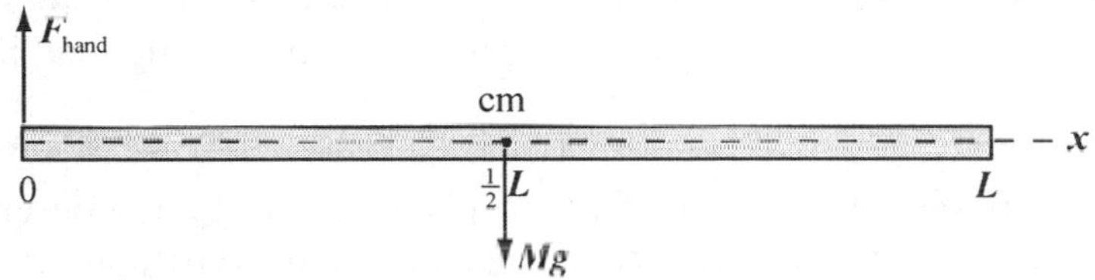

(*a*) Relate the acceleration of the far end of the meterstick to the angular acceleration of the meterstick:

$$a = L\alpha \qquad (1)$$

Apply $\sum \tau_P = I_P\alpha$ to the meterstick:

$$Mg\left(\frac{L}{2}\right) = I_P\alpha \Rightarrow \alpha = \frac{MgL}{2I_P}$$

From Table 9-1, for a rod pivoted at one end, we have:

$$I_P = \frac{1}{3}ML^2$$

Substitute for I_P in the expression for α to obtain:

$$\alpha = \frac{3MgL}{2ML^2} = \frac{3g}{2L}$$

Substitute for α in equation (1) to obtain:

$$a = \frac{3g}{2}$$

Substitute numerical values and evaluate a:

$$a = \frac{3(9.81\,\text{m/s}^2)}{2} = \boxed{14.7\,\text{m/s}^2}$$

(*b*) Express the acceleration of a point on the meterstick a distance x from the pivot point:

$$a = \alpha x = \frac{3g}{2L}x$$

Express the condition that the meterstick have an initial acceleration greater than g:

$$\frac{3g}{2L}x > g \Rightarrow x > \frac{2L}{3}$$

Substitute the numerical value of L and evaluate x:

$$x > \frac{2(100.0\,\text{cm})}{3} = \boxed{66.7\,\text{cm}}$$

125 ••• **[SSM]** Let's calculate the position y of the falling load attached to the winch in Example 9-8 as a function of time by numerical integration. Let the $+y$ direction be straight downward. Then, $v(y) = dy/dt$, or

$$t = \int_0^y \frac{1}{v(y')}dy' \approx \sum_{i=0}^{N} \frac{1}{v(y'_i)}\Delta y'$$

where t is the time taken for the bucket to fall a distance y, $\Delta y'$ is a small increment of y', and $y' = N\Delta y'$. Hence, we can calculate t as a function of d by numerical summation. Make a graph of y versus t between 0 s and 2.00 s. Assume $m_w = 10.0$ kg, $R = 0.50$ m, $m_b = 5.0$ kg, $L = 10.0$ m, and $m_c = 3.50$ kg. Use $\Delta y' = 0.10$ m. Compare this position to the position of the falling load if it were in free-fall.

Picture the Problem As the load falls, mechanical energy is conserved. As in Example 9-7, choose the initial potential energy to be zero and let the system include the winch, the bucket, and the earth. Apply conservation of mechanical energy to obtain an expression for the speed of the bucket as a function of its position and use the given expression for t to determine the time required for the bucket to travel a distance y.

Apply conservation of mechanical energy to the system to obtain:

$$U_f + K_f = U_i + K_i = 0 + 0 = 0 \quad (1)$$

Express the total potential energy when the bucket has fallen a distance y:

$$U_f = U_{bf} + U_{cf} + U_{wf} = -mgy - m_c'g\left(\frac{y}{2}\right)$$

where m_c' is the mass of the hanging part of the cable.

Assume the cable is uniform and express m_c' in terms of m_c, y, and L:

$$\frac{m_c'}{y} = \frac{m_c}{L} \text{ or } m_c' = \frac{m_c}{L}y$$

Substitute for m_c' to obtain:

$$U_f = -mgy - \frac{m_c gy^2}{2L}$$

Noting that bucket, cable, and rim of the winch have the same speed v, express the total kinetic energy when the bucket is falling with speed v:

$$K_f = K_{bf} + K_{cf} + K_{wf}$$
$$= \tfrac{1}{2}mv^2 + \tfrac{1}{2}m_c v^2 + \tfrac{1}{2}I\omega_f^2$$
$$= \tfrac{1}{2}mv^2 + \tfrac{1}{2}m_c v^2 + \tfrac{1}{2}\left(\tfrac{1}{2}MR^2\right)\frac{v^2}{R^2}$$
$$= \tfrac{1}{2}mv^2 + \tfrac{1}{2}m_c v^2 + \tfrac{1}{4}Mv^2$$

Substituting in equation (1) yields:

$$-mgy - \frac{m_c gy^2}{2L} + \tfrac{1}{2}mv^2 + \tfrac{1}{2}m_c v^2 + \tfrac{1}{4}Mv^2 = 0$$

Solving for v yields:

$$v = \sqrt{\frac{4mgy + \dfrac{2m_c gy^2}{L}}{M + 2m + 2m_c}}$$

A spreadsheet solution is shown below. The formulas used to calculate the quantities in the columns are as follows:

Cell	Formula/Content	Algebraic Form
D9	0	y_0
D10	D9+$B$8	$y + \Delta y$
E9	0	v_0
E10	((4*$B$3*$B$7*D10+2*$B$7*D10^2/(2*$B$5))/($B$1+2*$B$3+2*$B$4))^0.5	$\sqrt{\frac{4mgy + \frac{2m_c gy^2}{L}}{M + 2m + 2m_c}}$
F10	F9+$B$8/((E10+E9)/2)	$t_{n-1} + \left(\frac{v_{n-1} + v_n}{2}\right)\Delta y$
J9	0.5*$B$7*H9^2	$\tfrac{1}{2}gt^2$

	A	B	C	D	E	F	G	H	I	J
1	M=	10	kg							
2	R=	0.5	m							
3	m=	5	kg							
4	m_c=	3.5	kg							
5	L=	10	m							
6										

7	g=	9.81	m/s^2							
8	dy=	0.1	m	y	$v(y)$	$t(y)$		$t(y)$	y	$0.5gt^2$
9				0.0	0.00	0.00		0.00	0.0	0.00
10				0.1	0.85	0.23		0.23	0.1	0.27
11				0.2	1.21	0.33		0.33	0.2	0.54
12				0.3	1.48	0.41		0.41	0.3	0.81
13				0.4	1.71	0.47		0.47	0.4	1.08
15				0.5	1.91	0.52		0.52	0.5	1.35
105				9.6	9.03	2.24		2.24	9.6	24.61
106				9.7	9.08	2.25		2.25	9.7	24.85
107				9.8	9.13	2.26		2.26	9.8	25.09
108				9.9	9.19	2.27		2.27	9.9	25.34
109				10.0	9.24	2.28		2.28	10.0	25.58

The solid line on the following graph shows the position of the bucket as a function of time when it is in free fall and the dashed line shows its position as a function of time under the conditions modeled in this problem.

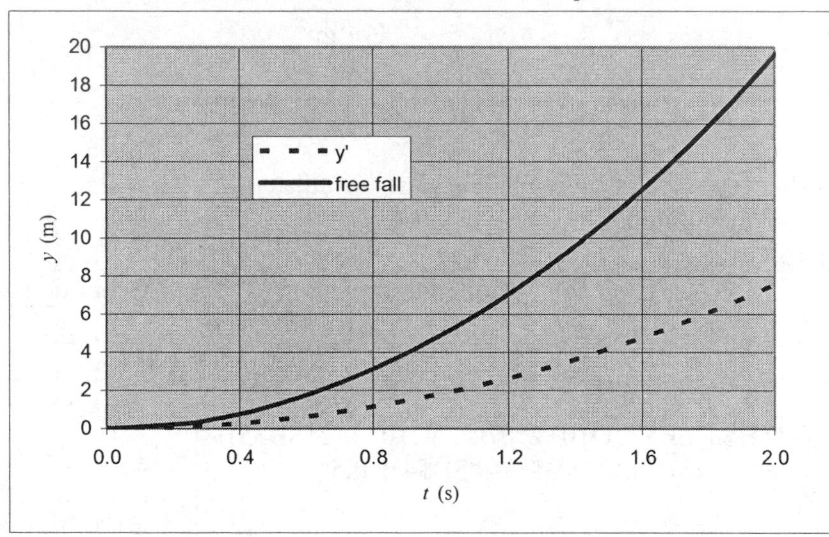

127 ••• [SSM] In problems dealing with a pulley with a nonzero moment of inertia, the magnitude of the tensions in the ropes hanging on either side of the pulley are not equal. The difference in the tension is due to the static frictional force between the rope and the pulley; however, the static frictional force cannot be made arbitrarily large. Consider a massless rope wrapped partly around a cylinder through an angle $\Delta\theta$ (measured in radians). It can be shown that if the tension on one side of the pulley is T, while the tension on the other side is T' ($T' > T$), the maximum value of T' that can be maintained without the rope slipping is $T'_{\text{max}} = Te^{\mu_s \Delta\theta}$, where μ_s is the coefficient of static friction. Consider the Atwood's machine in Figure 9-77: the pulley has a radius $R = 0.15$ m, the moment of inertia is $I = 0.35\ \text{kg}{\cdot}\text{m}^2$, and the coefficient of static friction between the wheel and the string is $\mu_s = 0.30$. (*a*) If the tension on one side of the pulley is 10 N, what is the maximum tension on the other side that will prevent the rope from slipping on the pulley? (*b*) What is the acceleration of the blocks in this case? (c) If the mass of one of the hanging blocks is 1.0 kg, what is the maximum mass of the other block if, after the blocks are released, the pulley is to rotate without slipping?

Picture the Problem Free-body diagrams for the pulley and the two blocks are shown to the right. Choose a coordinate system in which the direction of motion of the block whose mass is M (downward) is the positive y direction. We can use the given relationship $T'_{max} = Te^{\mu_s \Delta\theta}$ to relate the tensions in the rope on either side of the pulley and apply Newton's 2nd law in both rotational form (to the pulley) and translational form (to the blocks) to obtain a system of equations that we can solve simultaneously for a, T_1, T_2, and M.

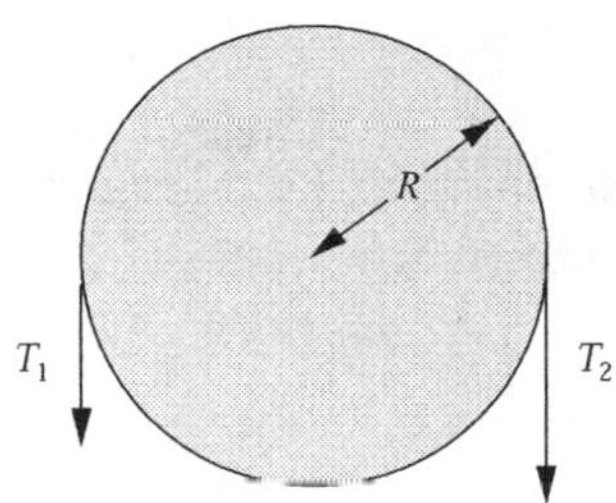

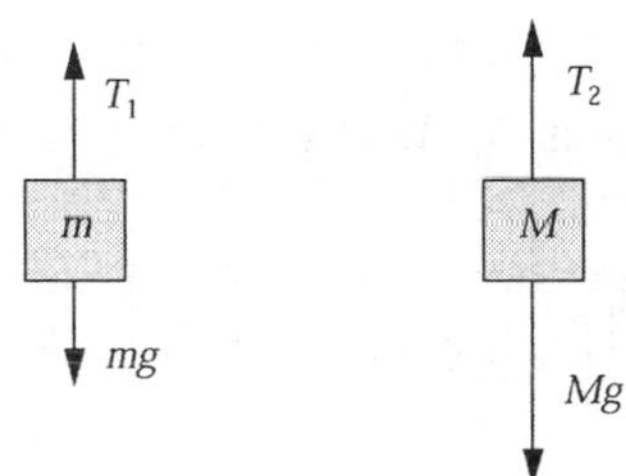

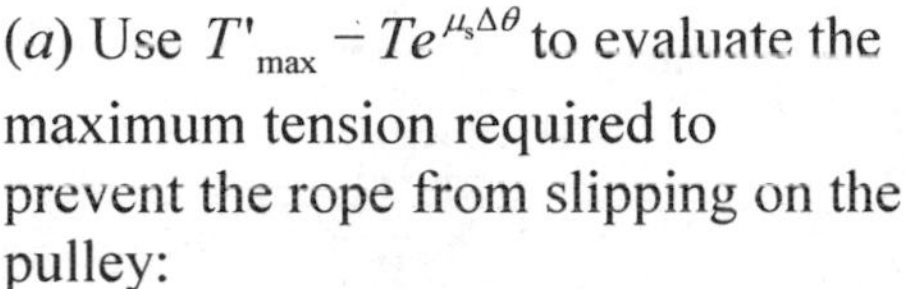

(*a*) Use $T'_{max} = Te^{\mu_s \Delta\theta}$ to evaluate the maximum tension required to prevent the rope from slipping on the pulley:

$$T'_{max} = (10\,\text{N})e^{(0.30)\pi} = 25.66\,\text{N} = \boxed{26\,\text{N}}$$

(*b*) Given that the angle of wrap is π radians, express T_2 in terms of T_1:

$$T_2 = T_1 e^{(0.30)\pi} \qquad (1)$$

Because the rope doesn't slip, we can relate the angular acceleration, α, of the pulley to the acceleration, a, of the hanging masses by:

$$\alpha = \frac{a}{R}$$

Apply $\sum \tau = I\alpha$ to the pulley to obtain:

$$(T_2 - T_1)R = I\frac{a}{R} \qquad (2)$$

Substitute for T_2 from equation (1) in equation (2) to obtain:

$$\left(T_1 e^{(0.30)\pi} - T_1\right)R = I\frac{a}{R}$$

Solving for T_1 yields:

$$T_1 = \frac{I}{\left(e^{(0.30)\pi} - 1\right)R^2}a$$

Apply $\sum F_y = ma_y$ to the block whose mass is m to obtain:

$$T_1 - mg = ma$$
and
$$T_1 = ma + mg \qquad (3)$$

Equating these two expressions for T_1 and solving for a yields:

$$a = \frac{g}{\dfrac{I}{\left(e^{(0.30)\pi} - 1\right)mR^2} - 1}$$

Substitute numerical values and evaluate a:

$$a = \frac{9.81\ \text{m/s}^2}{\dfrac{0.35\ \text{kg}\cdot\text{m}^2}{\left(e^{(0.30)\pi} - 1\right)(1.0\ \text{kg})(0.15\ \text{m})^2} - 1}$$
$$= 1.098\ \text{m/s}^2 = \boxed{1.1\ \text{m/s}^2}$$

(*c*) Apply $\sum F_y = ma_y$ to the block whose mass is M to obtain:

$$Mg - T_2 = Ma \Rightarrow M = \frac{T_2}{g - a}$$

Substitute for T_2 (from equation (1)) and T1 (from equation (3)) yields:

$$M = \frac{m(a + g)e^{(0.30)\pi}}{g - a}$$

Substitute numerical values and evaluate M:

$$M = \frac{(1.0\ \text{kg})\left(1.098\ \text{m/s}^2 + 9.81\ \text{m/s}^2\right)e^{(0.30)\pi}}{9.81\ \text{m/s}^2 - 1.098\ \text{m/s}^2} = \boxed{3.2\ \text{kg}}$$

Chapter 10
Angular Momentum

Conceptual Problems

5 • **[SSM]** A particle travels in a circular path and point *P* is at the center of the circle. (*a*) If the particle's linear momentum $\vec{p}$ is doubled without changing the radius of the circle, how is the magnitude of its angular momentum about *P* affected? (*b*) If the radius of the circle is doubled but the speed of the particle is unchanged, how is the magnitude of its angular momentum about *P* affected?

Determine the Concept $\vec{L}$ and $\vec{p}$ are related according to $\vec{L} = \vec{r} \times \vec{p}.$

(*a*) Because $\vec{L}$ is directly proportional to $\vec{p}$, *L* is doubled.

(*b*) Because $\vec{L}$ is directly proportional to $\vec{r}$, *L* is doubled.

11 •• **[SSM]** One way to tell if an egg is hardboiled or uncooked without breaking the egg is to lay the egg flat on a hard surface and try to spin it. A hardboiled egg will spin easily, while an uncooked egg will not. However, once spinning, the uncooked egg will do something unusual; if you stop it with your finger, it may start spinning again. Explain the difference in the behavior of the two types of eggs.

Determine the Concept The hardboiled egg is solid inside, so everything rotates with a uniform angular speed. By contrast, when you start an uncooked egg spinning, the yolk will not immediately spin with the shell, and when you stop it from spinning the yolk will initially continue to spin.

15 •• **[SSM]** The angular momentum of the propeller of a small single-engine airplane points forward. The propeller rotates clockwise if viewed from behind. (*a*) Just after liftoff, as the nose lifts and the airplane tends to veer to one side. To which side does it veer and why? (*b*) If the plane is flying horizontally and suddenly turns to the right, does the nose of the plane tend to move up or down? Why?

(*a*) The plane tends to veer to the right. The change in angular momentum $\Delta\vec{L}_{\text{prop}}$ for the propeller is up, so the net torque $\vec{\tau}$ on the propeller is up as well. The propeller must exert an equal but opposite torque on the plane. This downward torque exerted on the plane by the propeller tends to cause a downward change in the angular momentum of the plane. This means the plane tends to rotate clockwise as viewed from above.

(*b*) The plane tends to veer downward. The change in angular momentum $\Delta\vec{L}_{\text{prop}}$ for the propeller is to the right, so the net torque $\vec{\tau}$ on the propeller is toward the right as well. The propeller must exert an equal but opposite torque on the plane. This leftward directed torque exerted by the propeller on the plane tends to cause a leftward-directed change in angular momentum for the plane. This means the plane tends to rotate clockwise as viewed from the right.

17 •• **[SSM]** You are sitting on a spinning piano stool with your arms folded. (*a*) When you extend your arms out to the side, what happens to your kinetic energy? What is the cause of this change? (*b*) Explain what happens to your moment of inertia, angular speed and angular momentum as you extend your arms.

Determine the Concept The rotational kinetic energy of the you-stool system is given by $K_{\text{rot}} = \frac{1}{2}I\omega^2 = \frac{L^2}{2I}$. Because the net torque acting on the you-stool system is zero, its angular momentum $\vec{L}$ is conserved.

(*a*) Your kinetic energy decreases. Increasing your moment of inertia I while conserving your angular momentum L decreases your kinetic energy $K = L^2/(2I)$.

(*b*) Extending your arms out to the side increases your moment of inertia I s and decreases your angular speed. The angular momentum of the system is unchanged.

Estimation and Approximation

19 •• **[SSM]** An ice skater starts her pirouette with arms outstretched, rotating at 1.5 rev/s. Estimate her rotational speed (in revolutions per second) when she brings her arms tight against her body.

Picture the Problem Because we have no information regarding the mass of the skater, we'll assume that her body mass (not including her arms) is 50 kg and that each arm has a mass of 4.0 kg. Let's also assume that her arms are 1.0 m long and that her body is cylindrical with a radius of 20 cm. Because the net external torque acting on her is zero, her angular momentum will remain constant during her pirouette.

Because the net external torque acting on her is zero:

$$\Delta L = L_{\text{f}} - L_{\text{i}} = 0$$

or

$$I_{\text{arms in}}\omega_{\text{arms in}} - I_{\text{arms out}}\omega_{\text{arms out}} = 0 \quad (1)$$

Express her total moment of inertia with her arms out:

$$I_{\text{arms out}} = I_{\text{body}} + I_{\text{arms}}$$

Treating her body as though it is cylindrical, calculate the moment of inertia of her body, minus her arms:

$$I_{\text{body}} = \tfrac{1}{2}mr^2 = \tfrac{1}{2}(50\,\text{kg})(0.20\,\text{m})^2 = 1.00\,\text{kg}\cdot\text{m}^2$$

Modeling her arms as though they are rods, calculate their moment of inertia when she has them out:

$$I_{\text{arms}} = 2\left[\tfrac{1}{3}(4\,\text{kg})(1.0\,\text{m})^2\right] = 2.67\,\text{kg}\cdot\text{m}^2$$

Substitute to determine her total moment of inertia with her arms out:

$$I_{\text{arms out}} = 1.00\,\text{kg}\cdot\text{m}^2 + 2.67\,\text{kg}\cdot\text{m}^2 = 3.67\,\text{kg}\cdot\text{m}^2$$

Express her total moment of inertia with her arms in:

$$I_{\text{arms in}} = I_{\text{body}} + I_{\text{arms}} = 1.00\,\text{kg}\cdot\text{m}^2 + 2\left[(4.0\,\text{kg})(0.20\,\text{m})^2\right] = 1.32\,\text{kg}\cdot\text{m}^2$$

Solve equation (1) for $\omega_{\text{arms in}}$ to obtain:

$$\omega_{\text{arms in}} = \frac{I_{\text{arms out}}}{I_{\text{arms in}}}\omega_{\text{arms out}}$$

Substitute numerical values and evaluate $\omega_{\text{arms in}}$:

$$\omega_{\text{arms in}} = \frac{3.67\,\text{kg}\cdot\text{m}^2}{1.32\,\text{kg}\cdot\text{m}^2}(1.5\,\text{rev/s}) \approx \boxed{4\,\text{rev/s}}$$

23 •• **[SSM]** A 2.0-g particle moves at a constant speed of 3.0 mm/s around a circle of radius 4.0 mm. (*a*) Find the magnitude of the angular momentum of the particle. (*b*) If $L = \sqrt{\ell(\ell+1)}\hbar$, where ℓ is an integer, find the value of $\ell(\ell+1)$ and the approximate value of ℓ. (*c*) By how much does ℓ change if the particle's speed increases by one-millionth of a percent, nothing else changing? Use your result to explain why the quantization of angular momentum is not noticed in macroscopic physics.

Picture the Problem We can use $L = mvr$ to find the angular momentum of the particle. In (*b*) we can solve the equation $L = \sqrt{\ell(\ell+1)}\hbar$ for $\ell(\ell+1)$ and the approximate value of ℓ.

(*a*) Use the definition of angular momentum to obtain:

$$L = mvr = (2.0\times10^{-3}\text{ kg})(3.0\times10^{-3}\text{ m/s})(4.0\times10^{-3}\text{ m}) = 2.40\times10^{-8}\text{ kg}\cdot\text{m}^2/\text{s}$$
$$= \boxed{2.4\times10^{-8}\text{ kg}\cdot\text{m}^2/\text{s}}$$

(*b*) Solve the equation $L = \sqrt{\ell(\ell+1)}\hbar$ for $\ell(\ell+1)$:

$$\ell(\ell+1) = \frac{L^2}{\hbar^2} \qquad (1)$$

Substitute numerical values and evaluate $\ell(\ell+1)$:

$$\ell(\ell+1) = \left(\frac{2.40\times10^{-8}\text{ kg}\cdot\text{m}^2/\text{s}}{1.05\times01^{-34}\text{ J}\cdot\text{s}}\right)^2 = \boxed{5.2\times10^{52}}$$

Because $\ell >> 1$, approximate its value with the square root of $\ell(\ell+1)$:

$$\ell \approx \boxed{2.3\times10^{26}}$$

(*c*) The change in ℓ is:

$$\Delta\ell = \ell_{\text{new}} - \ell \qquad (2)$$

If the particle's speed increases by one-millionth of a percent while nothing else changes:

$$v \to v + 10^{-8}v = (1+10^{-8})v$$

and

$$\boldsymbol{L} \to \boldsymbol{L} + 10^{-8}\boldsymbol{L} = (1+10^{-8})\boldsymbol{L}$$

Equation (1) becomes:

$$\ell_{\text{new}}(\ell_{\text{new}}+1) = \frac{[(1+10^{-8})\boldsymbol{L}]^2}{\hbar^2}$$

and

$$\ell_{\text{new}} \approx \frac{(1+10^{-8})\boldsymbol{L}}{\hbar}$$

Substituting in equation (2) yields:

$$\Delta\ell = \ell_{\text{new}} - \ell \approx \frac{(1+10^{-8})\boldsymbol{L}}{\hbar} - \frac{\boldsymbol{L}}{\hbar} = 10^{-8}\frac{\boldsymbol{L}}{\hbar}$$

Substitute numerical values and evaluate $\Delta\ell$:

$$\Delta\ell = 10^{-8}\left(\frac{2.40\times10^{-8}\text{ kg}\cdot\text{m}^2/\text{s}}{1.05\times01^{-34}\text{ J}\cdot\text{s}}\right) = \boxed{2.3\times10^{18}}$$

and

$$\frac{\Delta\ell}{\ell} = \frac{2.3\times10^{18}}{2.3\times10^{26}} \approx 10^{-6}\%$$

The quantization of angular momentum is not noticed in macroscopic physics because no experiment can detect a fractional change in ℓ of $10^{-6}\%$.

The Cross Product and the Vector Nature of Torque and Rotation

27 • **[SSM]** A force of magnitude F is applied horizontally in the negative x direction to the rim of a disk of radius R as shown in Figure 10-42. Write $\vec{F}$ and $\vec{r}$ in terms of the unit vectors $\hat{i}$, $\hat{j}$, and $\hat{k}$, and compute the torque produced by this force about the origin at the center of the disk.

Picture the Problem We can express $\vec{F}$ and $\vec{r}$ in terms of the unit vectors $\hat{i}$ and $\hat{j}$ and then use the definition of the cross product to find $\vec{\tau}$.

Express $\vec{F}$ in terms of F and the unit vector $\hat{i}$:

$$\vec{F} = -F\hat{i}$$

Express $\vec{r}$ in terms of R and the unit vector $\hat{j}$:

$$\vec{r} = R\hat{j}$$

Calculate the cross product of $\vec{r}$ and $\vec{F}$:

$$\vec{\tau} = \vec{r} \times \vec{F} = FR\left(\hat{j} \times -\hat{i}\right) = FR\left(\hat{i} \times \hat{j}\right)$$
$$= \boxed{FR\hat{k}}$$

Torque and Angular Momentum

37 • **[SSM]** A 2.0-kg particle moves directly eastward at a constant speed of 4.5 m/s along an east-west line. (*a*) What is its angular momentum (including direction) about a point that lies 6.0 m north of the line? (*b*) What is its angular momentum (including direction) about a point that lies 6.0 m south of the line? (*c*) What is its angular momentum (including direction) about a point that lies 6.0 m directly east of the particle?

Picture the Problem The angular momentum of the particle is $\vec{L} = \vec{r} \times \vec{p}$ where $\vec{r}$ is the vector locating the particle relative to the reference point and $\vec{p}$ is the particle's linear momentum.

(*a*) The magnitude of the particle's angular momentum is given by:

$$L = rp\sin\phi = rmv\sin\phi = mv(r\sin\phi)$$

Substitute numerical values and evaluate L:

$$L = (2.0\,\text{kg})(4.5\,\text{m/s})(6.0\,\text{m})$$
$$= 54\,\text{kg}\cdot\text{m}^2/\text{s}$$

Use a right-hand rule to establish the direction of $\vec{L}$:

$$L = \boxed{54\,\text{kg}\cdot\text{m}^2/\text{s, upward}}$$

(*b*) Because the distance to the line along which the particle is moving is the same, only the direction of $\vec{L}$ differs:

$$L = \boxed{54\,\text{kg}\cdot\text{m}^2/\text{s, downward}}$$

(*c*) Because $\vec{r}\times\vec{p} = 0$ for a point on the line along which the particle is moving:

$$\vec{L} = \boxed{0}$$

45 •• **[SSM]** In Figure 10-46, the incline is frictionless and the string passes through the center of mass of each block. The pulley has a moment of inertia *I* and radius *R*. (*a*) Find the net torque acting on the system (the two masses, string, and pulley) about the center of the pulley. (*b*) Write an expression for the total angular momentum of the system about the center of the pulley. Assume the masses are moving with a speed *v*. (*c*) Find the acceleration of the masses by using your results for Parts (*a*) and (*b*) and by setting the net torque equal to the rate of change of the system's angular momentum.

Picture the Problem Let the system include the pulley, string, and the blocks and assume that the mass of the string is negligible. The angular momentum of this system changes because a *net* torque acts on it.

(*a*) Express the net torque about the center of mass of the pulley:

$$\tau_{\text{net}} = Rm_2 g\sin\theta - Rm_1 g = \boxed{Rg(m_2\sin\theta - m_1)}$$

where we have taken clockwise to be positive to be consistent with a positive upward velocity of the block whose mass is m_1 as indicated in the figure.

(*b*) Express the total angular momentum of the system about an axis through the center of the pulley:

$$L = I\omega + m_1 vR + m_2 vR = \boxed{vR\left(\frac{I}{R^2} + m_1 + m_2\right)}$$

(*c*) Express τ as the time derivative of the angular momentum:

$$\tau = \frac{dL}{dt} = \frac{d}{dt}\left[vR\left(\frac{I}{R^2} + m_1 + m_2\right)\right] = aR\left(\frac{I}{R^2} + m_1 + m_2\right)$$

Equate this result to that of Part (*a*) and solve for *a* to obtain:

$$\boxed{a = \frac{g(m_2 \sin\theta - m_1)}{\frac{I}{R^2} + m_1 + m_2}}$$

Conservation of Angular Momentum

49 •• **[SSM]** You stand on a frictionless platform that is rotating at an angular speed of 1.5 rev/s. Your arms are outstretched, and you hold a heavy weight in each hand. The moment of inertia of you, the extended weights, and the platform is 6.0 kg·m^2. When you pull the weights in toward your body, the moment of inertia decreases to 1.8 kg·m^2. (*a*) What is the resulting angular speed of the platform? (*b*) What is the change in kinetic energy of the system? (*c*) Where did this increase in energy come from?

Picture the Problem Let the system consist of you, the extended weights, and the platform. Because the net external torque acting on this system is zero, its angular momentum remains constant during the pulling in of the weights.

(*a*) Using conservation of angular momentum, relate the initial and final angular speeds of the system to its initial and final moments of inertia:

$$I_i\omega_i - I_f\omega_f = 0 \Rightarrow \omega_f = \frac{I_i}{I_f}\omega_i$$

Substitute numerical values and evaluate ω_f:

$$\omega_f = \frac{6.0\,\text{kg}\cdot\text{m}^2}{1.8\,\text{kg}\cdot\text{m}^2}(1.5\,\text{rev/s}) = \boxed{5.0\,\text{rev/s}}$$

(*b*) Express the change in the kinetic energy of the system:

$$\Delta K = K_f - K_i = \tfrac{1}{2}I_f\omega_f^2 - \tfrac{1}{2}I_i\omega_i^2$$

Substitute numerical values and evaluate ΔK:

$$\Delta K = \tfrac{1}{2}(1.8\,\text{kg}\cdot\text{m}^2)\left(5.0\frac{\text{rev}}{\text{s}}\times\frac{2\pi\,\text{rad}}{\text{rev}}\right)^2 - \tfrac{1}{2}(6.0\,\text{kg}\cdot\text{m}^2)\left(1.5\frac{\text{rev}}{\text{s}}\times\frac{2\pi\,\text{rad}}{\text{rev}}\right)^2$$

$$= \boxed{0.62\,\text{kJ}}$$

(*c*) Because no external agent does work on the system, the energy comes from your internal energy.

51 •• **[SSM]** A Lazy Susan consists of a heavy plastic cylinder mounted on a frictionless bearing resting on a vertical shaft. The cylinder has a radius

$R = 15$ cm and mass $M = 0.25$ kg. A cockroach (mass $m = 0.015$ kg) is on the Lazy Susan, at a distance of 8.0 cm from the center. Both the cockroach and the Lazy Susan are initially at rest. The cockroach then walks along a circular path concentric with the center of the Lazy Susan at a constant distance of 8.0 cm from the axis of the shaft. If the speed of the cockroach with respect to the Lazy Susan is 0.010 m/s, what is the speed of the cockroach with respect to the room?

Picture the Problem Because the net external torque acting on the Lazy Susan-cockroach system is zero, the net angular momentum of the system is constant (equal to zero because the Lazy Susan is initially at rest) and we can use conservation of angular momentum to find the angular velocity ω of the Lazy Susan. The speed of the cockroach relative to the floor v_f is the difference between its speed with respect to the Lazy Susan and the speed of the Lazy Susan at the location of the cockroach with respect to the floor.

Relate the speed of the cockroach with respect to the floor v_f to the speed of the Lazy Susan at the location of the cockroach:

$$v_f = v - \omega r \quad (1)$$

Use conservation of angular momentum to obtain:

$$L_{LS} - L_C = 0 \quad (2)$$

Express the angular momentum of the Lazy Susan:

$$L_{LS} = I_{LS}\omega = \tfrac{1}{2}MR^2\omega$$

Express the angular momentum of the cockroach:

$$L_C = I_C\omega_C = mr^2\left(\frac{v}{r} - \omega\right)$$

Substitute for L_{LS} and L_C in equation (2) to obtain:

$$\tfrac{1}{2}MR^2\omega - mr^2\left(\frac{v}{r} - \omega\right) = 0$$

Solving for ω yields:

$$\omega = \frac{2mrv}{MR^2 + 2mr^2}$$

Substitute for ω in equation (1) to obtain:

$$v_f = v - \frac{2mr^2v}{MR^2 + 2mr^2}$$

Substitute numerical values and evaluate v_f:

$$v_f = 0.010\,\text{m/s} - \frac{2(0.015\,\text{kg})(0.080\,\text{m})^2(0.010\,\text{m/s})}{(0.25\,\text{m})(0.15\,\text{m})^2 + 2(0.015\,\text{kg})(0.080\,\text{m})^2} = \boxed{10\,\text{mm/s}}$$

*Quantization of Angular Momentum

55 •• **[SSM]** The z component of the spin of an electron is $-\frac{1}{2}\hbar$, but the magnitude of the spin vector is $\sqrt{0.75}\hbar$. What is the angle between the electron's spin angular momentum vector and the positive z-axis?

Picture the Problem The electron's spin angular momentum vector is related to its z component as shown in the diagram. The angle between $\vec{s}$ and the positive z-axis is ϕ.

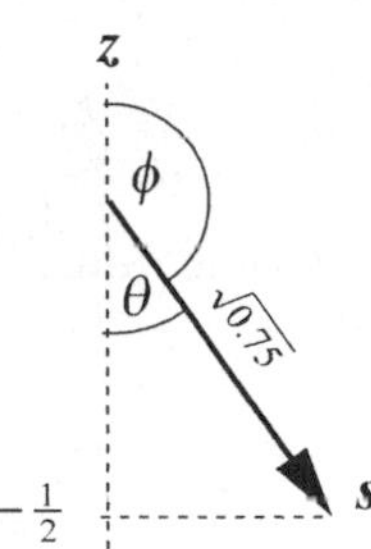

Express ϕ in terms of θ to obtain:

$$\phi = 180° - \theta$$

Using trigonometry, relate the magnitude of $\vec{s}$ to its $-z$ component:

$$\theta = \cos^{-1}\left(\frac{\frac{1}{2}\hbar}{\sqrt{0.75}\hbar}\right)$$

Substitute for θ in the expression for ϕ to obtain:

$$\theta = 180° - \cos^{-1}\left(\frac{\frac{1}{2}\hbar}{\sqrt{0.75}\hbar}\right) = \boxed{125°}$$

57 •• **[SSM]** You work in a bio-chemical research lab, where you are investigating the rotational energy levels of the HBr molecule. After consulting the periodic chart, you know that the mass of the bromine atom is 80 times that of the hydrogen atom. Consequently, in calculating the rotational motion of the molecule, you assume, to a good approximation, that the Br nucleus remains stationary as the H atom (mass 1.67×10^{-27} kg) revolves around it. You also know that the separation between the H atom and bromine nucleus is 0.144 nm. Calculate (*a*) the moment of inertia of the HBr molecule about the bromine nucleus, and (*b*) the rotational energies for the bromine nucleus's *ground state* (lowest energy) $\ell = 0$, and the next two states of higher energy (called the first and second *excited states*) described by $\ell = 1$, and $\ell = 2$.

Picture the Problem The rotational energies of HBr molecule are related to ℓ and E_{0r} according to $K_\ell = \ell(\ell+1)E_{0r}$ where $E_{0r} = \hbar^2/2I$.

(*a*) Neglecting the motion of the bromine molecule:

$$I_{HBr} \approx m_p r^2 = m_H r^2$$

Substitute numerical values and evaluate I_{HBr}:

$$I_{HBr} \approx (1.67\times10^{-27}\ \text{kg})(0.144\times10^{-9}\ \text{m})^2$$
$$= 3.463\times10^{-47}\ \text{kg}\cdot\text{m}^2$$
$$= \boxed{3.46\times10^{-47}\ \text{kg}\cdot\text{m}^2}$$

(*b*) Relate the rotational energies to ℓ and E_{0r}:

$$K_\ell = \ell(\ell+1)E_{0r} \text{ where } E_{0r} = \frac{\hbar^2}{2I_{HBr}}$$

Substitute numerical values and evaluate E_{0r}:

$$E_{0r} = \frac{\hbar^2}{2I} = \frac{(1.055\times10^{-34}\ \text{J}\cdot\text{s})^2}{2(3.463\times10^{-47}\ \text{kg}\cdot\text{m}^2)}$$
$$= 1.607\times10^{-22}\ \text{J}\times\frac{1\,\text{eV}}{1.602\times10^{-19}\ \text{J}}$$
$$= 1.003\,\text{meV}$$

Evaluate E_0 to obtain:

$$E_0 = K_0 = \boxed{1.00\,\text{meV}}$$

Evaluate E_1 to obtain:

$$E_1 = K_1 = (1+1)(1.003\,\text{meV})$$
$$= \boxed{2.01\,\text{meV}}$$

Evaluate E_2 to obtain:

$$E_2 = K_2 = 2(2+1)(1.003\,\text{meV})$$
$$= \boxed{6.02\,\text{meV}}$$

Collisions with Rotations

61 •• **[SSM]** Figure 10-52 shows a thin uniform bar of length L and mass M and a small blob of putty of mass m. The system is supported by a frictionless horizontal surface. The putty moves to the right with velocity $\vec{v}$, strikes the bar at a distance d from the center of the bar, and sticks to the bar at the point of contact. Obtain expressions for the velocity of the system's center of mass and for the angular speed following the collision.

Picture the Problem The velocity of the center of mass of the bar-blob system does not change during the collision and so we can calculate it before the collision using its definition. Because there are no external forces or torques acting on the bar-blob system, both linear and angular momentum are conserved in the collision. Let the direction the blob of putty is moving initially be the +*x* direction. Let lower-case letters refer to the blob of putty and upper-case letters refer to the bar. The diagram to the left shows the blob of putty approaching the bar and the diagram to the right shows the bar-blob system rotating about its center of mass and translating after the perfectly inelastic collision.

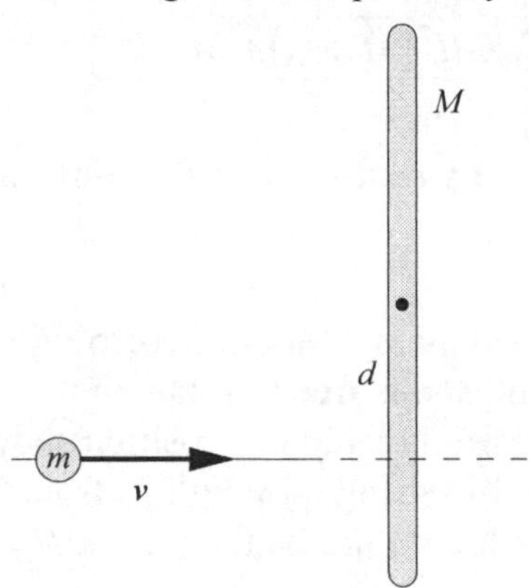

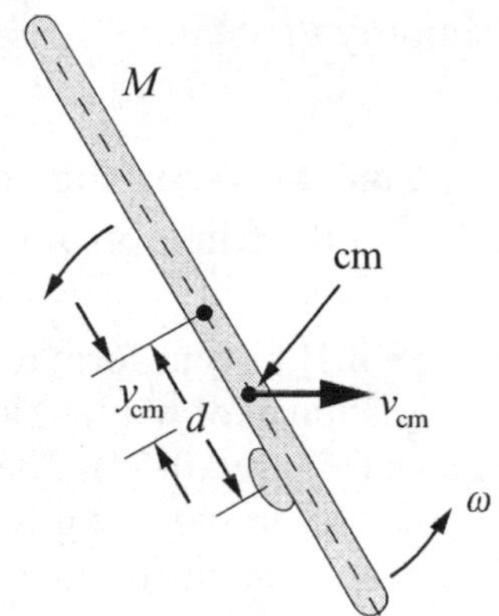

The velocity of the center of mass before the collision is given by:

$$(M+m)\vec{v}_{cm} = m\vec{v} + M\vec{V}$$

or, because $\vec{V} = 0$,

$$\boxed{\vec{v}_{cm} = \frac{m}{M+m}\vec{v}}$$

Using its definition, express the location of the center of mass relative to the center of the bar:

$$(M+m)y_{cm} = md \Rightarrow y_{cm} = \frac{md}{M+m}$$

below the center of the bar.

Express the angular momentum, relative to the center of mass, of the bar-blob system:

$$L_{cm} = I_{cm}\omega \Rightarrow \omega = \frac{L_{cm}}{I_{cm}} \qquad (1)$$

Express the angular momentum about the center of mass:

$$L_{cm} = mv(d - y_{cm}) = mv\left(d - \frac{md}{M+m}\right) = \frac{mMvd}{M+m}$$

Using the parallel axis theorem, express the moment of inertia of the system relative to its center of mass:

$$I_{cm} = \tfrac{1}{12}ML^2 + My_{cm}^2 + m(d - y_{cm})^2$$

Substitute for y_{cm} and simplify to obtain:

$$I_{cm} = \tfrac{1}{12}ML^2 + M\left(\frac{md}{M+m}\right)^2 + m\left(d - \frac{md}{M+m}\right)^2 = \tfrac{1}{12}ML^2 + \frac{mMd^2}{M+m}$$

Substitute for I_{cm} and L_{cm} in equation (1) and simplify to obtain:

$$\boxed{\omega = \frac{mMvd}{\frac{1}{12}ML^2(M+m)+Mmd^2}}$$

Remarks: You can verify the expression for I_{cm} by letting $m \to 0$ to obtain $I_{cm} = \frac{1}{12}ML^2$ and letting $M \to 0$ to obtain $I_{cm} = 0$.

67 •• **[SSM]** A uniform rod of length L_1 and mass M equal to 0.75 kg is supported by a hinge of negligible mass at one end and is free to rotate in the vertical plane (Figure 10-55). The rod is released from rest in the position shown. A particle of mass $m = 0.50$ kg is supported by a thin string of length L_2 from the hinge. The particle sticks to the rod on contact. What should be the ratio L_2/L_1 so that $\theta_{max} = 60°$ after the collision?

Picture the Problem Assume that there is no friction between the rod and the hinge. Because the net external torque acting on the system is zero, angular momentum is conserved in this perfectly inelastic collision. The rod, on its downward swing, acquires rotational kinetic energy. Angular momentum is conserved in the perfectly inelastic collision with the particle and the rotational kinetic of the after-collision system is then transformed into gravitational potential energy as the rod-plus-particle swing upward. Let the zero of gravitational potential energy be at a distance L_1 below the pivot and use both angular momentum and mechanical energy conservation to relate the distances L_1 and L_2 and the masses M and m.

Use conservation of energy to relate the initial and final potential energy of the rod to its rotational kinetic energy just before it collides with the particle:

$$K_f - K_i + U_f - U_i = 0$$

or, because $K_i = 0$,

$$K_f + U_f - U_i = 0$$

Substitute for K_f, U_f, and U_i to obtain:

$$\tfrac{1}{2}\left(\tfrac{1}{3}ML_1^2\right)\omega^2 + Mg\frac{L_1}{2} - MgL_1 = 0$$

Solving for ω yields:

$$\omega = \sqrt{\frac{3g}{L_1}}$$

Letting ω' represent the angular speed of the rod-and-particle system just after impact, use conservation of angular momentum to relate the angular momenta before and after the collision:

$$\Delta L = L_f - L_i = 0$$

or

$$\left(\tfrac{1}{3}ML_1^2 + mL_2^2\right)\omega' - \left(\tfrac{1}{3}ML_1^2\right)\omega = 0$$

Solve for ω' to obtain:

$$\omega' = \frac{\tfrac{1}{3}ML_1^2}{\tfrac{1}{3}ML_1^2 + mL_2^2}\omega$$

Use conservation of energy to relate the rotational kinetic energy of the rod-plus-particle just after their collision to their potential energy when they have swung through an angle θ_{max}:

$$K_f - K_i + U_f - U_i = 0$$

Because $K_f = 0$:

$$-\tfrac{1}{2}I\omega'^2 + Mg\left(\tfrac{1}{2}L_1\right)\left(1-\cos\theta_{max}\right) + mgL_2\left(1-\cos\theta_{max}\right) = 0 \qquad (1)$$

Express the moment of inertia of the system with respect to the pivot:

$$I = \tfrac{1}{3}ML_1^2 + mL_2^2$$

Substitute for θ_{max}, I and ω' in equation (1):

$$\frac{3\frac{g}{L_1}\left(\tfrac{1}{3}ML_1^2\right)^2}{\tfrac{1}{3}ML_1^2 + mL_2^2} = Mg\left(\tfrac{1}{2}L_1\right) + mgL_2$$

Simplify to obtain:

$$L_1^3 - 2\frac{m}{M}L_1^2L_2 + 3L_2^2L_1 + 6\frac{m}{M}L_2^3$$

Let $\alpha = m/M$ and $\beta = L_2/L_1$ to obtain:

$$6\alpha^2\beta^3 + 3\beta^2 + 2\alpha\beta - 1 = 0$$

Substitute for α and simplify to obtain the cubic equation in β:

$$8\beta^3 + 9\beta^2 + 4\beta - 3 = 0$$

Use the solver function* of your calculator to find the only real value of β:

$$\beta = \boxed{0.36}$$

Remarks: Most graphing calculators have a "solver" feature. One can solve the cubic equation using either the "graph" and "trace" capabilities or the "solver" feature. The root given above was found using SOLVER on a TI-85.

Precession

69 •• **[SSM]** A bicycle wheel that has a radius equal to 28 cm is mounted at the middle of an axle 50 cm long. The tire and rim weigh 30 N. The wheel is spun at 12 rev/s, and the axle is then placed in a horizontal position with one end resting on a pivot. (*a*) What is the angular momentum due to the spinning of the wheel? (Treat the wheel as a hoop.) (*b*) What is the angular velocity of precession? (*c*) How long does it take for the axle to swing through 360° around the pivot? (*d*) What is the angular momentum associated with the motion of the center of mass, that is, due to the precession? In what direction is this angular momentum?

Picture the Problem We can determine the angular momentum of the wheel and the angular velocity of its precession from their definitions. The period of the precessional motion can be found from its angular velocity and the angular momentum associated with the motion of the center of mass from its definition.

(*a*) Using the definition of angular momentum, express the angular momentum of the spinning wheel:

$$L = I\omega = MR^2\omega = \frac{w}{g}R^2\omega$$

Substitute numerical values and evaluate L:

$$L = \left(\frac{30\,\text{N}}{9.81\,\text{m/s}^2}\right)(0.28\,\text{m})^2 \times\left(12\frac{\text{rev}}{\text{s}}\times\frac{2\pi\,\text{rad}}{\text{rev}}\right) = 18.1\,\text{J}\cdot\text{s} = \boxed{18\,\text{J}\cdot\text{s}}$$

(*b*) Using its definition, express the angular velocity of precession:

$$\omega_{\text{p}} = \frac{d\phi}{dt} = \frac{MgD}{L}$$

Substitute numerical values and evaluate ω_{p}:

$$\omega_{\text{p}} = \frac{(30\,\text{N})(0.25\,\text{m})}{18.1\,\text{J}\cdot\text{s}} = 0.414\,\text{rad/s} = \boxed{0.41\,\text{rad/s}}$$

(*c*) Express the period of the precessional motion as a function of the angular velocity of precession:

$$T = \frac{2\pi}{\omega_p} = \frac{2\pi}{0.414\,\text{rad/s}} = \boxed{15\,\text{s}}$$

(*d*) Express the angular momentum of the center of mass due to the precession:

$$L_p = I_{cm}\omega_p = MD^2\omega_p$$

Substitute numerical values and evaluate L_p:

$$L_p = \left(\frac{30\,\text{N}}{9.81\,\text{m/s}^2}\right)(0.25\,\text{m})^2(0.414\,\text{rad/s}) = \boxed{0.079\,\text{J}\cdot\text{s}}$$

The direction of L_p is either up or down, depending on the direction of L.

General Problems

71 • **[SSM]** A particle whose mass is 3.0 kg moves in the *xy* plane with velocity $\vec{v} = (3.0\text{ m/s})\hat{i}$ along the line $y = 5.3$ m. (*a*) Find the angular momentum $\vec{L}$ about the origin when the particle is at (12 m, 5.3 m). (*b*) A force $\vec{F} = (-3.9\text{ N})\hat{i}$ is applied to the particle. Find the torque about the origin due to this force as the particle passes through the point (12 m, 5.3 m).

Picture the Problem While the 3-kg particle is moving in a straight line, it has angular momentum given by $\boldsymbol{L} = \vec{r} \times \vec{p}$ where $\vec{r}$ is its position vector and $\boldsymbol{p}$ is its linear momentum. The torque due to the applied force is given by $\vec{\tau} = \vec{r} \times \vec{F}$.

(*a*) The angular momentum of the particle is given by:

$$\vec{L} = \vec{r} \times \vec{p}$$

Express the vectors $\vec{r}$ and $\vec{p}$:

$$\vec{r} = (12\,\text{m})\hat{i} + (5.3\,\text{m})\hat{j}$$

and

$$\vec{p} = mv\hat{i} = (3.0\,\text{kg})(3.0\,\text{m/s})\hat{i} = (9.0\,\text{kg}\cdot\text{m/s})\hat{i}$$

Substitute for $\vec{r}$ and $\vec{p}$:and simplify to find $\vec{L}$:

$$\vec{L} = \left[(12\,\text{m})\hat{i} + (5.3\,\text{m})\hat{j}\right] \times (9.0\,\text{kg}\cdot\text{m/s})\hat{i} = (47.7\,\text{kg}\cdot\text{m}^2/\text{s})(\hat{j}\times\hat{i}) = \boxed{-(48\,\text{kg}\cdot\text{m}^2/\text{s})\hat{k}}$$

(*b*) Using its definition, express the torque due to the force:

$$\vec{\tau} = \vec{r} \times \vec{F}$$

Substitute for $\vec{r}$ and $\vec{F}$ and simplify to find $\vec{\tau}$:

$$\begin{aligned}\vec{\tau} &= \left[(12\,\text{m})\hat{i} + (5.3\,\text{m})\hat{j}\right] \times (-3.0\,\text{N})\hat{i} \\ &= -(15.9\,\text{N}\cdot\text{m})\left(\hat{j} \times \hat{i}\right) \\ &= \boxed{(16\,\text{N}\cdot\text{m})\hat{k}}\end{aligned}$$

77 •• **[SSM]** Repeat Problem 76, this time friction between the disks and the walls of the cylinder is not negligible. However, the coefficient of friction is not great enough to prevent the disks from reaching the ends of the cylinder. Can the final kinetic energy of the system be determined without knowing the coefficient of kinetic friction?

Determine the Concept Yes. The solution depends only upon conservation of angular momentum of the system, so it depends only upon the initial and final moments of inertia.

79 •• **[SSM]** Kepler's second law states: *The line from the center of the Sun to the center of a planet sweeps out equal areas in equal times.* Show that this law follows directly from the law of conservation of angular momentum and the fact that the force of gravitational attraction between a planet and the Sun acts along the line joining the centers of the two celestial objects.

Picture the Problem The pictorial representation shows an elliptical orbit. The triangular element of the area is $dA = \frac{1}{2}r(rd\theta) = \frac{1}{2}r^2 d\theta$.

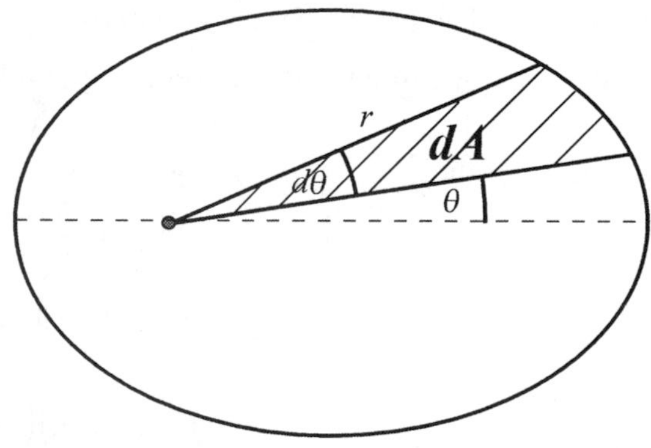

Differentiate dA with respect to t to obtain:

$$\frac{dA}{dt} = \frac{1}{2}r^2\frac{d\theta}{dt} = \frac{1}{2}r^2\omega \qquad (1)$$

Because the gravitational force acts along the line joining the two objects, $\tau = 0$. Hence:

$$L = mr^2\omega = \text{constant} \qquad (2)$$

Eliminate $r^2\omega$ between equations (1) and (2) to obtain:

$$\frac{dA}{dt} = \boxed{\frac{L}{2m} = \text{constant}}$$

83 •• **[SSM]** The term precession of the equinoxes refers to the fact that the Earth's spin axis does not stay fixed but moves with a period of about 26,000 y. (This explains why our pole star, Polaris, will not remain the pole star forever.) The reason for this instability is that Earth is a giant gyroscope. The spin axis of Earth precesses because of the torques exerted on it by the gravitational forces of the Sun and Moon. The angle between the direction of Earth's spin axis and the normal to the ecliptic plane (the plane of Earth's orbit) is 22.5 degrees. Calculate an approximate value for this torque, given that the period of rotation of the earth is 1.00 d and its moment of inertia is 8.03×10^{37} kg·m^2.

Picture the Problem Let ω_P be the angular velocity of precession of the earth-as-gyroscope, ω_S its angular velocity about its spin axis, and I its moment of inertia with respect to an axis through its poles, and relate ω_P to ω_S and I using its definition.

Use its definition to express the precession rate of the earth as a giant gyroscope:

$$\omega_P = \frac{\tau}{L}$$

Substitute for I and solve for τ to obtain:

$$\tau = L\omega_P = I\omega\omega_P$$

The angular velocity ω_s of the earth about its spin axis is given by:

$\omega = \dfrac{2\pi}{T}$ where T is the period of rotation of the earth.

Substitute for ω to obtain:

$$\tau = \frac{2\pi I\omega_P}{T}$$

Substitute numerical values and evaluate τ:

$$\tau = \frac{2\pi\left(8.03\times10^{37}\ \text{kg}\cdot\text{m}^2\right)\left(7.66\times10^{-12}\ \text{s}^{-1}\right)}{1\,\text{d}\times\dfrac{24\,\text{h}}{\text{d}}\times\dfrac{3600\,\text{s}}{\text{h}}} = \boxed{4.47\times10^{22}\ \text{N}\cdot\text{m}}$$

Chapter R
Special Relativity

Conceptual Problems

3 • [SSM] If event A occurs at a different location than event B in some reference frame, might it be possible for there to be a second reference frame in which they occur at the same location? If so, give an example. If not, explain why not.

Determine the Concept Yes. Let the initial frame of reference be frame 1. In frame 1 let L be the distance between the events, let T be the time between the events, and let the $+x$ direction be the direction of event B relative to event A. Next, calculate the value of L/T. If L/T is less than c, then consider the two events in a reference frame 2, a frame moving at speed $v = L/T$ in the $+x$ direction. In frame 2 both events occur at the same location.

5 • [SSM] Two events are simultaneous in a frame in which they also occur at the same location. Are they simultaneous in all other reference frames?

Determine the Concept Yes. If two events occur at the same time *and* place in one reference frame they occur at the same time *and* place in all reference frames. (Any pair of events that occur at the same time *and* at the same place in one reference frame are called a spacetime coincidence.)

11 •• [SSM] Many nuclei of atoms are unstable; for example, ^{14}C, an isotope of carbon, has a half-life of 5700 years. (By definition, the *half-life* is the time it takes for any given number of unstable particles to decay to half that number of particles.) This fact is used extensively for archeological and biological dating of old artifacts. Such unstable nuclei decay into several decay products, each with significant kinetic energy. Which of the following is true? (*a*) The mass of the unstable nucleus is larger than the sum of the masses of the decay products. (*b*) The mass of the unstable nucleus is smaller than the sum of the masses of the decay products. (*c*) The mass of the unstable nucleus is the same as the sum of the masses of the decay products. Explain your choice.

Determine the Concept Because mass is converted into the kinetic energy of the fragments, the mass of the unstable nucleus is larger than the sum of the masses of the decay products. $\boxed{(a)}$ is correct.

Estimation and Approximation

13 •• [SSM] In 1975, an airplane carrying an atomic clock flew back and forth at low altitude for 15 hours at an average speed of 140 m/s as part of a time-dilation experiment. The time on the clock was compared to the time on an atomic clock kept on the ground. What is the time difference between the atomic clock on the airplane and the atomic clock on the ground? (Ignore any effects that accelerations of the airplane have on the atomic clock which is on the airplane. Also assume that the airplane travels at constant speed.)

Picture the Problem We can use the time dilation equation to relate the elapsed time in the frame of reference of the airborne clock to the elapsed time in the frame of reference of the clock kept on the ground.

Use the time dilation equation to relate the elapsed time Δt according to the clock on the ground to the elapsed time Δt_0 according to the airborne atomic clock:

$$\Delta t = \frac{\Delta t_0}{\sqrt{1-\left(\frac{v}{c}\right)^2}} \quad (1)$$

Because $v \ll c$, we can use the approximation $\frac{1}{\sqrt{1-x}} \approx 1+\frac{1}{2}x$ to obtain:

$$\frac{1}{\sqrt{1-\left(\frac{v}{c}\right)^2}} \approx 1+\frac{1}{2}\left(\frac{v}{c}\right)^2$$

Substitute in equation (1) to obtain:

$$\Delta t = \left[1+\frac{1}{2}\left(\frac{v}{c}\right)^2\right]\Delta t_0 = \Delta t_0 + \frac{1}{2}\left(\frac{v}{c}\right)^2 \Delta t_0 \quad (2)$$

where the second term represents the additional time measured by the clock on the ground.

Evaluate the proper elapsed time according to the clock on the airplane:

$$\Delta t_0 = (15\,\text{h})\left(3600\,\frac{\text{s}}{\text{h}}\right) = 5.40\times10^4\,\text{s}$$

Substitute numerical values and evaluate the second term in equation (2):

$$\Delta t' = \frac{1}{2}\left(\frac{140\,\text{m/s}}{2.998\times10^8\,\text{m/s}}\right)^2 (5.40\times10^4\,\text{s}) = 5.89\times10^{-9}\,\text{s} \approx \boxed{5.9\,\text{ns}}$$

Length Contraction and Time Dilation

17 • **[SSM]** In the reference frame of a pion in Problem 18, how far does the laboratory travel in 2.6×10^{-8} s?

Picture the Problem We can use $\Delta x = v\Delta t_\pi$, where Δt_π is the proper mean lifetime of the pions, to find the distance traveled by the laboratory frame in a typical pion lifetime.

The average distance the laboratory will travel before the pions decay is the product of the speed of the pions and their proper mean lifetime:

$$\Delta x = v\Delta t_\pi$$

Substitute numerical values and evaluate Δx:

$$\Delta x = (0.85c)(2.6\times10^{-8}\ \text{s}) = 6.63\ \text{m}$$
$$= \boxed{6.6\ \text{m}}$$

21 • **[SSM]** A spaceship travels from Earth to a star 95 light-years away at a speed of 2.2×10^8 m/s. How long does the spaceship take to get to the star (*a*) as measured on Earth and (*b*) as measured by a passenger on the spaceship?

Picture the Problem We can use $\Delta x = v\Delta t$ to find the time for the trip as measured on earth and $\Delta t_0 = \Delta t\sqrt{1-(v/c)^2}$ to find the time measured by a passenger on the spaceship.

(*a*) Express the elapsed time, as measured on earth, in terms of the distance traveled and the speed of the spaceship:

$$\Delta t = \frac{\Delta x}{v}$$

Substitute numerical values and evaluate Δt:

$$\Delta t = \frac{95c\cdot\text{y}}{2.2\times10^8\ \text{m/s}}\times\frac{9.461\times10^{15}\ \text{m}}{c\cdot\text{y}}$$
$$= 4.09\times10^9\ \text{s}\times\frac{1\ \text{y}}{31.56\ \text{Ms}} = 129\ \text{y}$$
$$= \boxed{1.3\times10^2\ \text{y}}$$

(*b*) A passenger on the spaceship will measure the proper time:

$$\Delta t_0 = \Delta t\sqrt{1-\left(\frac{v}{c}\right)^2}$$

Substitute numerical values and evaluate the proper time:

$$\Delta t_0 = (129\ \text{y})\sqrt{1-\frac{(2.2\times10^8\ \text{m/s})^2}{(2.998\times10^8\ \text{m/s})^2}}$$
$$= \boxed{88\ \text{y}}$$

25 •• **[SSM]** Your friend, who is the same age as you, travels to the star Alpha Centauri, which is 4.0 light-years away, and returns immediately. He claims that the entire trip took just 6.0 y. What was his speed? Ignore any accelerations of your friend's spaceship and assume the spaceship traveled at the same speed during the entire trip.

Picture the Problem To calculate the speed in the reference frame of the friend, who is named Ed, we consider each leg of the trip separately. Consider an imaginary stick extending from Earth to Alpha Centauri that is at rest relative to Earth. In Ed's frame the length of the stick, and thus the distance between Earth and Alpha Centauri, is shortened in accord with the length contraction formula. As three years pass on Ed's watch Alpha Centauri travels at speed v from its initial location to him.

Sketch the situation as it is in your reference frame. The distance between Earth and Alpha C. is the rest length L_0 of the stick discussed in Picture the Problem:

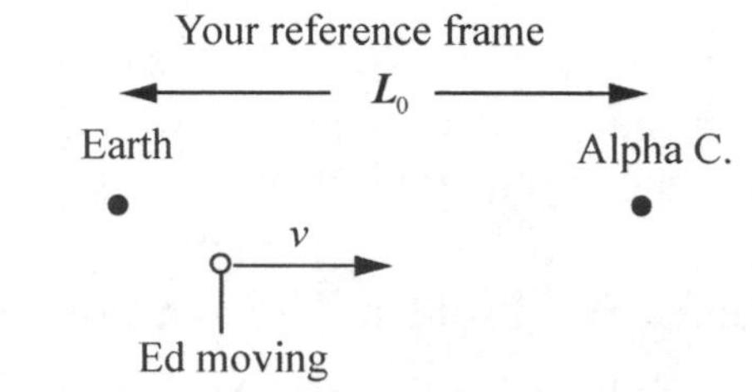

Sketch the situation as it is in Ed's reference frame. The distance between Earth and Alpha C. is the length L of the moving stick discussed in Picture the Problem:

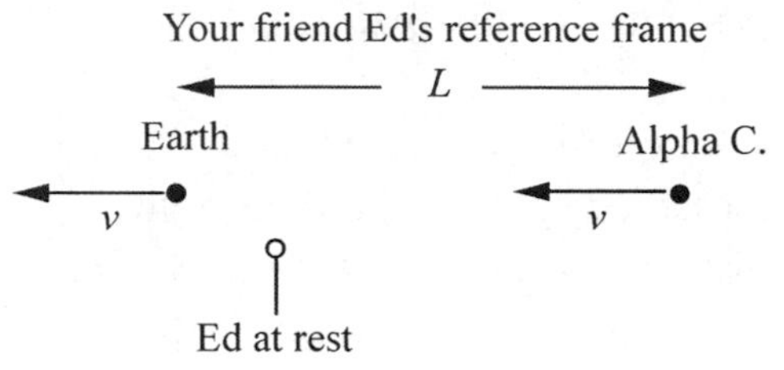

The two events are Ed leaves Earth and Ed arrive at Alpha Centauri. In Ed's frame these two events occur at the same place (next to Ed). Thus, the time between those two events Δt_0 is the proper time between the two events.

$$\Delta t_0 = 3\text{ y}$$

The distance L traveled by Alpha Centauri in Ed's frame during the first three years equals the speed multiplied by the time in Ed's frame:

$$L = v\Delta t_0$$

The distance between Earth and Alpha Centauri in Ed's frame is the contracted length of the imaginary stick:

$$L = L_0\sqrt{1-\frac{v^2}{c^2}}$$

Equate these expressions for L to obtain:

$$L_0\sqrt{1-\frac{v^2}{c^2}} = v\Delta t_0$$

Substituting numerical values yields:

$$(4.0\,c\cdot\text{y})\sqrt{1-\frac{v^2}{c^2}} = v(3.0\text{ y})$$

or

$$\sqrt{1-\frac{v^2}{c^2}} = 0.75\frac{v}{c}$$

Solve the quadratic equation in v/c ot obtain:

$$\frac{v}{c} = 0.80 \Rightarrow v = \boxed{0.80c}$$

The Relativity of Simultaneity

35 •• **[SSM]** In an inertial reference frame *S*, event B occurs 2.00 μs after event A and 1.50 km distant from event A. How fast must an observer be moving along the line joining the two events so that the two events occur simultaneously? For an observer traveling fast enough is it possible for event B to precede event A?

Picture the Problem Because event A is ahead of event B by $L_p v/c^2$ where v is the speed of the observer moving along the line joining the two events, we can use this expression and the given time between the events to find v.

Express the time Δt between events A and B in terms of L_p, v, and c:

$$\Delta t = \frac{L_p v}{c^2} \Rightarrow v = \frac{c^2\Delta t}{L_p}$$

Substitute numerical values and evaluate v:

$$v = \frac{(2.998\times10^8\text{ m/s})^2(2.00\,\mu\text{s})}{1.50\text{ km}}$$

$$= (1.20\times10^8\text{ m/s})\left(\frac{c}{2.998\times10^8\text{ m/s}}\right)$$

$$= \boxed{0.400c}$$

Rewrite the expression for v in terms of t_A and t_B yields:

$$v = \frac{c^2(t_B - t_A)}{L_p}$$

Express the condition on $t_B - t_A$ if event B is to precede event A:

$$t_B - t_A < 0 \text{ and } v > \frac{c^2(t_B - t_A)}{L_p}$$

Substitute numerical values and evaluate v:

$$v > \frac{\left(2.998\times10^{8}\ \text{m/s}\right)^{2}\left(2.00\ \mu\text{s}\right)}{1.50\ \text{km}}$$

$$= 1.20\times10^{8}\ \text{m/s} = 0.400c$$

Event B can precede event A provided $v > 0.400c$.

Relativistic Energy and Momentum

41 • **[SSM]** How much energy would be required to accelerate a particle of mass m from rest to (*a*) $0.500c$, (*b*) $0.900c$, and (*c*) $0.990c$? Express your answers as multiples of the rest energy, mc^2.

Picture the Problem We can use Equation R-14 to find the energy required to accelerate this particle from rest to the given speeds.

From Equation R-14 we have:

$$K = \frac{mc^2}{\sqrt{1-(v/c)^2}} - mc^2$$

$$= \left(\frac{1}{\sqrt{1-(v/c)^2}} - 1\right)mc^2$$

(*a*)Substitute numerical values and evaluate $K(0.500c)$:

$$K(0.500c) = \left(\frac{1}{\sqrt{1-(0.500c/c)^2}} - 1\right)mc^2$$

$$= \boxed{0.155mc^2}$$

(*b*) Substitute numerical values and evaluate $K(0.900c)$:

$$K(0.900c) = \left(\frac{1}{\sqrt{1-(0.900c/c)^2}} - 1\right)mc^2$$

$$= \boxed{1.29mc^2}$$

(*c*) Substitute numerical values and evaluate $K(0.990c)$:

$$K(0.990c) = \left(\frac{1}{\sqrt{1-(0.990c/c)^2}} - 1\right)mc^2$$

$$= \boxed{6.09mc^2}$$

45 •• **[SSM]** (*a*) Show that the speed *v* of a particle of mass *m* and total energy *E* is given by $\frac{v}{c}=\left[1-\frac{(mc^2)^2}{E^2}\right]^{1/2}$ and that when *E* is much greater than mc^2, this can be approximated by $\frac{v}{c}\approx 1-\frac{(mc^2)^2}{2E^2}$. Find the speed of an electron with kinetic energy of (*b*) 0.510 MeV and (*c*) 10.0 MeV.

Picture the Problem We can solve the equation for the relativistic energy of a particle to obtain the first result and then use the binomial expansion subject to $E >> mc^2$ to obtain the second result. In parts (*b*) and (*c*) we can use the first expression obtained in (*a*), with $E = E_0 + K$, to find the speeds of electrons with the given kinetic energies. See Table 39-1 for the rest energy of an electron.

(*a*) The relativistic energy of a particle is given by Equation R-15:

$$E=\frac{mc^2}{\sqrt{1-\frac{v^2}{c^2}}}$$

Solving for *v*/*c* yields:

$$\boxed{\frac{v}{c}=\left[1-\frac{(mc^2)^2}{E^2}\right]^{1/2}} \qquad (1)$$

Expand the radical expression binomially to obtain:

$$\frac{v}{c}=\sqrt{1-\frac{(mc^2)^2}{E^2}}=1-\frac{1}{2}\frac{(mc^2)^2}{E^2}+\text{higher-order terms}$$

Because the higher-order terms are much smaller than the 2nd-degree term when $E >> mc^2$:

$$\frac{v}{c}\approx\boxed{1-\frac{(mc^2)^2}{2E^2}}$$

(*b*) Solve equation (1) for *v*:

$$v=c\sqrt{1-\frac{(mc^2)^2}{E^2}}$$

Because $E = E_0 + K$:

$$v=c\sqrt{1-\frac{E_0^2}{(E_0+K)^2}}=c\sqrt{1-\frac{1}{\left(1+\frac{K}{E_0}\right)^2}}$$

For an electron whose kinetic energy is 0.510 MeV:

$$v(0.510\,\text{MeV}) = c\sqrt{1-\frac{1}{\left(1+\frac{0.510\,\text{MeV}}{0.511\,\text{MeV}}\right)^2}} = \boxed{0.866c}$$

(*c*) For an electron whose kinetic energy is 10.0 MeV:

$$v(10.0\,\text{MeV}) = c\sqrt{1-\frac{1}{\left(1+\frac{10.0\,\text{MeV}}{0.511\,\text{MeV}}\right)^2}} = \boxed{0.999c}$$

General Problems

53 •• **[SSM]** Particles called muons traveling at 0.99995*c* are detected at the surface of Earth. One of your fellow students claims that the muons might have originated from the Sun. Prove him wrong. (The proper mean lifetime of the muon is 2.20 μs.)

Picture the Problem Your fellow student is thinking that the time dilation factor might allow muons to travel the 150,000,000,000 m from the Sun to the Earth. You can discredit your classmate's assertion by considering the mean lifetime of the muon from the Earth's reference frame. Doing so will demonstrate that the distance traveled during as many as 5 proper mean lifetimes is consistent with the origination of muons within the Earth's atmosphere.

The distance, in the earth frame of reference, a muon can travel in n mean lifetimes τ is given by:

$$d = n\frac{d_0}{\sqrt{1-(v/c)^2}} = \frac{nv\tau}{\sqrt{1-(v/c)^2}} = \frac{n(v/c)c\tau}{\sqrt{1-(v/c)^2}}$$

Substitute numerical values for v, c, and τ and simplify to obtain:

$$d = n\frac{(0.99995)(2.998\times10^8\ \text{m/s})(2.20\ \mu\text{s})}{\sqrt{1-(0.99995)^2}} = (66.0\ \text{km})n$$

In 5 lifetimes a muon would travel a distance: $d = (66.0\ \text{km})(5) = 330\ \text{km}$, a distance approximating a low Earth orbit.

In 100 lifetimes, $d \approx 6600$ km, or approximately one Earth radius. This relatively short distance should convince your classmate that the origin of the muons that are observed on Earth is within our atmosphere and that they certainly are not from the Sun.

Chapter 11
Gravity

Conceptual Problems

1 • **[SSM]** True or false:
(*a*) For Kepler's law of equal areas to be valid, the force of gravity must vary inversely with the square of the distance between a given planet and the Sun.
(*b*) The planet closest to the Sun has the shortest orbital period.
(*c*) Venus's orbital speed is larger than the orbital speed of Earth.
(*d*) The orbital period of a planet allows accurate determination of that planet's mass.

(*a*) False. Kepler's law of equal areas is a consequence of the fact that the gravitational force acts along the line joining two bodies but is independent of the manner in which the force varies with distance.

(*b*) True. The periods of the planets vary with the three-halves power of their distances from the sun. So the shorter the distance from the sun, the shorter the period of the planet's motion.

(*c*) True. Setting up a proportion involving the orbital speeds of the two planets in terms of their orbital periods and mean distances from the Sun (see Table 11-1) shows that $v_{\text{Venus}} = 1.17v_{\text{Earth}}$.

(*d*) False. The orbital period of a planet is independent of the planet's mass.

3 • **[SSM]** During what season in the northern hemisphere does Earth attain its maximum orbital speed about the Sun? What season is related to its minimum orbital speed? HINT: The major factor determining the seasons on Earth is *not* the variation in distance from the Sun.

Determine the Concept Earth is closest to the Sun during winter in the northern hemisphere. This is the time of fastest orbital speed. Summer would be the time for minimum orbital speed.

7 • **[SSM]** At the surface of the moon, the acceleration due to the gravity of the moon is *a*. At a distance from the center of the moon equal to four times the radius of the moon, the acceleration due to the gravity of the moon is (*a*) 16*a*, (*b*) *a*/4, (*c*) *a*/3, (*d*) *a*/16, (*e*) None of the above.

Picture the Problem The acceleration due to gravity varies inversely with the square of the distance from the center of the moon.

Express the dependence of the acceleration due to the gravity of the moon on the distance from its center:

$$a' \propto \frac{1}{r^2}$$

Express the dependence of the acceleration due to the gravity of the moon at its surface on its radius:

$$a \propto \frac{1}{R_M^2}$$

Divide the first of these expressions by the second to obtain:

$$\frac{a'}{a} = \frac{R_M^2}{r^2}$$

Solving for a' and simplifying yields:

$$a' = \frac{R_M^2}{r^2}a = \frac{R_M^2}{(4R_M)^2}a = \tfrac{1}{16}a$$

and $\boxed{(d)}$ is correct.

11 •• **[SSM]** Suppose the escape speed from a planet was only slightly larger than the escape speed from Earth, yet it was considerably larger than Earth. How would the planet's (average) density compare to Earth's (average) density? (*a*) It must be more dense. (*b*) It must be less dense. (*c*) It must be the same density. (*d*) You cannot determine the answer based on the data given.

Picture the Problem The densities of the planets are related to the escape speeds from their surfaces through $v_e = \sqrt{2GM/R}$.

The escape speed from the planet is given by:

$$v_{\text{planet}} = \sqrt{\frac{2GM_{\text{planet}}}{R_{\text{planet}}}}$$

The escape speed from Earth is given by:

$$v_{\text{Earth}} = \sqrt{\frac{2GM_{\text{Earth}}}{R_{\text{Earth}}}}$$

Expressing the ratio of the escape speed from the planet to the escape speed from Earth and simplifying yields:

$$\frac{v_{\text{planet}}}{v_{\text{Earth}}} = \frac{\sqrt{\dfrac{2GM_{\text{planet}}}{R_{\text{planet}}}}}{\sqrt{\dfrac{2GM_{\text{Earth}}}{R_{\text{Earth}}}}} = \sqrt{\frac{R_{\text{Earth}}}{R_{\text{planet}}}\frac{M_{\text{planet}}}{M_{\text{Earth}}}}$$

Because $v_{\text{planet}} \approx v_{\text{Earth}}$:

$$1 \approx \sqrt{\frac{R_{\text{Earth}}}{R_{\text{planet}}}\frac{M_{\text{planet}}}{M_{\text{Earth}}}}$$

Squaring both sides of the equation yields:

$$1 \approx \frac{R_{\text{Earth}}}{R_{\text{planet}}} \frac{M_{\text{planet}}}{M_{\text{Earth}}}$$

Express M_{planet} and M_{Earth} in terms of their densities and simplify to obtain:

$$1 \approx \frac{R_{\text{Earth}}}{R_{\text{planet}}} \frac{\rho_{\text{planet}} V_{\text{planet}}}{\rho_{\text{Earth}} V_{\text{Earth}}} = \frac{R_{\text{Earth}}}{R_{\text{planet}}} \frac{\rho_{\text{planet}} V_{\text{planet}}}{\rho_{\text{Earth}} V_{\text{Earth}}} = \frac{R_{\text{Earth}}}{R_{\text{planet}}} \frac{\rho_{\text{planet}} \frac{4}{3}\pi R_{\text{planet}}^3}{\rho_{\text{Earth}} \frac{4}{3}\pi R_{\text{Earth}}^3} = \frac{\rho_{\text{planet}} R_{\text{planet}}^2}{\rho_{\text{Earth}} R_{\text{Earth}}^2}$$

Solving for the ratio of the densities yields:

$$\frac{\rho_{\text{planet}}}{\rho_{\text{Earth}}} \approx \frac{R_{\text{Earth}}^2}{R_{\text{planet}}^2}$$

Because the planet is considerably larger than Earth:

$$\frac{\rho_{\text{planet}}}{\rho_{\text{Earth}}} << 1$$

and $\boxed{(b)}$ is correct.

13 •• **[SSM]** Near the end of their useful lives, several large Earth-orbiting satellites have been maneuvered so as to burn up as they enter Earth's atmosphere. These maneuvers have to be done carefully so large fragments do not impact populated land areas. You are in charge of such a project. Assuming the satellite of interest has on-board propulsion, in what direction would you fire the rockets for a short burn time to start this downward spiral? What would happen to the kinetic energy, gravitational potential energy and total mechanical energy following the burn as the satellite came closer and closer to Earth?

Determine the Concept You should fire the rocket in a direction to oppose the orbital motion of the satellite. As the satellite gets closer to Earth after the burn, the potential energy will decrease as the satellite gets closer to Earth. However, the total mechanical energy will decrease due to the frictional drag forces transforming mechanical energy into thermal energy. The kinetic energy will increase until the satellite enters the atmosphere where the drag forces slow its motion.

Estimation and Approximation

17 • **[SSM]** Estimate the mass of our galaxy (the Milky Way) if the Sun orbits the center of the galaxy with a period of 250 million years at a mean distance of 30,000 $c{\cdot}y$. Express the mass in terms of multiples of the solar mass M_S. (Neglect the mass farther from the center than the Sun, and assume that the mass closer to the center than the Sun exerts the same force on the Sun as would a point particle of the same mass located at the center of the galaxy.)

Picture the Problem To approximate the mass of the galaxy we'll assume the galactic center to be a point mass with the sun in orbit about it and apply Kepler's 3rd law.

Using Kepler's 3rd law, relate the period of the sun T to its mean distance r from the center of the galaxy:

$$T^2 = \frac{4\pi^2}{GM_{\text{galaxy}}}r^3 = \frac{\dfrac{4\pi^2}{M_s}}{G\dfrac{M_{\text{galaxy}}}{M_s}}r^3$$

Solve for $\frac{r^3}{T^2}$ and simplify to obtain:

$$\frac{r^3}{T^2} = \frac{G\dfrac{M_{\text{galaxy}}}{M_s}}{\dfrac{4\pi^2}{M_s}} = \frac{\dfrac{M_{\text{galaxy}}}{M_s}}{\dfrac{4\pi^2}{GM_s}}$$

If we measure distances in AU and times in years:

$$\frac{4\pi^2}{GM_s} = 1\frac{\text{y}^2}{(\text{AU})^3}$$

and

$$\frac{r^3}{T^2} = \frac{M_{\text{galaxy}}}{M_s}\frac{(\text{AU})^3}{\text{y}^2}$$

Substitute numerical values and evaluate M_{galaxy}/M_s:

$$\frac{M_{\text{galaxy}}}{M_s} = \frac{\left(3.00\times10^4\ c\cdot\text{y}\times\dfrac{9.461\times10^{15}\ \text{m}}{c\cdot\text{y}}\times\dfrac{1\ \text{AU}}{1.50\times10^{11}\ \text{m}}\right)^3\dfrac{\text{y}^2}{(\text{AU})^3}}{\left(250\times10^6\ \text{y}\right)^2} = 1.08\times10^{11}$$

or

$$M_{\text{galaxy}} = \boxed{1.08\times10^{11}M_s}$$

Kepler's Laws

25 •• **[SSM]** One of the so-called "Kirkwood gaps" in the asteroid belt occurs at an orbital radius at which the period of the orbit is half that of Jupiter's. The reason there is a gap for orbits of this radius is because of the periodic pulling (by Jupiter) that an asteroid experiences at the same place in its orbit every *other* orbit around the sun. Repeated tugs from Jupiter of this kind would eventually change the orbit of such an asteroid – therefore all asteroids that would otherwise have orbited at this radius have presumably been cleared away from the area due to this resonance phenomenon. How far from the sun is this particular 2:1 "Kirkwood" gap?

Picture the Problem The period of an orbit is related to its semi-major axis (for circular orbits this distance is the orbital radius). Because we know the orbital periods of Jupiter and a hypothetical asteroid in the Kirkwood gap, we can use

Kepler's 3rd law to set up a proportion relating the orbital periods and average distances of Jupiter and the asteroid from the Sun from which we can obtain an expression for the orbital radius of an asteroid in the Kirkwood gap.

Use Kepler's 3rd law to relate Jupiter's orbital period to its mean distance from the Sun:

$$T_{\text{Jupiter}}^2 = Cr_{\text{Jupiter}}^3$$

Use Kepler's 3rd law to relate the orbital period of an asteroid in the Kirkwood gap to its mean distance from the Sun:

$$T_{\text{Kirkwood}}^2 = Cr_{\text{Kirkwood}}^3$$

Dividing the second of these equations by the first yields:

$$\frac{T_{\text{Kirkwood}}^2}{T_{\text{Jupiter}}^2} = \frac{Cr_{\text{Kirkwood}}^3}{Cr_{\text{Jupiter}}^3} = \frac{r_{\text{Kirkwood}}^3}{r_{\text{Jupiter}}^3}$$

Solving for r_{Kirkwood} yields:

$$r_{\text{Kirkwood}} = \sqrt[3]{\left(\frac{T_{\text{Kirkwood}}}{T_{\text{Jupiter}}}\right)^2}\, r_{\text{Jupiter}}$$

Because the period of the orbit of an asteroid in the Kirkwood gap is half that of Jupiter's:

$$r_{\text{Kirkwood}} = \sqrt[3]{\left(\frac{\frac{1}{2}T_{\text{Jupiter}}}{T_{\text{Jupiter}}}\right)^2}\left(77.8\times10^{10}\ \text{m}\right) = \boxed{4.90\times10^{11}\ \text{m}}$$

29 •• **[SSM]** Kepler determined distances in the Solar System from his data. For example, he found the relative distance from the Sun to Venus (as compared to the distance from the Sun to Earth) as follows. Because Venus's orbit is closer to the Sun than is Earth's, Venus is a morning or evening star—its position in the sky is never very far from the Sun (see Figure 11-24). If we consider the orbit of Venus as a perfect circle, then consider the relative orientation of Venus, Earth, and the Sun at maximum extension—when Venus is farthest from the Sun in the sky. (*a*) Under this condition, show that angle *b* in Figure 11-24 is 90°. (*b*) If the maximum elongation angle *a* between Venus and the Sun is 47°, what is the distance between Venus and the Sun in AU? (*c*) Use this result to estimate the length of a Venusian "year."

Picture the Problem We can use a property of lines tangent to a circle and radii drawn to the point of contact to show that $b = 90°$. Once we've established that *b* is a right angle we can use the definition of the sine function to relate the distance from the Sun to Venus to the distance from the Sun to Earth.

(*a*) The line from Earth to Venus′ orbit is tangent to the orbit of Venus at the point of maximum extension. Venus will appear closer to the sun in earth's sky when it passes the line drawn from Earth and tangent to its orbit. Hence $b = \boxed{90°}$

(*b*) Using trigonometry, relate the distance from the sun to Venus d_{SV} to the angle a:

$$\sin a = \frac{d_{\text{SV}}}{d_{\text{SE}}} \Rightarrow d_{\text{SV}} = d_{\text{SE}} \sin a$$

Substitute numerical values and evaluate d_{SV}:

$$d_{\text{SV}} = (1.00\,\text{AU})\sin 47° = 0.731\,\text{AU}$$
$$= \boxed{0.73\,\text{AU}}$$

(*c*) Use Kepler's 3rd law to relate Venus's orbital period to its mean distance from the Sun:

$$T_{\text{Venus}}^2 = Cr_{\text{Venus}}^3$$

Use Kepler's 3rd law to relate Earth's orbital period to its mean distance from the Sun:

$$T_{\text{Earth}}^2 = Cr_{\text{Earth}}^3$$

Dividing the first of these equations by the second yields:

$$\frac{T_{\text{Venus}}^2}{T_{\text{Earth}}^2} = \frac{Cr_{\text{Venus}}^3}{Cr_{\text{Earth}}^3} = \frac{r_{\text{Venus}}^3}{r_{\text{Earth}}^3}$$

Solving for T_{Venus} yields:

$$T_{\text{Venus}} = \sqrt{\left(\frac{r_{\text{Venus}}}{r_{\text{Earth}}}\right)^3}\, T_{\text{Earth}}$$

Using the result from part (*b*) yields:

$$T_{\text{Venus}} = \sqrt{\left(\frac{0.731\,\text{AU}}{1.00\,\text{AU}}\right)^3}\,(1.00\,\text{y})$$
$$= \boxed{0.63\,\text{y}}$$

Remarks: The correct distance from the sun to Venus is closer to 0.723 AU.

Newton's Law of Gravity

31 • **[SSM]** Jupiter's satellite Europa orbits Jupiter with a period of 3.55 d at an average orbital radius of 6.71×10^8 m. (*a*) Assuming that the orbit is circular, determine the mass of Jupiter from the data given. (*b*) Another satellite of Jupiter, Callisto, orbits at an average radius of 18.8×10^8 m with an orbital period of 16.7 d. Show that this data is consistent with an inverse square force law for gravity (*Note*: DO NOT use the value of *G* anywhere in Part (*b*)).

Picture the Problem While we could apply Newton's Law of Gravitation and 2nd Law of Motion to solve this problem from first principles, we'll use Kepler's 3rd law (derived from these laws) to find the mass of Jupiter in Part (*a*). In Part (*b*) we can compare the ratio of the centripetal accelerations of Europa and Callisto to show that they are consistent with an inverse square law for gravity.

(*a*) Assuming a circular orbit, apply Kepler's 3rd law to the motion of Europa to obtain:

$$T_E^2 = \frac{4\pi^2}{GM_J}R_E^3 \Rightarrow M_J = \frac{4\pi^2}{GT_E^2}R_E^3$$

Substitute numerical values and evaluate M_J:

$$M_J = \frac{4\pi^2\left(6.71\times10^8\ \text{m}\right)^3}{\left(6.673\times10^{-11}\ \text{N}\cdot\text{m}^2/\text{kg}^2\right)\left(3.55\,\text{d}\times\frac{24\,\text{h}}{\text{d}}\times\frac{3600\,\text{s}}{\text{h}}\right)^2} = \boxed{1.90\times10^{27}\ \text{kg}}$$

Note that this result is in excellent agreement with the accepted value of 1.902×10^{27} kg.

(*b*) Express the centripetal acceleration of both of the moons to obtain:

$$a_{\text{centripetal}} = \frac{v^2}{R} = \frac{\left(\frac{2\pi R}{T}\right)^2}{R} = \frac{4\pi^2 R}{T^2}$$

where R and T are the radii and periods of their motion.

Using this result, express the centripetal accelerations of Europa and Callisto:

$$a_E = \frac{4\pi^2 R_E}{T_E^2} \text{ and } a_C = \frac{4\pi^2 R_C}{T_C^2}$$

Divide the first of these equations by the second and simplify to obtain:

$$\frac{a_E}{a_C} = \frac{\frac{4\pi^2 R_E}{T_E^2}}{\frac{4\pi^2 R_C}{T_C^2}} = \frac{T_C^2}{T_E^2}\frac{R_E}{R_C}$$

Substitute for the periods of Callisto and Europa using Kepler's 3rd law to obtain:

$$\frac{a_E}{a_C} = \frac{CR_C^3}{CR_E^3}\frac{R_E}{R_C} = \frac{R_C^2}{R_E^2}$$

This result, together with the fact that the gravitational force is directly proportional to the acceleration of the moons, demonstrates that the gravitational force varies inversely with the square of the distance.

33 • **[SSM]** The mass of Saturn is 5.69×10^{26} kg. (*a*) Find the period of its moon Mimas, whose mean orbital radius is 1.86×10^8 m. (*b*) Find the mean orbital radius of its moon Titan, whose period is 1.38×10^6 s.

Picture the Problem While we could apply Newton's Law of Gravitation and 2nd Law of Motion to solve this problem from first principles, we'll use Kepler's 3rd law (derived from these laws) to find the period of Mimas and to relate the periods of the moons of Saturn to their mean distances from its center.

(*a*) Using Kepler's 3rd law, relate the period of Mimas to its mean distance from the center of Saturn:

$$T_M^2 = \frac{4\pi^2}{GM_S} r_M^3 \Rightarrow T_M = \sqrt{\frac{4\pi^2}{GM_S} r_M^3}$$

Substitute numerical values and evaluate T_M:

$$T_M = \sqrt{\frac{4\pi^2 (1.86\times10^8\ \text{m})^3}{(6.6726\times10^{-11}\ \text{N}\cdot\text{m}^2/\text{kg}^2)(5.69\times10^{26}\ \text{kg})}} = 8.18\times10^4\ \text{s} \approx \boxed{22.7\ \text{h}}$$

(*b*) Using Kepler's 3rd law, relate the period of Titan to its mean distance from the center of Saturn:

$$T_T^2 = \frac{4\pi^2}{GM_S} r_T^3 \Rightarrow r_T = \sqrt[3]{\frac{T_T^2 GM_S}{4\pi^2}}$$

Substitute numerical values and evaluate r_T:

$$r_T = \sqrt[3]{\frac{(1.38\times10^6\ \text{s})^2 (6.6726\times10^{-11}\ \text{N}\cdot\text{m}^2/\text{kg}^2)(5.69\times10^{26}\ \text{kg})}{4\pi^2}} = \boxed{1.22\times10^9\ \text{m}}$$

41 •• **[SSM]** A superconducting gravity meter can measure changes in gravity of the order $\Delta g/g = 1.00 \times 10^{-11}$. (*a*) You are hiding behind a tree holding the meter, and your 80-kg friend approaches the tree from the other side. How close to you can your friend get before the meter detects a change in *g* due to his presence? (*b*) You are in a hot air balloon and are using the meter to determine the rate of ascent (assume the balloon has constant acceleration). What is the smallest change in altitude that results in a detectable change in the gravitational field of Earth?

Picture the Problem We can determine the maximum range at which an object with a given mass can be detected by substituting the equation for the gravitational field in the expression for the resolution of the meter and solving for the distance. Differentiating $g(r)$ with respect to r, separating variables to obtain dg/g, and approximating Δr with dr will allow us to determine the vertical change in the position of the gravity meter in Earth's gravitational field is detectable.

(*a*) Express the gravitational field of Earth:

$$g_E = \frac{GM_E}{R_E^2}$$

Express the gravitational field due to the mass m (assumed to be a point mass) of your friend and relate it to the resolution of the meter:

$$g(r)=\frac{Gm}{r^2}=1.00\times10^{-11}g_E$$
$$=1.00\times10^{-11}\frac{GM_E}{R_E^2}$$

Solving for r yields:

$$r=R_E\sqrt{\frac{1.00\times10^{11}m}{M_E}}$$

Substitute numerical values and evaluate r:

$$r=(6.37\times10^6\ \text{m})\sqrt{\frac{1.00\times10^{11}(80\ \text{kg})}{5.98\times10^{24}\ \text{kg}}}$$
$$=\boxed{7.37\ \text{m}}$$

(b) Differentiate $g(r)$ and simplify to obtain:

$$\frac{dg}{dr}=\frac{-2Gm}{r^3}=-\frac{2}{r}\left(\frac{Gm}{r^2}\right)=-\frac{2}{r}g$$

Separate variables to obtain:

$$\frac{dg}{g}=-2\frac{dr}{r}=10^{-11}$$

Approximating dr with Δr, evaluate Δr with $r=R_E$:

$$\Delta r=\left|-\tfrac{1}{2}(1.00\times10^{-11})(6.37\times10^6\ \text{m})\right|$$
$$=\boxed{31.9\ \mu\text{m}}$$

43 •• **[SSM]** Earth's radius is 6370 km and the moon's radius is 1738 km. The acceleration of gravity at the surface of the moon is 1.62 m/s^2. What is the ratio of the average density of the moon to that of Earth?

Picture the Problem We can use the definitions of the gravitational fields at the surfaces of Earth and the moon to express the accelerations due to gravity at these locations in terms of the average densities of Earth and the moon. Expressing the ratio of these accelerations will lead us to the ratio of the densities.

Express the acceleration due to gravity at the surface of Earth in terms of Earth's average density:

$$g_E=\frac{GM_E}{R_E^2}=\frac{G\rho_E V_E}{R_E^2}=\frac{G\rho_E\frac{4}{3}\pi R_E^3}{R_E^2}$$
$$=\tfrac{4}{3}G\rho_E\pi R_E$$

The acceleration due to gravity at the surface of the moon in terms of the moon's average density is:

$$g_M=\tfrac{4}{3}G\rho_M\pi R_M$$

Divide the second of these equations by the first to obtain:

$$\frac{g_M}{g_E} = \frac{\rho_M R_M}{\rho_E R_E} \Rightarrow \frac{\rho_M}{\rho_E} = \frac{g_M R_E}{g_E R_M}$$

Substitute numerical values and evaluate $\frac{\rho_M}{\rho_E}$:

$$\frac{\rho_M}{\rho_E} = \frac{(1.62\,\text{m/s}^2)(6.37\times10^6\,\text{m})}{(9.81\,\text{m/s}^2)(1.738\times10^6\,\text{m})} = \boxed{0.605}$$

Gravitational Potential Energy

47 • **[SSM]** Find the escape speed for a projectile leaving the surface of the moon. The acceleration of gravity on the moon is 0.166 times that on Earth and the moon's radius is 0.273 R_E.

Picture the Problem The escape speed from the moon or Earth is given by $v_e = \sqrt{2GM/R}$, where M and R represent the masses and radii of the moon or Earth.

Express the escape speed from the moon:

$$v_{e.m} = \sqrt{\frac{2GM_m}{R_m}} = \sqrt{2g_m R_m} \qquad (1)$$

Express the escape speed from Earth:

$$v_{e.E} = \sqrt{\frac{2GM_E}{R_E}} = \sqrt{2g_E R_E} \qquad (2)$$

Divide equation (1) by equation (2) to obtain:

$$\frac{v_{e.m}}{v_{e.E}} = \frac{\sqrt{g_m R_m}}{\sqrt{g_E R_E}} = \sqrt{\frac{g_m R_m}{g_E R_E}}$$

Solving for $v_{e,m}$ yields:

$$v_{e.m} = \sqrt{\frac{g_m R_m}{g_E R_E}}\, v_{e.E}$$

Substitute numerical values and evaluate $v_{e,m}$:

$$v_{e.m} = \sqrt{(0.166)(0.273)}(11.2\,\text{km/s}) = \boxed{2.38\,\text{km/s}}$$

51 •• **[SSM]** An object is dropped from rest from a height of 4.0×10^6 m above the surface of Earth. If there is no air resistance, what is its speed when it strikes Earth?

Picture the Problem Let the zero of gravitational potential energy be at infinity and let m represent the mass of the object. We'll use conservation of energy to relate the initial potential energy of the object-Earth system to the final potential and kinetic energies.

Use conservation of energy to relate the initial potential energy of the system to its energy as the object is about to strike Earth:

$$K_f - K_i + U_f - U_i = 0$$

or, because $K_i = 0$,

$$K(R_E) + U(R_E) - U(R_E + h) = 0 \quad (1)$$

where h is the initial height above Earth's surface.

Express the potential energy of the object-Earth system when the object is at a distance r from the surface of Earth:

$$U(r) = -\frac{GM_E m}{r}$$

Substitute in equation (1) to obtain:

$$\tfrac{1}{2}mv^2 - \frac{GM_E m}{R_E} + \frac{GM_E m}{R_E + h} = 0$$

Solving for v yields:

$$v = \sqrt{2\left(\frac{GM_E}{R_E} - \frac{GM_E}{R_E + h}\right)} = \sqrt{2gR_E\left(\frac{h}{R_E + h}\right)}$$

Substitute numerical values and evaluate v:

$$v = \sqrt{\frac{2(9.81\,\text{m/s}^2)(6.37\times10^6\,\text{m})(4.0\times10^6\,\text{m})}{6.37\times10^6\,\text{m} + 4.0\times10^6\,\text{m}}} = \boxed{6.9\,\text{km/s}}$$

Gravitational Orbits

59 •• **[SSM]** Many satellites orbit Earth with maximum altitudes of 1000 km or less. *Geosynchronous* satellites, however, orbit at an altitude of 3579 km above Earth's surface. How much more energy is required to launch a 500-kg satellite into a geosynchronous orbit than into an orbit 1000 km above the surface of Earth?

Picture the Problem We can express the energy difference between these two orbits in terms of the total energy of a satellite at each elevation. The application of Newton's 2nd law to the force acting on a satellite will allow us to express the total energy of each satellite as function of its mass, the radius of Earth, and its orbital radius.

Express the energy difference:

$$\Delta E = E_{geo} - E_{1000} \quad (1)$$

Express the total energy of an orbiting satellite:

$$E_{tot} = K + U = \tfrac{1}{2}mv^2 - \frac{GM_E m}{R} \quad (2)$$

where R is the orbital radius.

Apply Newton's 2nd law to a satellite to relate the gravitational force to the orbital speed:

$$F_{radial} = \frac{GM_E m}{R^2} = m\frac{v^2}{R}$$

Solving for v^2 yields:

$$v^2 = \frac{gR_E^2}{R}$$

Substitute in equation (2) to obtain:

$$E_{tot} = \tfrac{1}{2}m\frac{gR_E^2}{R} - \frac{gR_E^2 m}{R} = -\frac{mgR_E^2}{2R}$$

Substituting in equation (1) and simplifying yields:

$$\Delta E = -\frac{mgR_E^2}{2R_{geo}} + \frac{mgR_E^2}{2R_{1000}} = \frac{mgR_E^2}{2}\left(\frac{1}{R_{1000}} - \frac{1}{R_{geo}}\right)$$
$$= \frac{mgR_E^2}{2}\left(\frac{1}{R_E + 1000\text{ km}} - \frac{1}{R_E + 3579\text{ km}}\right)$$

Substitute numerical values and evaluate ΔE:

$$\Delta E = \tfrac{1}{2}(500\text{ kg})(9.81\text{ N/kg})(6.37\times10^6\text{ m})^2\left(\frac{1}{7.37\times10^6\text{ m}} - \frac{1}{9.95\times10^6\text{ m}}\right) = \boxed{3.50\text{ GJ}}$$

The Gravitational Field ($\vec{g}$)

63 •• **[SSM]** A point particle of mass m is on the x axis at $x = L$ and an identical point particle is on the y axis at $y = L$. (*a*) What is the gravitational field at the origin? (*b*) What is the magnitude of this field?

Picture the Problem We can use the definition of the gravitational field due to a point mass to find the x and y components of the field at the origin and then add these components to find the resultant field. We can find the magnitude of the field from its components using the Pythagorean theorem.

(*a*) The gravitational field at the origin is the sum of its x and y components:

$$\vec{g} = \vec{g}_x + \vec{g}_y \quad (1)$$

Express the gravitational field due to the point mass at $x = L$:

$$\vec{g}_x = \frac{Gm}{L^2}\hat{i}$$

Express the gravitational field due to the point mass at $y = L$:

$$\vec{g}_y = \frac{Gm}{L^2}\hat{j}$$

Substitute in equation (1) to obtain:

$$\vec{g} = \vec{g}_x + \vec{g}_y = \boxed{\frac{Gm}{L^2}\hat{i} + \frac{Gm}{L^2}\hat{j}}$$

(*b*) The magnitude of $\vec{g}$ is given by:

$$|\vec{g}| = \sqrt{g_x^2 + g_y^2}$$

Substitute for g_x and g_y and simplify to obtain:

$$|\vec{g}| = \sqrt{\left(\frac{Gm}{L^2}\right)^2 + \left(\frac{Gm}{L^2}\right)^2} = \boxed{\sqrt{2}\frac{Gm}{L^2}}$$

67 ••• **[SSM]** A nonuniform thin rod of length L lies on the x axis. One end of the rod is at the origin, and the other end is at $x = L$. The rod's mass per unit length λ varies as $\lambda = Cx$, where C is a constant. (Thus, an element of the rod has mass $dm = \lambda\,dx$.) (*a*) What is the total mass of the rod? (*b*) Find the gravitational field due to the rod on the x axis at $x = x_0$, where $x_0 > L$.

Picture the Problem We can find the mass of the rod by integrating dm over its length. The gravitational field at $x_0 > L$ can be found by integrating $d\vec{g}$ at x_0 over the length of the rod.

(*a*) The total mass of the stick is given by:

$$M = \int_0^L \lambda\,dx$$

Substitute for λ and evaluate the integral to obtain"

$$M = C\int_0^L x\,dx = \boxed{\tfrac{1}{2}CL^2}$$

(*b*) Express the gravitational field due to an element of the stick of mass *dm*:

$$d\vec{g} = -\frac{Gdm}{(x_0 - x)^2}\hat{i} = -\frac{G\lambda\, dx}{(x_0 - x)^2}\hat{i} = -\frac{GCxdx}{(x_0 - x)^2}\hat{i}$$

Integrate this expression over the length of the stick to obtain:

$$\vec{g} = -GC\int_0^L \frac{xdx}{(x_0 - x)^2}\hat{i} = \boxed{\frac{2GM}{L^2}\left[\ln\left(\frac{x_0}{x_0 - L}\right) - \left(\frac{L}{x_0 - L}\right)\right]\hat{i}}$$

The Gravitational Field ($\vec{g}$) due to Spherical Objects

71 •• **[SSM]** Two widely separated solid spheres, S_1 and S_2, each have radius *R* and mass *M*. Sphere S_1 is uniform, whereas the density of sphere S_2 is given by $\rho(r) = C/r$, where *r* is the distance from its center. If the gravitational field strength at the surface of S_1 is g_1, what is the gravitational field strength at the surface of S_2?

Picture the Problem The gravitational field strength at the surface of a sphere is given by $g = GM/R^2$, where *R* is the radius of the sphere and *M* is its mass.

Express the gravitational field strength on the surface of S_1:

$$g_1 = \frac{GM}{R^2}$$

Express the gravitational field strength on the surface of S_2:

$$g_2 = \frac{GM}{R^2}$$

Divide the second of these equations by the first and simplify to obtain:

$$\frac{g_2}{g_1} = \frac{\frac{GM}{R^2}}{\frac{GM}{R^2}} = 1 \Rightarrow \boxed{g_1 = g_2}$$

75 •• **[SSM]** Suppose you are standing on a spring scale in an elevator that is descending at constant speed in a mine shaft located on the equator. (*a*) Show that the force on you due to Earth's gravity alone is proportional to your distance from the center of the planet. (*b*) Assume that the mine shaft located on the equator and is vertical. Do not neglect Earth's rotational motion. Show that the reading on the spring scale is proportional to your distance from the center of the planet.

Picture the Problem There are two forces acting on you as you descend in the elevator and are at a distance r from the center of Earth; an upward normal force (F_N) exerted by the scale, and a downward gravitational force (mg) exerted by Earth. Because you are in equilibrium (you are descending at constant speed) under the influence of these forces, the normal force exerted by the scale is equal in magnitude to the gravitational force acting on you. We can use Newton's law of gravity to express this gravitational force.

(*a*) Express the force of gravity acting on you when you are a distance r from the center of Earth:

$$F_g = \frac{GM(r)m}{r^2} \qquad (1)$$

Using the definition of density, express the density of Earth between you and the center of Earth and the density of Earth as a whole:

$$\rho = \frac{M(r)}{V(r)} = \frac{M(r)}{\frac{4}{3}\pi r^3}$$

The density of Earth is also given by:

$$\rho = \frac{M_E}{V_E} = \frac{M_E}{\frac{4}{3}\pi R^3}$$

Equating these two expressions for ρ and solving for $M(r)$ yields:

$$M(r) = M_E\left(\frac{r}{R}\right)^3$$

Substitute for $M(r)$ in equation (1) and simplify to obtain:

$$F_g = \frac{GM_E\left(\frac{r}{R}\right)^3 m}{r^2} = \frac{GM_E m}{R^2}\frac{r}{R} \qquad (2)$$

Apply Newton's law of gravity to yourself at the surface of Earth to obtain:

$$mg = \frac{GM_E m}{R^2} \Rightarrow g = \frac{GM_E}{R^2}$$

where g is the magnitude of the gravitational field at the surface of Earth.

Substitute for g in equation (2) to obtain:

$$\boxed{F_g = \left(\frac{mg}{R}\right)r}$$

That is, the force of gravity on you is proportional to your distance from the center of Earth.

(*b*) Apply Newton's 2nd law to your body to obtain:

$$F_N - mg\frac{r}{R} = -mr\omega^2$$

where the net force $(-mr\omega^2)$, directed toward the center of Earth, is the centripetal force acting on your body.

Solving for F_N yields:

$$F_N = \left(\frac{mg}{R}\right)r - mr\omega^2$$

Note that this equation tells us that your effective weight increases linearly with distance from the center of Earth. However, due just to the effect of rotation, as you approach the center the centripetal force decreases linearly and, doing so, increases your effective weight.

77 •• **[SSM]** A solid sphere of radius R has its center at the origin. It has a uniform mass density ρ_0, except that there is a spherical cavity in it of radius $r = \frac{1}{2}R$ centered at $x = \frac{1}{2}R$ as in Figure 11-27. Find the gravitational field at points on the x axis for $|x| > R$. *Hint: The cavity may be thought of as a sphere of mass* $m = (4/3)\pi r^3\rho_0$ *plus a sphere of "negative" mass* $-m$.

Picture the Problem We can use the hint to find the gravitational field along the x axis.

Using the hint, express $g(x)$:

$$g(x) = g_{\text{solid sphere}} + g_{\text{hollow sphere}}$$

Substitute for $g_{\text{solid sphere}}$ and $g_{\text{hollow sphere}}$ and simplify to obtain:

$$g(x) = \frac{GM_{\text{solid sphere}}}{x^2} + \frac{GM_{\text{hollow sphere}}}{\left(x - \frac{1}{2}R\right)^2} = \frac{G\rho_0\left(\frac{4}{3}\pi R^3\right)}{x^2} + \frac{G\rho_0\left[-\frac{4}{3}\pi\left(\frac{1}{2}R\right)^3\right]}{\left(x - \frac{1}{2}R\right)^2}$$

$$= \boxed{G\left(\frac{4\pi\rho_0 R^3}{3}\right)\left[\frac{1}{x^2} - \frac{1}{8\left(x - \frac{1}{2}R\right)^2}\right]}$$

81 ••• **[SSM]** A small diameter hole is drilled into the sphere of Problem 80 toward the center of the sphere to a depth of 2.0 m below the sphere's surface. A small mass is dropped from the surface into the hole. Determine the speed of the small mass as it strikes the bottom of the hole.

Picture the Problem We can use conservation of energy to relate the work done by the gravitational field to the speed of the small object as it strikes the bottom of the hole. Because we're given the mass of the sphere, we can find C by expressing the mass of the sphere in terms of C. We can then use the definition of the gravitational field to find the gravitational field of the sphere inside its surface. The work done by the field equals the negative of the change in the potential energy of the system as the small object falls in the hole.

Use conservation of energy to relate the work done by the gravitational field to the speed of the small object as it strikes the bottom of the hole:

$K_f - K_i + \Delta U = 0$

or, because $K_i = 0$ and $W = -\Delta U$,

$$W = \tfrac{1}{2}mv^2 \Rightarrow v = \sqrt{\frac{2W}{m}} \qquad (1)$$

where v is the speed with which the object strikes the bottom of the hole and W is the work done by the gravitational field.

Express the mass of a differential element of the sphere:

$$dm = \rho dV = \rho\left(4\pi r^2 dr\right)$$

Integrate to express the mass of the sphere in terms of C:

$$M = 4\pi C \int_0^{5.0\,\text{m}} r dr = \left(50\,\text{m}^2\right)\pi C$$

Solving for C yields:

$$C = \frac{M}{\left(50\,\text{m}^2\right)\pi}$$

Substitute numerical values and evaluate C:

$$\boldsymbol{C} = \frac{1.0\times10^{11}\,\text{kg}}{\left(50\,\text{m}^2\right)\boldsymbol{\pi}} = 6.37\times10^8\,\text{kg/m}^2$$

Use its definition to express the gravitational field of the sphere at a distance from its center less than its radius:

$$g = \frac{F_g}{m} = \frac{GM}{r^2} = G\frac{\int_0^r 4\pi r^2 \rho\, dr}{r^2} = G\frac{\int_0^r 4\pi r^2 \frac{C}{r} dr}{r^2} = G\frac{4\pi C\int_0^r r\, dr}{r^2} = 2\pi GC$$

Express the work done on the small object by the gravitational force acting on it:

$$W = -\int_{5.0\,\text{m}}^{3.0\,\text{m}} mgdr = (2\,\text{m})mg$$

Substitute in equation (1) and simplify to obtain:

$$v = \sqrt{\frac{2(2.0\,\text{m})m(2\pi GC)}{m}} = \sqrt{(8.0\,\text{m})\pi GC}$$

Substitute numerical values and evaluate v:

$$\boldsymbol{v} = \sqrt{(8.0\,\text{m})\boldsymbol{\pi}\left(6.673\times10^{-11}\,\text{N}\cdot\text{m}^2/\text{kg}^2\right)\left(6.37\times10^8\,\text{kg/m}^2\right)} = \boxed{1.0\,\text{m/s}}$$

83 ••• **[SSM]** Two identical spherical cavities are made in a lead sphere of radius R. The cavities each have a radius $R/2$. They touch the outside surface of the sphere and its center as in Figure 11-28. The mass of a solid uniform lead sphere of radius R is M. Find the force of attraction on a point particle of mass m located at a distance d from the center of the lead sphere.

Picture the Problem The force of attraction of the small sphere of mass m to the lead sphere of mass M is the sum of the forces due to the solid sphere ($\vec{F}_S$) and the cavities ($\vec{F}_C$) of negative mass.

Express the force of attraction:

$$\vec{F} = \vec{F}_S + \vec{F}_C \qquad (1)$$

Use the law of gravity to express the force due to the solid sphere:

$$\vec{F}_S = -\frac{GMm}{d^2}\hat{i}$$

Express the magnitude of the force acting on the small sphere due to one cavity:

$$F_C = \frac{GM'm}{d^2 + \left(\frac{R}{2}\right)^2}$$

where M' is the negative mass of a cavity.

Relate the negative mass of a cavity to the mass of the sphere before hollowing:

$$M' = -\rho V = -\rho\left[\tfrac{4}{3}\pi\left(\frac{R}{2}\right)^3\right] = -\tfrac{1}{8}\left(\tfrac{4}{3}\pi\rho R^3\right) = -\tfrac{1}{8}M$$

Letting θ be the angle between the x axis and the line joining the center of the small sphere to the center of either cavity, use the law of gravity to express the force due to the two cavities:

$$\vec{F}_C = 2\frac{GMm}{8\left(d^2 + \frac{R^2}{4}\right)}\cos\theta\,\hat{i}$$

because, by symmetry, the y components add to zero.

Use the figure to express $\cos\theta$:

$$\cos\theta = \frac{d}{\sqrt{d^2 + \frac{R^2}{4}}}$$

Substitute for $\cos\theta$ and simplify to obtain:

$$\vec{F}_C = \frac{GMm}{4\left(d^2 + \frac{R^2}{4}\right)} \frac{d}{\sqrt{d^2 + \frac{R^2}{4}}}\hat{i} = \frac{GMmd}{4\left(d^2 + \frac{R^2}{4}\right)^{3/2}}\hat{i}$$

Substitute in equation (1) and simplify:

$$\vec{F} = -\frac{GMm}{d^2}\hat{i} + \frac{GMmd}{4\left(d^2 + \frac{R^2}{4}\right)^{3/2}}\hat{i}$$

$$\boxed{= -\frac{GMm}{d^2}\left[1 - \frac{\frac{d^3}{4}}{\left\{d^2 + \frac{R^2}{4}\right\}^{3/2}}\right]\hat{i}}$$

General Problems

89 •• **[SSM]** A *neutron star* is a highly condensed remnant of a massive star in the last phase of its evolution. It is composed of neutrons (hence the name) because the star's gravitational force causes electrons and protons to "coalesce" into the neutrons. Suppose at the end of its current phase, the Sun collapsed into a neutron star (it can't in actuality because it does not have enough mass) of radius 12.0 km, without losing any mass in the process. (*a*) Calculate the ratio of the gravitational acceleration at the surface of the Sun following its collapse compared to its value at the surface of the Sun today. (*b*) Calculate the ratio of the escape speed from the surface of the neutron-Sun to its value today.

Picture the Problem We can apply Newton's 2nd law and the law of gravity to an object of mass *m* at the surface of the Sun and the neutron-Sun to find the ratio of the gravitational accelerations at their surfaces. Similarly, we can express the ratio of the corresponding expressions for the escape speeds from the two suns to determine their ratio.

(*a*) Express the gravitational force acting on an object of mass *m* at the surface of the Sun:

$$F_g = ma_g = \frac{GM_{Sun}m}{R_{Sun}^2}$$

Solving for a_g yields:

$$a_g = \frac{GM_{\text{Sun}}}{R_{\text{Sun}}^2} \quad (1)$$

The gravitational force acting on an object of mass m at the surface of a neutron-Sun is:

$$F_g = ma'_g = \frac{GM_{\text{neutron-Sun}}m}{R_{\text{neutron-Sun}}^2}$$

Solving for a'_g yields:

$$a'_g = \frac{GM_{\text{neutron-Sun}}}{R_{\text{neutron-Sun}}^2} \quad (2)$$

Divide equation (2) by equation (1) to obtain:

$$\frac{a'_g}{a_g} = \frac{\dfrac{GM_{\text{neutron-Sun}}}{R_{\text{neutron-Sun}}^2}}{\dfrac{GM_{\text{Sun}}}{R_{\text{Sun}}^2}} = \frac{\dfrac{M_{\text{neutron-Sun}}}{R_{\text{neutron-Sun}}^2}}{\dfrac{M_{\text{Sun}}}{R_{\text{Sun}}^2}}$$

Because $M_{\text{neutron-Sun}} = M_{\text{Sun}}$:

$$\frac{a'_g}{a_g} = \frac{R_{\text{Sun}}^2}{R_{\text{neutron-Sun}}^2}$$

Substitute numerical values and evaluate the ratio a'_g/a_g:

$$\frac{a'_g}{a_g} = \left(\frac{6.96\times10^8\text{ m}}{12.0\times10^3\text{ m}}\right)^2 = \boxed{3.36\times10^9}$$

(*b*) The escape speed from the neutron-Sun is given by:

$$v'_e = \sqrt{\frac{GM_{\text{neutron-Sun}}}{R_{\text{neutron-Sun}}}}$$

The escape speed from the Sun is given by:

$$v_e = \sqrt{\frac{GM_{\text{Sun}}}{R_{\text{Sun}}}}$$

Dividing the first of these equations by the second and simplifying yields:

$$\frac{v'_e}{v_e} = \sqrt{\frac{R_{\text{Sun}}}{R_{\text{neutron-Sun}}}}$$

Substitute numerical values and evaluate v'_e/v_e:

$$\frac{v'_e}{v_e} = \sqrt{\frac{6.96\times10^8\text{ m}}{12.0\times10^3\text{ m}}} = \boxed{241}$$

95 •• **[SSM]** Uranus, the seventh planet in the Solar System, was first observed in 1781 by William Herschel. Its orbit was then analyzed in terms of Kepler's Laws. By the 1840's, observations of Uranus clearly indicated that its true orbit was different from the Keplerian calculation by an amount that could not be accounted for by observational uncertainty. The conclusion was that there must be another influence other than the Sun and the known planets lying inside Uranus's orbit. This influence was hypothesized to be due to an eighth planet, whose predicted orbit was described in 1845 independently by two astronomers: John Adams (no relation to our president) and Urbain LeVerrier. In September of 1846, John Galle, searching in the sky at the place predicted by Adams and LeVerrier, made the first observation of Neptune. Uranus and Neptune are in orbit about the Sun with periods of 84.0 and 164.8 years, respectively. To see the effect that Neptune had on Uranus, determine the ratio of the gravitational force between Neptune and Uranus to that between Uranus and the Sun, when Neptune and Uranus are at their closest approach to one another (i.e. when aligned with the Sun). The masses of the Sun, Uranus and Neptune are 333,000, 14.5 and 17.1 times that of Earth, respectively.

Picture the Problem We can use the law of gravity and Kepler's 3rd law to express the ratio of the gravitational force between Neptune and Uranus to that between Uranus and the Sun, when Neptune and Uranus are at their closest approach to one another.

The ratio of the gravitational force between Neptune and Uranus to that between Uranus and the Sun, when Neptune and Uranus are at their closest approach to one another is given by:

$$\frac{F_{\text{g,N-U}}}{F_{\text{g,U-S}}} = \frac{\dfrac{GM_{\text{N}}M_{\text{U}}}{(r_{\text{N}} - r_{\text{U}})^2}}{\dfrac{GM_{\text{U}}M_{\text{S}}}{r_{\text{U}}^2}} = \frac{M_{\text{N}}r_{\text{U}}^2}{M_{\text{S}}(r_{\text{N}} - r_{\text{U}})^2} \quad (1)$$

Applying Kepler's 3rd law to Uranus yields:

$$T_{\text{U}}^2 = Cr_{\text{U}}^3 \quad (2)$$

Applying Kepler's 3rd law to Neptune yields:

$$T_{\text{N}}^2 = Cr_{\text{N}}^3 \quad (3)$$

Divide equation (3) by equation (2) to obtain:

$$\frac{T_{\text{N}}^2}{T_{\text{U}}^2} = \frac{Cr_{\text{N}}^3}{Cr_{\text{U}}^3} = \frac{r_{\text{N}}^3}{r_{\text{U}}^3} \Rightarrow r_{\text{N}} = r_{\text{U}}\left(\frac{T_{\text{N}}}{T_{\text{U}}}\right)^{2/3}$$

Substitute for r_{N} in equation (1) to obtain:

$$\frac{F_{\text{g,N-U}}}{F_{\text{g,U-S}}} = \frac{M_{\text{N}}r_{\text{U}}^2}{M_{\text{S}}\left(r_{\text{U}}\left(\dfrac{T_{\text{N}}}{T_{\text{U}}}\right)^{2/3} - r_{\text{U}}\right)^2}$$

Simplifying this expression yields:

$$\frac{F_{\text{g,N-U}}}{F_{\text{g,U-S}}} = \frac{M_{\text{N}}}{M_{\text{S}}\left(\left(\dfrac{T_{\text{N}}}{T_{\text{U}}}\right)^{2/3} - 1\right)^2}$$

Because $M_N = 17.1M_E$ and $M_S = 333{,}000M_E$:

$$\frac{F_{g,N\text{-}U}}{F_{g,U\text{-}S}} = \frac{17.1M_E}{3.33\times10^5 M_E\left(\left(\frac{T_N}{T_U}\right)^{2/3}-1\right)^2} = \frac{17.1}{3.33\times10^5\left(\left(\frac{T_N}{T_U}\right)^{2/3}-1\right)^2}$$

Substitute numerical values and evaluate $\frac{F_{g,N\text{-}U}}{F_{g,U\text{-}S}}$:

$$\frac{F_{g,N\text{-}U}}{F_{g,U\text{-}S}} = \frac{17.1}{3.33\times10^5\left(\left(\frac{164.8\text{ y}}{84.0\text{ y}}\right)^{2/3}-1\right)^2} \approx 2\times10^{-4}$$

Because this ratio is so small, during the time at which Neptune is closest to Uranus, the force exerted on Uranus by Neptune is much less than the force exerted on Uranus by the Sun.

97 •• **[SSM]** Four identical planets are arranged in a square as shown in Figure 11-29. If the mass of each planet is M and the edge length of the square is a, what must be their speed if they are to orbit their common center under the influence of their mutual attraction?

Picture the Problem We can find the orbital speeds of the planets from their distance from the center of mass of the system and the period of their motion. Application of Kepler's 3rd law will allow us to express the period of their motion T in terms of the effective mass of the system; which we can find from its definition.

Express the orbital speeds of the planets in terms of their period T:

$$v = \frac{2\pi R}{T} \qquad (1)$$

where R is the distance to the center of mass of the four-planet system.

Apply Kepler's 3rd law to express the period of the planets:

$$T = \sqrt{\frac{4\pi^2}{GM_{eff}}R^3}$$

where M_{eff} is the effective mass of the four planets.

Substitute for T in equation (1) to obtain:

$$v = \frac{2\pi R}{\sqrt{\frac{4\pi^2}{GM_{eff}}R^3}} = \sqrt{\frac{GM_{eff}}{R}} \qquad (2)$$

The distance of each planet from the effective mass is:

$$R = \frac{a}{\sqrt{2}}$$

Find M_{eff} from its definition:

$$\frac{1}{M_{eff}} = \frac{1}{M} + \frac{1}{M} + \frac{1}{M} + \frac{1}{M}$$

and

$$M_{eff} = \tfrac{1}{4}M$$

Substitute for R and M_{eff} in equation (2) and simplify to obtain:

$$\boxed{v = \sqrt{\frac{\sqrt{2}GM}{4a}}}$$

103 ••• **[SSM]** In this problem you are to find the gravitational potential energy of the thin rod in Example 11-8 and a point particle of mass m_0 that is on the x axis at $x = x_0$. (*a*) Show that the potential energy shared by an element of the rod of mass dm (shown in Figure 11-14) and the point particle of mass m_0 located at $x_0 \geq \frac{1}{2}L$ is given by

$$dU = -\frac{Gm_0dm}{x_0 - x_s} = \frac{GMm_0}{L(x_0 - x_s)}dx_s$$

where $U = 0$ at $x_0 = \infty$. (*b*) Integrate your result for Part (*a*) over the length of the rod to find the total potential energy for the system. Generalize your function $U(x_0)$ to any place on the x axis in the region $x > L/2$ by replacing x_0 by a general coordinate x and write it as $U(x)$. (*c*) Compute the force on m_0 at a general point x using $F_x = -dU/dx$ and compare your result with m_0g, where g is the field at x_0 calculated in Example 11-8.

Picture the Problem Let $U = 0$ at $x = \infty$. The potential energy of an element of the stick dm and the point mass m_0 is given by the definition of gravitational potential energy: $dU = -Gm_0dm/r$ where r is the separation of dm and m_0.

(*a*) Express the potential energy of the masses m_0 and dm:

$$dU = -\frac{Gm_0dm}{x_0 - x_s}$$

The mass dm is proportional to the size of the element dx_s:

$$dm = \lambda\, dx_s$$

where $\lambda = \dfrac{M}{L}$.

Substitute for dm and λ to express dU in terms of x_s:

$$dU = -\frac{Gm_0\lambda\, dx_s}{x_0 - x_s} = \boxed{-\frac{GMm_0\, dx_s}{L(x_0 - x_s)}}$$

(*b*) Integrate dU to find the total potential energy of the system:

$$U = -\frac{GMm_0}{L}\int_{-L/2}^{L/2}\frac{dx_s}{x_0 - x_s} = \frac{GMm_0}{L}\left[\ln\left(x_0 - \frac{L}{2}\right) - \ln\left(x_0 + \frac{L}{2}\right)\right]$$

$$= \boxed{-\frac{GMm_0}{L}\ln\left(\frac{x_0 + L/2}{x_0 - L/2}\right)}$$

(*c*) Because x_0 is a general point along the x axis:

$$F(x_0) = -\frac{dU}{dx_0} = \frac{GMm_0}{L}\left[\frac{1}{x_0 + \frac{L}{2}} - \frac{1}{x_0 - \frac{L}{2}}\right]$$

Further simplification yields:

$$F(x_0) = -\frac{Gmm_0}{x^2 - L^2/4}$$

This answer and the answer given in Example 11-8 are the same.

Chapter 12
Static Equilibrium and Elasticity

Conceptual Problems

1 • **[SSM]** True or false:

(*a*) $\sum_i \vec{F}_i = 0$ is sufficient for static equilibrium to exist.

(*b*) $\sum_i \vec{F}_i = 0$ is necessary for static equilibrium to exist.

(*c*) In static equilibrium, the net torque about any point is zero.

(*d*) An object in equilibrium cannot be moving.

(*a*) False. The conditions $\sum \vec{F} = 0$ and $\sum \vec{\tau} = 0$ must be satisfied.

(*b*) True. The necessary and sufficient conditions for static equilibrium are $\sum \vec{F} = 0$ and $\sum \vec{\tau} = 0$.

(*c*) True. The conditions $\sum \vec{F} = 0$ and $\sum \vec{\tau} = 0$ must be satisfied.

(*d*) False. An object can be moving with constant speed (translational or rotational) when the conditions $\sum \vec{F} = 0$ and $\sum \vec{\tau} = 0$ are satisfied.

9 •• **[SSM]** An aluminum wire and a steel wire of the same length L and diameter D are joined end-to-end to form a wire of length $2L$. One end of the wire is then fastened to the ceiling and an object of mass M is attached to the other end. Neglecting the mass of the wires, which of the following statements is true? (*a*) The aluminum portion will stretch by the same amount as the steel portion. (*b*) The tensions in the aluminum portion and the steel portion are equal. (*c*) The tension in the aluminum portion is greater than that in the steel portion. (*d*) None of the above

Determine the Concept We know that equal lengths of aluminum and steel wire of the same diameter will stretch different amounts when subjected to the same tension. Also, because we are neglecting the mass of the wires, the tension in them is independent of which one is closer to the roof and depends only on Mg. $\boxed{(b)}$ is correct.

Estimation and Approximation

11 •• **[SSM]** Consider an atomic model for Young's modulus. Assume that a large number of atoms are arranged in a cubic array, with each atom at a corner of a cube and each atom at a distance a from its six nearest neighbors. Imagine that each atom is attached to its 6 nearest neighbors by little springs each with force constant k. (a) Show that this material, if stretched, will have a Young's modulus $Y = k/a$. (b) Using Table 12-1 and assuming that $a \approx 1.0$ nm, estimate a typical value for the "atomic force constant" k in a metal.

Picture the Problem We can derive this expression by imagining that we pull on an area A of the given material, expressing the force each spring will experience, finding the fractional change in length of the springs, and substituting in the definition of Young's modulus.

(a) The definition of Young's modulus is:

$$Y = \frac{F/A}{\Delta L/L} \quad (1)$$

Express the elongation ΔL of each spring:

$$\Delta L = \frac{F_s}{k} \quad (2)$$

The force F_s each spring will experience as a result of a force F acting on the area A is:

$$F_s = \frac{F}{N}$$

Express the number of springs N in the area A:

$$N = \frac{A}{a^2}$$

Substituting for N yields:

$$F_s = \frac{Fa^2}{A}$$

Substitute F_s in equation (2) to obtain, for the extension of one spring:

$$\Delta L = \frac{Fa^2}{kA}$$

Assuming that the springs extend/compress linearly, the fractional extension of the springs is:

$$\frac{\Delta L_{tot}}{L} = \frac{\Delta L}{a} = \frac{1}{a}\frac{Fa^2}{kA} = \frac{Fa}{kA}$$

Substitute in equation (1) and simplify to obtain:

$$Y = \frac{\frac{F}{A}}{\frac{Fa}{kA}} = \boxed{\frac{k}{a}}$$

(b) From our result in Part (a):

$$k = Ya$$

From Table 12-1:

$$Y = 200\text{GN/m}^2 = 2.00\times10^{11}\ \text{N/m}^2$$

Substitute numerical values and evaluate k:

$$k = (2.00\times10^{11}\ \text{N/m}^2)(1.0\times10^{-9}\ \text{m}) = \boxed{2.0\ \text{N/cm}}$$

Static Equilibrium

17 • **[SSM]** Figure 12-31 shows a 25-foot sailboat. The mast is a uniform 120-kg pole that is supported on the deck and held fore and aft by wires as shown. The tension in the *forestay* (wire leading to the bow) is 1000 N. Determine the tension in the *backstay* (wire leading aft) and the normal force that the deck exerts on the mast. (Assume that the frictional force the deck exerts on the mast to be negligible.)

Picture the Problem The force diagram shows the forces acting on the mast. Let the origin of the coordinate system be at the foot of the mast with the $+x$ direction to the right and the $+y$ direction upward. Because the mast is in equilibrium, we can apply the conditions for translational and rotational equilibrium to find the tension in the backstay, T_B, and the normal force, F_D, that the deck exerts on the mast.

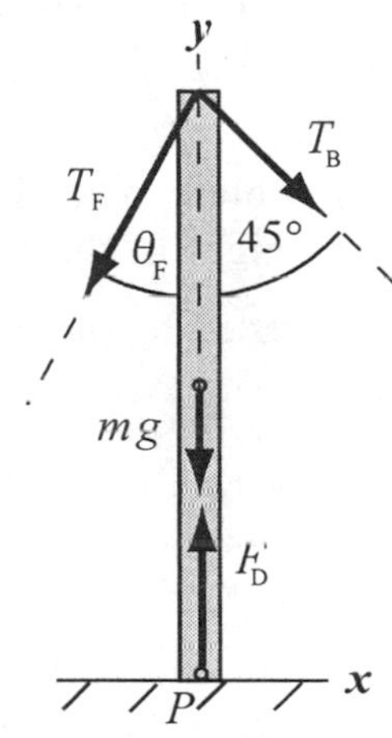

Apply $\sum\vec{\tau} = 0$ to the mast about an axis through point P:

$$(4.88\,\text{m})(1000\,\text{N})\sin\theta_F - (4.88\,\text{m})T_B\sin 45.0° = 0$$

Solve for T_B to obtain:

$$T_B = \frac{(1000\,\text{N})\sin\theta_F}{\sin 45.0°} \quad (1)$$

Find θ_F, the angle of the forestay with the vertical:

$$\theta_F = \tan^{-1}\left(\frac{2.74\,\text{m}}{4.88\,\text{m}}\right) = 29.3°$$

Substitute numerical values in equation (1) and evaluate T_B:

$$T_B = \frac{(1000\,\text{N})\sin 29.3°}{\sin 45.0°} = \boxed{692\,\text{N}}$$

Apply the condition for translational equilibrium in the y direction to the mast:

$$\sum F_y = F_D - T_F\cos\theta_F - T_B\cos 45° - mg = 0$$

Solving for F_D yields:

$$F_D = T_F\cos\theta_F + T_B\cos 45° + mg$$

Substitute numerical values and evaluate F_D:

$$F_D = (1000\text{ N})\cos 29.3° + (692\text{ N})\cos 45° + (120\text{ kg})(9.81\text{ m/s}^2) = \boxed{2.54\text{ kN}}$$

19 •• **[SSM]** A *gravity board* is a convenient and quick way to determine the location of the center of gravity of a person. It consists of a horizontal board supported by a fulcrum at one end and a scale at the other end. To demonstrate this in class, your physics professor calls on you to lie horizontally on the board with the top of your head directly above the fulcrum point as shown in Figure 12-33. The scale is 2.00 m from the fulcrum. In preparation for this experiment, you had accurately weighed yourself and determined your mass to be 70.0 kg. When you are at rest on the gravity board, the scale advances 250 N beyond its reading when the board is there by itself. Use this data to determine the location of your center of gravity relative to your feet.

Picture the Problem The diagram shows $\vec{w}$, the weight of the student, $\vec{F}_P$, the force exerted by the board at the pivot, and $\vec{F}_s$, the force exerted by the scale, acting on the student. Because the student is in equilibrium, we can apply the condition for rotational equilibrium to the student to find the location of his center of gravity.

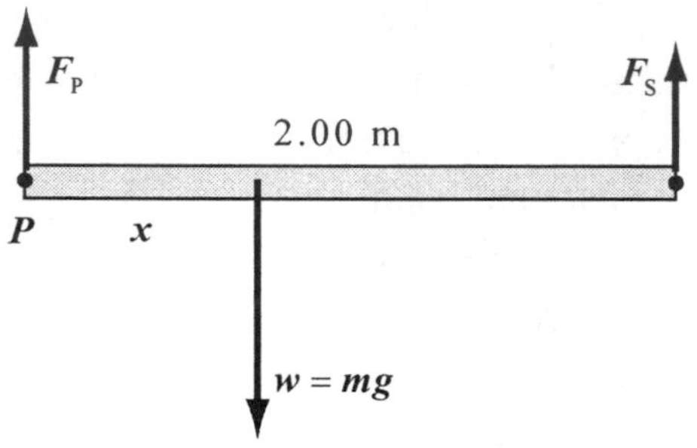

Apply $\sum \vec{\tau} = 0$ about an axis through the pivot point P:

$$F_s(2.00\text{ m}) - wx = 0$$

Solving for x yields:

$$x = \frac{(2.00\text{ m})F_s}{w}$$

Substitute numerical values and evaluate x:

$$x = \frac{(2.00\text{ m})(250\text{ N})}{(70.0\text{ kg})(9.81\text{ m/s}^2)} = \boxed{0.728\text{ m}}$$

25 •• **[SSM]** A cylinder of mass M and radius R rolls against a step of height h as shown in Figure 12-37. When a horizontal force of magnitude F is applied to the top of the cylinder, the cylinder remains at rest. (*a*) Find an expression for the normal force exerted by the floor on the cylinder. (*b*) Find an expression for the horizontal force exerted by the edge of the step on the cylinder. (*c*) Find an expression for the vertical component of the force exerted by the edge of the step on the cylinder.

Picture the Problem The figure to the right shows the forces acting on the cylinder. Choose a coordinate system in which the positive x direction is to the right and the positive y direction is upward. Because the cylinder is in equilibrium, we can apply the conditions for translational and rotational equilibrium to find F_n and the horizontal and vertical components of the force the corner of the step exerts on the cylinder.

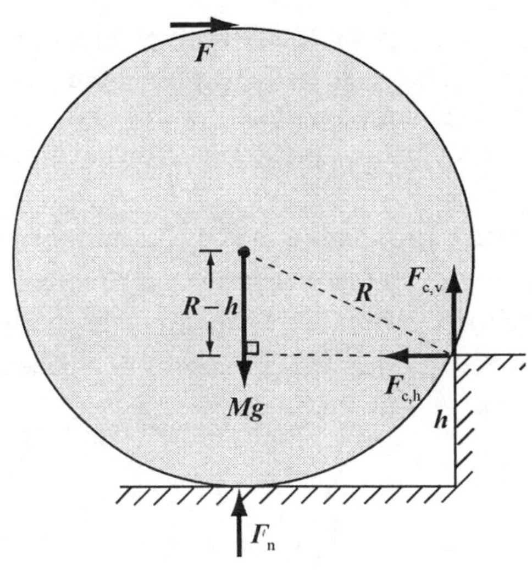

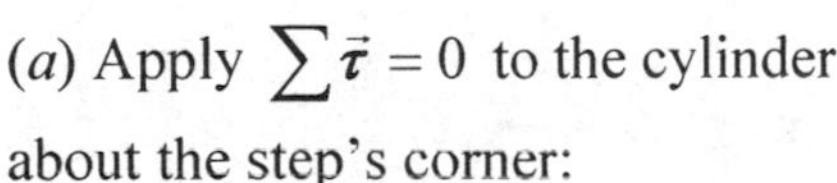

(*a*) Apply $\sum \vec{\tau} = 0$ to the cylinder about the step's corner:

$$Mg\ell - F_n\ell - F(2R-h) = 0$$

Solving for F_n yields:

$$F_n = Mg - \frac{F(2R-h)}{\ell}$$

Express ℓ as a function of R and h:

$$\ell = \sqrt{R^2-(R-h)^2} = \sqrt{2Rh-h^2}$$

Substitute for ℓ in the expression for F_n and simplify to obtain:

$$F_n = Mg - \frac{F(2R-h)}{\sqrt{2Rh-h^2}}$$

$$= \boxed{Mg - F\sqrt{\frac{2R-h}{h}}}$$

(*b*) Apply $\sum F_x = 0$ to the cylinder:

$$-F_{c,h} + F = 0$$

Solve for $F_{c,h}$:

$$F_{c,h} = \boxed{F}$$

(*c*) Apply $\sum F_y = 0$ to the cylinder:

$$F_n - Mg + F_{c,v} = 0 \Rightarrow F_{c,v} = Mg - F_n$$

Substitute the result from Part (*a*) and simplify to obtain:

$$F_{c,v} = Mg - \left\{Mg - F\sqrt{\frac{2R-h}{h}}\right\}$$

$$= \boxed{F\sqrt{\frac{2R-h}{h}}}$$

31 •• **[SSM]** Two 80-N forces are applied to opposite corners of a rectangular plate as shown in Figure 12-41. (*a*) Find the torque produced by this couple using Equation 12-6. (*b*) Show that the result is the same as if you determine the torque about the lower left-hand corner.

Picture the Problem The forces shown in the figure constitute a couple and will cause the plate to experience a counterclockwise angular acceleration. The couple equation is $\boldsymbol{\tau} = \boldsymbol{FD}$. The following diagram shows the geometric relationships between the variables in terms of a generalized angle θ.

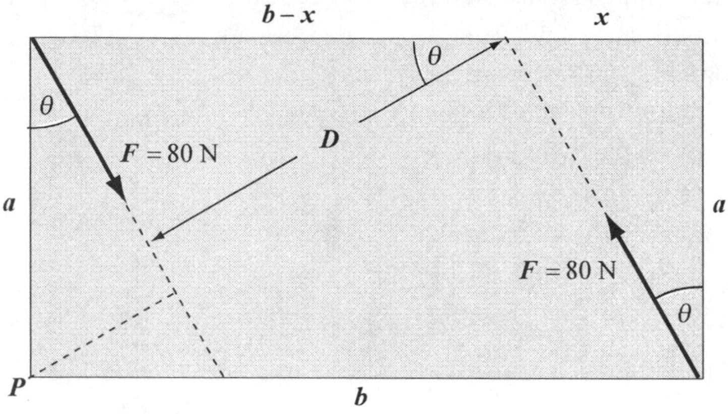

(*a*) The couple equation is: $\tau = FD$ (1)

From the diagram, *D* is given by: $D = (b - x)\cos\theta$ (2)

Again, referring to the diagram: $x = a\tan\theta$

Substituting for *x* in equation (2) and simplifying yields:

$$D = (b - a\tan\theta)\cos\theta = b\cos\theta - a\sin\theta$$

Substituting for *D* in equation (1) yields: $\tau = F(b\cos\theta - a\sin\theta)$ (3)

Substitute numerical values and evaluate τ:

$$\tau = (80\text{ N})(b\cos 30° - a\sin 30°) = \boxed{(69\text{ N})b - (40\text{ N})a}$$

(*b*) Letting the counterclockwise direction be the positive direction, apply $\sum\vec{\tau} = 0$ about an axis normal to the plane of the rectangle and passing through point *P*:

$$F(\ell + D) - F\ell = 0$$

Substituting for D yields: $F(\ell + b\cos\theta - a\sin\theta) - F\ell = 0$

Solve for τ to obtain: $\tau = \boxed{F(b\cos\theta - a\sin\theta)}$, in agreement with equation (3).

33 •• **[SSM]** A ladder of negligible mass and of length L leans against a slick wall making an angle of θ with the horizontal floor. The coefficient of friction between the ladder and the floor is μ_s. A man climbs the ladder. What height h can he reach before the ladder slips?

Picture the Problem Let the mass of the man be M. The ladder and the forces acting on it are shown in the diagram. Because the wall is slick, the force the wall exerts on the ladder must be horizontal. Because the ladder is in equilibrium, we can apply the conditions for translational and rotational equilibrium to it.

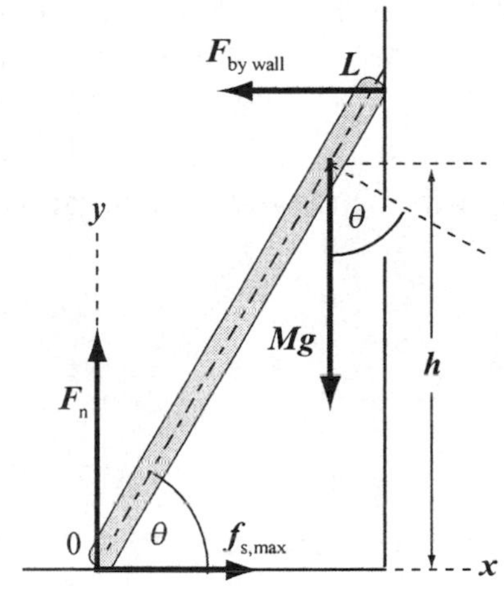

Apply $\sum F_y = 0$ to the ladder and solve for F_n: $F_n - Mg = 0 \Rightarrow F_n = Mg$

Apply $\sum F_x = 0$ to the ladder and solve for $f_{s,max}$: $f_{s,max} - F_W = 0 \Rightarrow f_{s,max} = F_W$

Apply $\sum \vec{\tau} = 0$ about the bottom of the ladder to obtain: $Mg\ell\cos\theta - F_W L\sin\theta = 0$

Solving for ℓ and simplifying yields:

$$\ell = \frac{F_W L\sin\theta}{Mg\cos\theta} = \frac{f_{s,max}L}{Mg}\tan\theta$$
$$= \frac{\mu_s F_n L}{Mg}\tan\theta = \mu_s L\tan\theta$$

Referring to the figure, relate ℓ to h: $h = \ell\sin\theta$

Substituting for ℓ yields: $h = \boxed{\mu_s L \tan\theta \sin\theta}$

39 ••• **[SSM]** A tall, uniform, rectangular block sits on an inclined plane as shown in Figure 12-45. A cord is attached to the top of the block to prevent it from falling down the incline. What is the maximum angle θ for which the block will not slide on the incline? Assume the block has a height-to-width ratio, *b*/*a*, of 4.0 and the coefficient of static friction between it and the incline is $\mu_s = 0.80$.

Picture the Problem Consider what happens just as θ increases beyond θ_{max}. Because the top of the block is fixed by the cord, the block will in fact rotate with only the lower right edge of the block remaining in contact with the plane. It follows that just prior to this slipping, F_n and $f_s = \mu_s F_n$ act at the lower right edge of the block. Choose a coordinate system in which up the incline is the $+x$ direction and the direction of $\vec{F}_n$ is the $+y$ direction. Because the block is in equilibrium, we can apply the conditions for translational and rotational equilibrium.

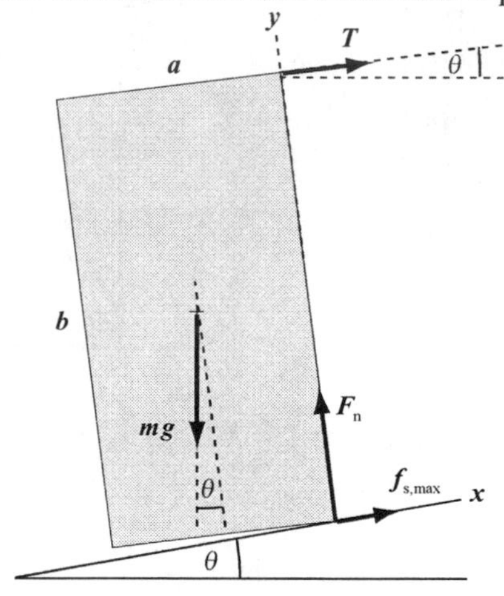

Apply $\sum F_x = 0$ to the block: $T + \mu_s F_n - mg\sin\theta = 0$ (1)

Apply $\sum F_y = 0$ to the block: $F_n - mg\cos\theta = 0$ (2)

Apply $\sum \vec{\tau} = 0$ about an axis through the lower right edge of the block: $\frac{1}{2}a(mg\cos\theta) + \frac{1}{2}b(mg\sin\theta) - bT = 0$ (3)

Eliminate F_n between equations (1) and (2) and solve for T: $T = mg(\sin\theta - \mu_s \cos\theta)$

Substitute for T in equation (3):

$$\frac{1}{2}a(mg\cos\theta) + \frac{1}{2}b(mg\sin\theta) - b[mg(\sin\theta - \mu_s\cos\theta)] = 0$$

Substitute $4a$ for b:

$$\tfrac{1}{2}a(mg\cos\theta)+\tfrac{1}{2}(4.0a)(mg\sin\theta) - (4.0a)[mg(\sin\theta-\mu_s\cos\theta)]=0$$

Simplify to obtain:

$$(1+8.0\mu_s)\cos\theta-4.0\sin\theta=0$$

Solving for θ yields:

$$\theta=\tan^{-1}\left[\frac{1+8.0\mu_s}{4.0}\right]$$

Substitute numerical values and evaluate θ:

$$\theta=\tan^{-1}\left[\frac{1+(8.0)(0.80)}{4.0}\right]=\boxed{62°}$$

Stress and Strain

41 • **[SSM]** Copper has a tensile strength of about 3.0×10^8 N/m^2. (*a*) What is the maximum load that can be hung from a copper wire of diameter 0.42 mm? (*b*) If half this maximum load is hung from the copper wire, by what percentage of its length will it stretch?

Picture the Problem L is the unstretched length of the wire, F is the force acting on it, and A is its cross-sectional area. The stretch in the wire ΔL is related to Young's modulus by $Y = \text{stress/strain} = (F/A)/(\Delta L/L)$.

(*a*) Express the maximum load in terms of the wire's tensile strength:

$$F_{\text{max}} = \text{tensile strength} \times A = \text{tensile strength} \times \pi r^2$$

Substitute numerical values and evaluate F_{max}:

$$F_{\text{max}} = (3.0\times10^8\ \text{N/m}^2)\pi(0.21\times10^{-3}\ \text{m})^2 = 41.6\ \text{N} = \boxed{42\ \text{N}}$$

(*b*) Using the definition of Young's modulus, express the fractional change in length of the copper wire:

$$\frac{\Delta L}{L}=\frac{F}{AY}=\frac{\tfrac{1}{2}F_{\text{max}}}{AY}$$

Substitute numerical values and evaluate $\frac{\Delta L}{L}$:

$$\frac{\Delta L}{L}=\frac{\tfrac{1}{2}(41.6\ \text{N})}{\pi(0.21\ \text{mm})^2(1.10\times10^{11}\ \text{N/m}^2)} = \boxed{0.14\%}$$

43 • **[SSM]** As a runner's foot pushes off on the ground, the shearing force acting on an 8.0-mm-thick sole is shown in Figure 12-46. If the force of 25 N is distributed over an area of 15 cm^2, find the angle of shear θ, given that the shear modulus of the sole is 1.9×10^5 N/m^2.

Picture the Problem The shear stress, defined as the ratio of the shearing force to the area over which it is applied, is related to the shear strain through the definition of the shear modulus; $M_\text{s} = \dfrac{\text{shear stress}}{\text{shear strain}} = \dfrac{F_\text{s}/A}{\tan\theta}$.

Using the definition of shear modulus, relate the angle of shear, θ to the shear force and shear modulus:

$$\tan\theta = \frac{F_\text{s}}{M_\text{s}A} \Rightarrow \theta = \tan^{-1}\left(\frac{F_\text{s}}{M_\text{s}A}\right)$$

Substitute numerical values and evaluate θ:

$$\theta = \tan^{-1}\left[\frac{25\ \text{N}}{(1.9\times10^5\ \text{N/m}^2)(15\times10^{-4}\ \text{m}^2)}\right] = \boxed{5.0°}$$

45 •• **[SSM]** Equal but opposite forces of magnitude F are applied to both ends of a thin wire of length L and cross-sectional area A. Show that if the wire is modeled as a spring, the force constant k is given by $k = AY/L$ and the potential energy stored in the wire is $U = \frac{1}{2}F\Delta L$, where Y is Young's modulus and ΔL is the amount the wire has stretched.

Picture the Problem We can use Hooke's law and Young's modulus to show that, if the wire is considered to be a spring, the force constant k is given by $k = AY/L$. By treating the wire as a spring we can show the energy stored in the wire is $U = \frac{1}{2}F\Delta L$.

Express the relationship between the stretching force, the force constant, and the elongation of a spring:

$$F = k\Delta L \Rightarrow k = \frac{F}{\Delta L}$$

Using the definition of Young's modulus, express the ratio of the stretching force to the elongation of the wire:

$$\frac{F}{\Delta L} = \frac{AY}{L} \qquad (1)$$

Equate these two expressions for $F/\Delta L$ to obtain:

$$k = \boxed{\frac{AY}{L}}$$

Treating the wire as a spring, express its stored energy:

$$U = \tfrac{1}{2}k(\Delta L)^2 = \tfrac{1}{2}\frac{AY}{L}(\Delta L)^2 = \tfrac{1}{2}\left(\frac{AY\Delta L}{L}\right)\Delta L$$

Solving equation (1) for F yields:

$$F = \frac{AY\Delta L}{L}$$

Substitute for F in the expression for U to obtain:

$$U = \boxed{\tfrac{1}{2}F\Delta L}$$

51 •• **[SSM]** An elevator cable is to be made of a new type of composite developed by Acme Laboratories. In the lab, a sample of the cable that is 2.00 m long and has a cross-sectional area of 0.200 mm^2 fails under a load of 1000 N. The actual cable used to support the elevator will be 20.0 m long and have a cross-sectional area of 1.20 mm^2. It will need to support a load of 20,000 N safely. Will it?

Picture the Problem We can use the definition of stress to calculate the failing stress of the cable and the stress on the elevator cable. Note that the failing stress of the composite cable is the same as the failing stress of the test sample.

The stress on the elevator cable is:

$$\text{Stress}_{\text{cable}} = \frac{F}{A} = \frac{20.0\,\text{kN}}{1.20\times10^{-6}\,\text{m}^2} = 1.67\times10^{10}\,\text{N/m}^2$$

The failing stress of the sample is:

$$\text{Stress}_{\text{failing}} = \frac{F}{A} = \frac{1000\,\text{N}}{0.2\times10^{-6}\,\text{m}^2} = 0.500\times10^{10}\,\text{N/m}^2$$

Because $\text{Stress}_{\text{failing}} < \text{Stress}_{\text{cable}}$, the cable will not support the elevator.

53 •• **[SSM]** You are given a wire with a circular cross-section of radius r and a length L. If the wire is made from a material whose density remains constant when it is stretched in one direction, then show that $\Delta r/r = -\tfrac{1}{2}\Delta L/L$, assuming that $\Delta L << L$. (See Problem 52.)

Picture the Problem We can evaluate the differential of the volume of the wire and, using the assumptions that the volume of the wire does not change under stretching and that the change in its length is small compared to its length, show that $\Delta r/r = -\tfrac{1}{2}\Delta L/L\,L$.

Express the volume of the wire: $V = \pi r^2 L$

Evaluate the differential of V to obtain: $dV = \pi r^2 dL + 2\pi rLdr$

Because $dV = 0$: $0 = rdL + 2Ldr \Rightarrow \frac{dr}{r} = -\frac{1}{2}\frac{dL}{L}$

Because $\Delta L << L$, we can approximate the differential changes dr and dL with small changes Δr and ΔL to obtain: $\frac{\Delta r}{r} = \boxed{-\frac{1}{2}\frac{\Delta L}{L}}$

General Problems

55 • **[SSM]** A standard bowling ball weighs 16 pounds. You wish to hold a bowling ball in front of you, with the elbow bent at a right angle. Assume that your biceps attaches to your forearm at 2.5 cm out from the elbow joint, and that your biceps muscle pulls vertically upward, that is, it acts at right angles to the forearm. Also assume that the ball is held 38 cm out from the elbow joint. Let the mass of your forearm be 5.0 kg and assume its center of gravity is located 19 cm out from the elbow joint. How much force must your biceps muscle apply to forearm in order to hold out the bowling ball at the desired angle?

Picture the Problem We can model the forearm as a cylinder of length L = 38 cm with the forces shown in the pictorial representation acting on it. Because the forearm is in both translational and rotational equilibrium under the influence of these forces, the forces in the diagram must add (vectorially) to zero and the net torque with respect to any axis must also be zero.

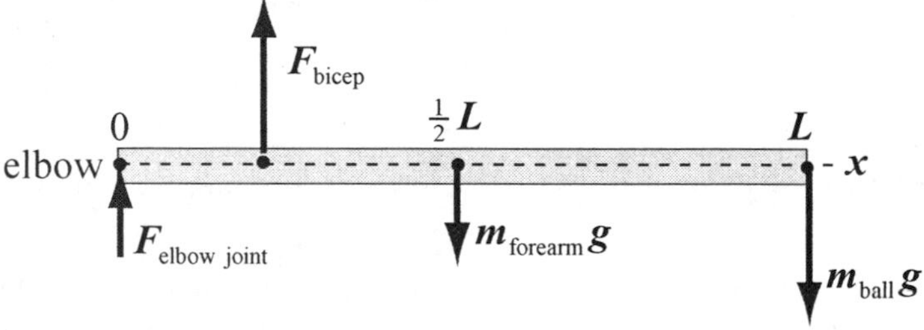

Apply $\sum \vec{\tau} = 0$ about an axis through the elbow and perpendicular to the plane of the diagram: $\ell F_{bicep} - \frac{1}{2}Lm_{forearm}g - Lm_{ball}g = 0$

Solving for F_{bicep} and simplifying yields:

$$F_{\text{bicep}} = \frac{\frac{1}{2}Lm_{\text{forearm}}g + Lm_{\text{ball}}g}{\ell} = \frac{\left(\frac{1}{2}m_{\text{forearm}} + m_{\text{ball}}\right)Lg}{\ell}$$

Substitute numerical values and evaluate F_{bicep}:

$$F_{\text{bicep}} = \frac{\left(\frac{1}{2}(5.0\text{ kg}) + 16\text{ lb}\times\frac{1\text{ kg}}{2.205\text{ lb}}\right)(38\text{ cm})(9.81\text{ m/s}^2)}{2.5\text{ cm}} = \boxed{1.5\text{ kN}}$$

59 •• **[SSM]** Consider a rigid 2.5-m-long beam (Figure 12-51) that is supported by a fixed 1.25-m-high post through its center and pivots on a frictionless bearing at its center atop the vertical 1.25-m-high post. One end of the beam is connected to the floor by a spring that has a force constant $k = 1250$ N/m. When the beam is horizontal, the spring is vertical and unstressed. If an object is hung from the opposite end of the beam, the beam settles into an equilibrium position where it makes an angle of 17.5° with the horizontal. What is the mass of the object?

Picture the Problem Because the beam is in rotational equilibrium, we can apply $\sum\vec{\tau} = 0$ to it to determine the mass of the object suspended from its left end.

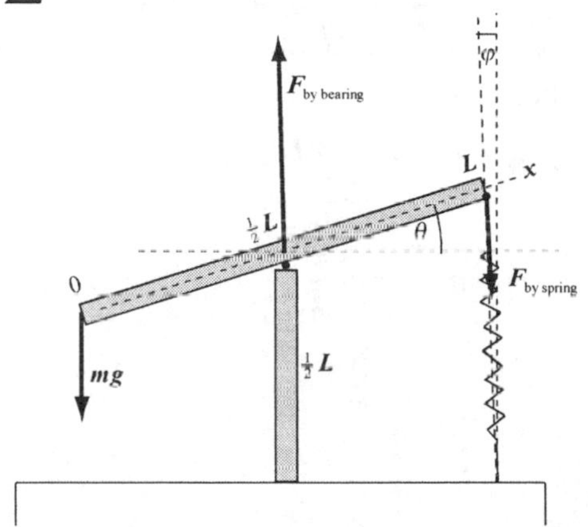

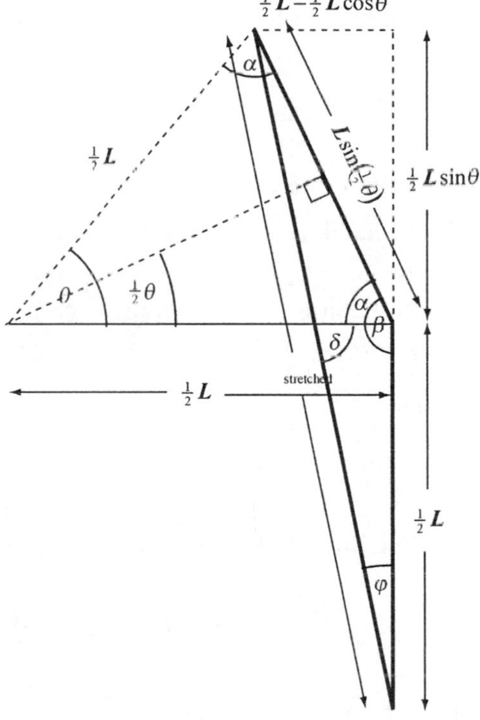

The pictorial representation directly above shows the forces acting on the beam when it is in static equilibrium. The pictorial representation to the right is an enlarged view of the right end of the beam. We'll use this diagram to determine the length of the stretched spring.

Apply $\sum \vec{\tau} = 0$ to the beam about an axis through the bearing point to obtain:

$$\boldsymbol{mg}\left(\tfrac{1}{2}\boldsymbol{L}\cos\boldsymbol{\theta}\right) - \boldsymbol{F}_{\text{by spring}}\left(\tfrac{1}{2}\boldsymbol{L}\sin\boldsymbol{\varphi}\right) = 0$$

or, because $\boldsymbol{F}_{\text{by spring}} = \boldsymbol{k}\Delta\ell_{\text{spring}}$,

$$\boldsymbol{mg}\cos\boldsymbol{\theta} - \boldsymbol{k}\Delta\ell_{\text{spring}}\sin\boldsymbol{\varphi} = 0 \qquad (1)$$

Use the right-hand diagram above to relate the angles φ and θ:

$$\boldsymbol{\varphi} = \tan^{-1}\left(\frac{\frac{1}{2}\boldsymbol{L} - \frac{1}{2}\boldsymbol{L}\cos\boldsymbol{\theta}}{\frac{1}{2}\boldsymbol{L} + \frac{1}{2}\boldsymbol{L}\sin\boldsymbol{\theta}}\right) = \tan^{-1}\left(\frac{1-\cos\boldsymbol{\theta}}{1+\sin\boldsymbol{\theta}}\right)$$

Substitute for φ in equation (1) to obtain:

$$mg\cos\theta - k\Delta\ell_{\text{spring}}\sin\left(\tan^{-1}\left(\frac{1-\cos\theta}{1+\sin\theta}\right)\right) = 0$$

Solve for m to obtain:

$$m = \frac{k\Delta\ell_{\text{spring}}\sin\left(\tan^{-1}\left(\frac{1-\cos\theta}{1+\sin\theta}\right)\right)}{g\cos\theta} \qquad (2)$$

$\Delta\ell_{\text{spring}}$ is given by:

$$\Delta\ell_{\text{spring}} = \ell_{\text{stretched}} - \ell_{\text{unstretched}}$$

or, because $\ell_{\text{unstretched}} = \tfrac{1}{2}\boldsymbol{L}$,

$$\Delta\ell_{\text{spring}} = \ell_{\text{stretched}} - \tfrac{1}{2}\boldsymbol{L} \qquad (3)$$

To find the value of $\ell_{\text{stretched}}$, refer to the right-hand diagram and note that:

$$2\boldsymbol{\alpha} + \boldsymbol{\theta} = \boldsymbol{\pi} \Rightarrow \boldsymbol{\alpha} = \frac{\boldsymbol{\pi}}{2} - \frac{1}{2}\boldsymbol{\theta}$$

Again, referring to the diagram, relate β to α:

$$\boldsymbol{\beta} = \boldsymbol{\alpha} + \frac{\boldsymbol{\pi}}{2}$$

Substituting for α yields:

$$\boldsymbol{\beta} = \frac{\boldsymbol{\pi}}{2} - \frac{1}{2}\boldsymbol{\theta} + \frac{\boldsymbol{\pi}}{2} = \boldsymbol{\pi} - \frac{1}{2}\boldsymbol{\theta}$$

Apply the law of cosines to the triangle defined with bold sides:

$$\ell^2_{\text{stretched}} = \left(\tfrac{1}{2}L\right)^2 + \left(L\sin\tfrac{1}{2}\theta\right)^2 - 2\left(\tfrac{1}{2}L\right)\left(L\sin\tfrac{1}{2}\theta\right)\cos\left(\pi - \tfrac{1}{2}\theta\right)$$

Use the formula for the cosine of the difference of two angles to obtain:

$$\cos\left(\boldsymbol{\pi} - \tfrac{1}{2}\boldsymbol{\theta}\right) = -\cos\tfrac{1}{2}\boldsymbol{\theta}$$

Substituting for $\cos\left(\pi - \frac{1}{2}\theta\right)$ yields:

$$\begin{aligned}\ell^2_{\text{stretched}} &= \left(\tfrac{1}{2}L\right)^2 + \left(L\sin\tfrac{1}{2}\theta\right)^2 + 2\left(\tfrac{1}{2}L\right)\left(L\sin\tfrac{1}{2}\theta\right)\cos\tfrac{1}{2}\theta \\ &= \tfrac{1}{4}L^2 + L^2\sin^2\tfrac{1}{2}\theta + \tfrac{1}{2}L^2\left(2\sin\tfrac{1}{2}\theta\cos\tfrac{1}{2}\theta\right)\end{aligned}$$

Use the trigonometric identities $\sin\theta = 2\sin\frac{1}{2}\theta\cos\frac{1}{2}\theta$ and $\sin^2\frac{1}{2}\theta = \dfrac{1-\cos\theta}{2}$ to obtain:

$$\ell^2_{\text{stretched}} = \tfrac{1}{4}L^2 + L^2\left(\frac{1-\cos\theta}{2}\right) + \tfrac{1}{2}L^2(\sin\theta)$$

Simplifying yields:

$$\ell^2_{\text{stretched}} = \tfrac{1}{4}L^2\left[3 - 2(\cos\theta - \sin\theta)\right]$$

or

$$\ell_{\text{stretched}} = \tfrac{1}{2}L\sqrt{3 - 2(\cos\theta - \sin\theta)}$$

Substituting for $\ell_{\text{stretched}}$ in equation (3) yields:

$$\Delta\ell_{\text{spring}} = \tfrac{1}{2}L\sqrt{3 - 2(\cos\theta - \sin\theta)} - \tfrac{1}{2}L = \tfrac{1}{2}L\left(\sqrt{3 - 2(\cos\theta - \sin\theta)} - 1\right)$$

Substitute for $\Delta\ell_{\text{spring}}$ in equation (2) to obtain:

$$m = \frac{\frac{1}{2}kL\left(\sqrt{3 - 2(\cos\theta - \sin\theta)} - 1\right)\sin\left(\tan^{-1}\left(\frac{1-\cos\theta}{1+\sin\theta}\right)\right)}{g\cos\theta}$$

Substitute numerical values and evaluate m:

$$m = \frac{\frac{1}{2}(1250\ \text{N/m})(2.5\ \text{m})\left(\sqrt{3 - 2(\cos 17.5° - \sin 17.5°)} - 1\right)\sin\left(\tan^{-1}\left(\frac{1-\cos 17.5°}{1+\sin 17.5°}\right)\right)}{\left(9.81\ \text{m/s}^2\right)\cos 17.5°}$$

$$= \boxed{1.8\ \text{kg}}$$

65 •• **[SSM]** A cube leans against a frictionless wall making an angle of θ with the floor as shown in Figure 12-55. Find the minimum coefficient of static friction μ_s between the cube and the floor that is needed to keep the cube from slipping.

Picture the Problem Let the mass of the cube be *M.* The figure shows the location of the cube's center of mass and the forces acting on the cube. The opposing couple is formed by the friction force $f_{s,max}$ and the force exerted by the wall. Because the cube is in equilibrium, we can use the condition for translational equilibrium to establish that $f_{s,max} = F_W$ and $F_n = Mg$ and the condition for rotational equilibrium to relate the opposing couples.

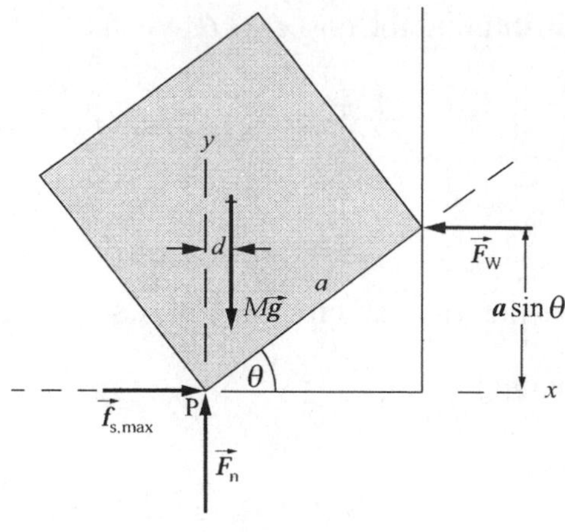

Apply $\sum\vec{F} = 0$ to the cube:

$$\sum F_y = F_n - Mg = 0 \Rightarrow F_n = Mg$$

and

$$\sum F_x = f_s - F_W = 0 \Rightarrow F_W = f_s$$

Noting that $\vec{f}_{s,max}$ and $\vec{F}_W$ form a couple (their magnitudes are equal), as do $\vec{F}_n$ and $M\vec{g}$, apply $\sum\vec{\tau} = 0$ about an axis though point P to obtain:

$$f_{s,max}a\sin\theta - Mgd = 0$$

Referring to the diagram to the right, note that $d = \frac{a}{\sqrt{2}}\sin(45° + \theta)$.

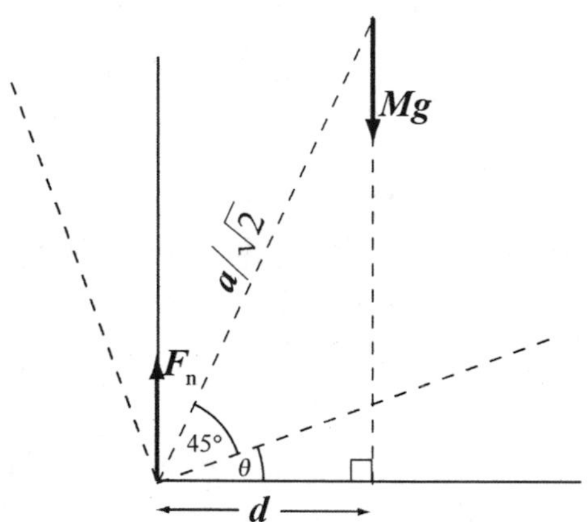

Substitute for d and $f_{s,max}$ to obtain:

$$\mu_s Mga\sin\theta - Mg\frac{a}{\sqrt{2}}\sin(45° + \theta) = 0$$

or

$$\mu_s \sin\theta - \frac{1}{\sqrt{2}}\sin(45° + \theta) = 0$$

Solve for μ_s and simplify to obtain:

$$\mu_s = \frac{1}{\sqrt{2}\sin\theta}\sin(45° + \theta) = \frac{1}{\sqrt{2}\sin\theta}(\sin 45°\cos\theta + \cos 45°\sin\theta)$$
$$= \frac{1}{\sqrt{2}\sin\theta}\left(\frac{1}{\sqrt{2}}\cos\theta + \frac{1}{\sqrt{2}}\sin\theta\right) = \boxed{\frac{1}{2}(\cot\theta + 1)}$$

67 •• **[SSM]** Figure 12-57 shows a 20.0-kg ladder leaning against a frictionless wall and resting on a frictionless horizontal surface. To keep the ladder from slipping, the bottom of the ladder is tied to the wall with a thin wire. When no one is on the ladder, the tension in the wire is 29.4 N. (The wire will break if the tension exceeds 200 N.) (*a*) If an 80.0-kg person climbs halfway up the ladder, what force will be exerted by the ladder against the wall? (*b*) How far from the bottom end of the ladder can an 80.0-kg person climb?

Picture the Problem Let m represent the mass of the ladder and M the mass of the person. The force diagram shows the forces acting on the ladder for Part (*b*). From the condition for translational equilibrium, we can conclude that $T = F_{\text{by wall}}$, a result we'll need in Part (*b*). Because the ladder is also in rotational equilibrium, summing the torques about the bottom of the ladder will eliminate both F_n and T.

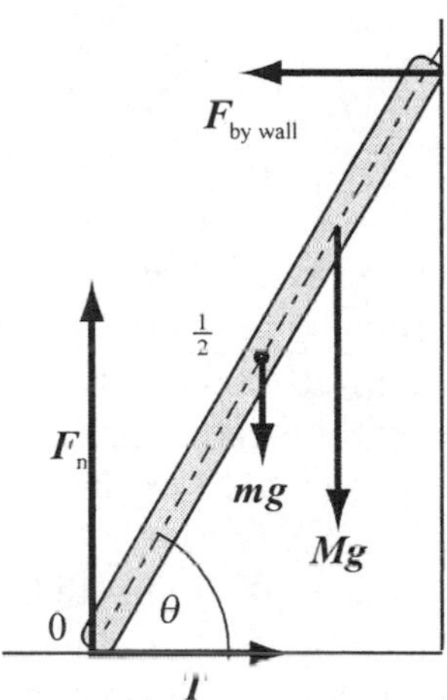

(*a*) Apply $\sum\vec{\tau} = 0$ about an axis through the bottom of the ladder:

$$F_{\text{by wall}}(\ell\sin\theta) - mg\left(\tfrac{1}{2}\ell\cos\theta\right) - Mg\left(\tfrac{1}{2}\ell\cos\theta\right) = 0$$

Solve for $F_{\text{by wall}}$ and simplify to obtain:

$$F_{\text{by wall}} = \frac{(m+M)g\cos\theta}{2\sin\theta} = \frac{(m+M)g}{2\tan\theta}$$

Refer to Figure 12-57 to determine θ:

$$\theta = \tan^{-1}\left(\frac{5.0\text{ m}}{1.5\text{ m}}\right) = 73.3°$$

Substitute numerical values and evaluate $F_{\text{by wall}}$:

$$F_{\text{by wall}} = \frac{(20\text{ kg} + 80\text{ kg})(9.81\text{ m/s}^2)}{2\tan 73.3°} = \boxed{0.15\text{ kN}}$$

(*b*) Apply $\sum F_x = 0$ to the ladder to obtain:

$$T - F_{\text{by wall}} = 0 \Rightarrow T = F_{\text{by wall}}$$

Apply $\sum \vec{\tau} = 0$ about an axis through the bottom of the ladder subject to the condition that $F_{\text{by wall}} = T_{\max}$:

$$T_{\max}(\ell \sin\theta) - mg\left(\tfrac{1}{2}\ell\cos\theta\right) - Mg(L\cos\theta) = 0$$

where L is the maximum distance along the ladder that the person can climb without exceeding the maximum tension in the wire.

Solving for L and simplifying yields:

$$L = \frac{T_{\max}\ell\sin\theta - \frac{1}{2}mg\ell\cos\theta}{Mg\cos\theta}$$

Substitute numerical values and evaluate L:

$$L = \frac{(200\text{ N})(5.0\text{ m}) - \frac{1}{2}(20\text{ kg})(9.81\text{ m/s}^2)(1.5\text{ m})}{(80\text{ kg})(9.81\text{ m/s}^2)\cos 73.3^\circ} = \boxed{3.8\text{ m}}$$

71 ••• **[SSM]** There are a large number of identical uniform bricks, each of length L. If they are stacked one on top of another lengthwise (see Figure 12-60), the maximum offset that will allow the top brick to rest on the bottom brick is $L/2$. (*a*) Show that if this two-brick stack is placed on top of a third brick, the maximum offset of the second brick on the third brick is $L/4$. (*b*) Show that, in general, if you build a stack of N bricks, the maximum overhang of the $(n-1)$th brick (counting down from the top) on the nth brick is $L/2n$. (*c*) Write a **spreadsheet** program to calculate total offset (the sum of the individual offsets) for a stack of N bricks, and calculate this for $L = 20$ cm and $N = 5$, 10, and 100. (*d*) Does the sum of the individual offsets approach a finite limit as $N \to \infty$? If so, what is that limit?

Picture the Problem Let the weight of each uniform brick be w. The downward force of all the bricks above the nth brick *must act at its corner,* because the upward reaction force points through the center of mass of all the bricks above the nth one. Because there is no vertical acceleration, the upward force exerted by the $(n + 1)$th brick on the nth brick must equal the total weight of the bricks above it. Thus this force is just nw. Note that it is convenient to develop the general relationship of Part (*b*) initially and then extract the answer for Part (*a*) from this general result.

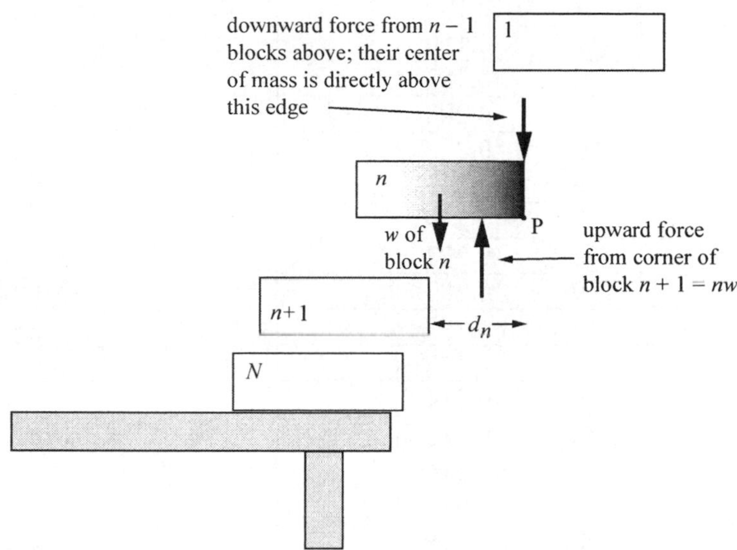

(*a*) and (*b*) Noting that the line of action of the downward forces exerted by the blocks above the *n*th block passes through the point P, resulting in a lever arm of zero, apply $\sum \tau_P = 0$ to the *n*th brick to obtain:

$$w\left(\tfrac{1}{2}L\right) - nwd_n = 0$$

where d_n is the overhang of the *n*th brick beyond the edge of the $(n+1)$th brick.

Solving for d_n yields:

$$d_n = \boxed{\frac{L}{2n}} \text{ where } n = 1,2,3,\ldots$$

For block number 2:

$$d_2 = \boxed{\frac{L}{4}}$$

(*c*) A spreadsheet program to calculate the sum of the offsets as a function of *n* is shown below. The formulas used to calculate the quantities in the columns are as follows:

Cell	Formula/Content	Algebraic Form
B5	B4+1	$n+1$
C5	C4+$B$1/(2*B5)	$d_n + \dfrac{L}{2n}$

	A	B	C	D
1	$L=$	0.20	m	
2				
3		n	offset	

4		1	0.100	
5		2	0.150	
6		3	0.183	
7		4	0.208	
8		5	0.228	
9		6	0.245	
10		7	0.259	
11		8	0.272	
12		9	0.283	
13		10	0.293	
98		95	0.514	
99		96	0.515	
100		97	0.516	
101		98	0.517	
102		99	0.518	
103		100	0.519	

From the table we see that $d_5 = \boxed{15\,\text{cm},}$ $d_{10} = \boxed{26\,\text{cm},}$ and $d_{100} = \boxed{0.52\,\text{cm}.}$

(*d*) The sum of the individual offsets *S* is given by:

$$S = \sum_{n=1}^{N} d_n = \frac{L}{2}\sum_{n=1}^{N}\frac{1}{n}$$

Because this series is a harmonic series, *S* approaches infinity as the number of blocks *N* grows without bound. The following graph, plotted using a spreadsheet program, suggests that *S* has no limit.

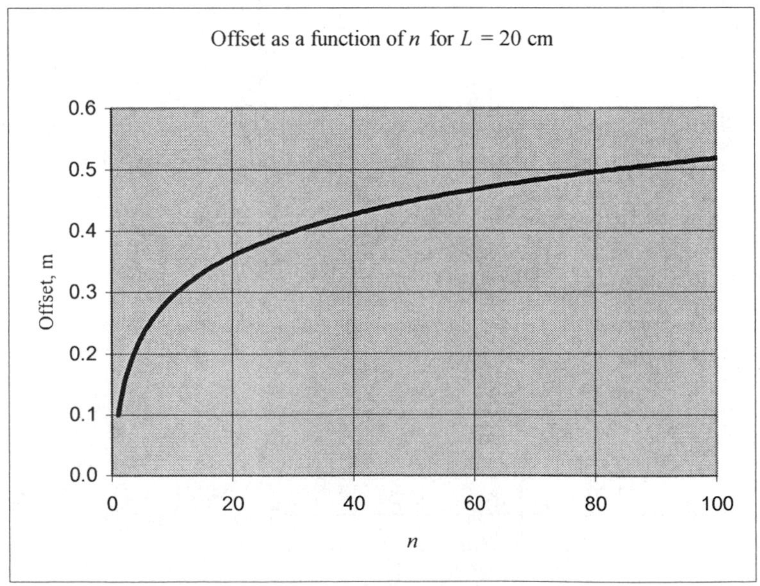

75 ••• **[SSM]** Two solid smooth (frictionless) spheres of radius r are placed inside a cylinder of radius R, as in Figure 12-62. The mass of each sphere is m. Find the force exerted by the bottom of the cylinder on the bottom sphere, the force exerted by the wall of the cylinder on each sphere, and the force exerted by one sphere on the other. All forces should be expressed in terms of m, R, and r.

Picture the Problem The geometry of the system is shown in the drawing. Let upward be the positive y direction and to the right be the positive x direction. Let the angle between the vertical center line and the line joining the two centers be θ. Then $\sin\theta = \dfrac{R-r}{r}$ and $\tan\theta = \dfrac{R-r}{\sqrt{R(2r-R)}}$.

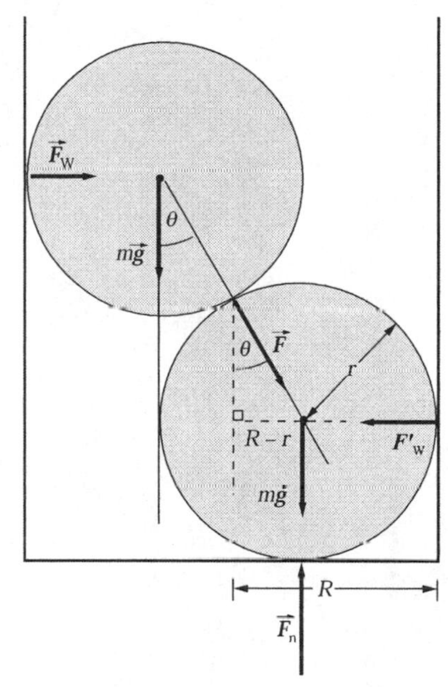

The force exerted by the bottom of the cylinder is just $2mg$. Let F be the force that the top sphere exerts on the lower sphere. Because the spheres are in equilibrium, we can apply the condition for translational equilibrium.

Apply $\sum F_y = 0$ to the spheres:

$$F_n - mg - mg = 0 \Rightarrow F_n = \boxed{2mg}$$

Because the cylinder wall is smooth, $F\cos\theta = mg$, and:

$$F = \frac{mg}{\cos\theta} = \boxed{mg\frac{r}{R(2r-R)}}$$

Express the x component of F:

$$F_x = F\sin\theta = mg\tan\theta$$

Express the force that the wall of the cylinder exerts:

$$F_W = \boxed{mg\frac{R-r}{\sqrt{R(2r-R)}}}$$

Remarks: Note that as r approaches $R/2$, $F_W \to \infty$.

Chapter 13
Fluids

Conceptual Problems

3 • **[SSM]** Two objects differ in density and mass. Object A has a mass that is eight times the mass of object B. The density of object A is four times the density of object B. How do their volumes compare? *(a)* $V_A = \frac{1}{2}V_B$, *(b)* $V_A = V_B$, *(c)* $V_A = 2V_B$, *(d)* not enough information is given to compare their volumes.

Determine the Concept The density of an object is its mass per unit volume. We can determine the relationship between the volumes of A and B by examining their ratio.

Express the volumes of the two objects:

$$V_A = \frac{m_A}{\rho_A} \text{ and } V_B = \frac{m_B}{\rho_B}$$

Divide the first of these equations by the second and simplify to obtain:

$$\frac{V_A}{V_B} = \frac{\frac{m_A}{\rho_A}}{\frac{m_B}{\rho_B}} = \frac{\rho_B}{\rho_A}\frac{m_A}{m_B}$$

Substituting for the masses and densities and simplifying yields:

$$\frac{V_A}{V_B} = \frac{\rho_B}{4\rho_B}\frac{8m_B}{m_B} = 2 \Rightarrow V_A = 2V_B$$

and $\boxed{(c)}$ is correct.

7 •• **[SSM]** A solid 200-g block of lead and a solid 200-g block of copper are completely submerged in an aquarium filled with water. Each block is suspended just above the bottom of the aquarium by a thread. Which of the following is true?

(a) The buoyant force on the lead block is greater than the buoyant force on the copper block.
(b) The buoyant force on the copper block is greater than the buoyant force on the lead block.
(c) The buoyant force is the same on both blocks.
(d) More information is needed to choose the correct answer.

Determine the Concept The buoyant forces acting on these submerged objects are equal to the weight of the water each displaces. The weight of the displaced water, in turn, is directly proportional to the volume of the submerged object. Because $\rho_{Pb} > \rho_{Cu}$, the volume of the copper must be greater than that of the lead and, hence, the buoyant force on the copper is greater than that on the lead. $\boxed{(b)}$ is correct.

13 •• [SSM] An upright glass of water is accelerating to the right along a flat, horizontal surface. What is the origin of the force that produces the acceleration on a small element of water in the middle of the glass? Explain by using a diagram. *Hint: The water surface will not remain level as long as the glass of water is accelerating. Draw a free body diagram of the small element of water.*

Determine the Concept The pictorial representation shows the glass and an element of water in the middle of the glass. As is readily established by a simple demonstration, the surface of the water is not level while the glass is accelerated, showing that there is a pressure gradient (a difference in pressure) due to the differing depths ($h_1 > h_2$ and hence $F_1 > F_2$) of water on the two sides of the element of water. This pressure gradient results in a net force on the element as shown in the figure. The upward buoyant force is equal in magnitude to the downward gravitational force.

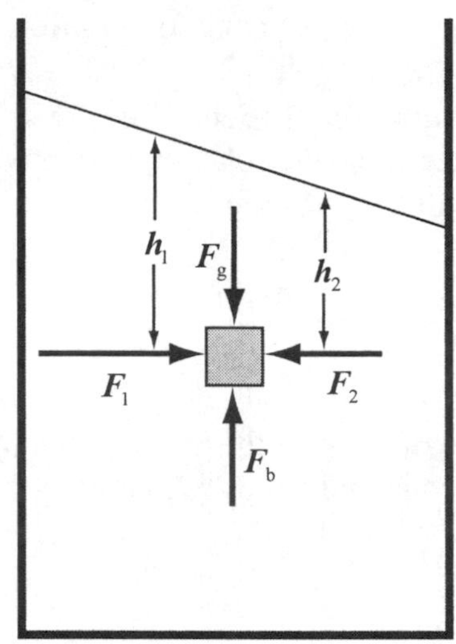

17 •• [SSM] Figure 13-30 is a diagram of a prairie dog tunnel. The geometry of the two entrances are such that entrance 1 is surrounded by a mound and entrance 2 is surrounded by flat ground. Explain how the tunnel remains ventilated, and indicate in which direction air will flow through the tunnel.

Determine the Concept The opening above which the air flows faster will be at a lower pressure than the opening above which the air flows slower. The mounding around entrance 1 will cause the air to flow faster over the opening than at entrance 2. This difference in pressure will cause the pressure to be lower at opening 1 and a circulation of air in the tunnel from entrance 2 toward entrance 1. It has been demonstrated that enough air will circulate inside the tunnel even with the slightest breeze outside.

Density

21 • [SSM] Consider a room measuring 4.0 m × 5.0 m × 4.0 m. Under normal atmospheric conditions at Earth's surface, what would be the mass of the air in the room?

Picture the Problem The mass of the air in the room is the product of its density and volume. The density of air can be found in Figure 13-1.

Use the definition of density to express the mass of the air in the room:

$$m = \rho V = \rho LWH$$

Substitute numerical values and evaluate m:

$$m = (1.293\,\text{kg/m}^3)(4.0\,\text{m})(5.0\,\text{m})(4.0\,\text{m}) = \boxed{1.0\times10^2\ \text{kg}}$$

Pressure

31 • **[SSM]** A hydraulic lift is used to raise an automobile of mass 1500 kg. The radius of the shaft of the lift is 8.00 cm and that of the piston is 1.00 cm. How much force must be applied to the compressor's piston to raise the automobile? *Hint*: *The shaft of the lift is the other piston.*

Picture the Problem The pressure applied to an enclosed liquid is transmitted undiminished to every point in the fluid and to the walls of the container. Hence we can equate the pressure produced by the force applied to the piston to the pressure due to the weight of the automobile and solve for F.

Express the pressure the weight of the automobile exerts on the shaft of the lift:

$$P_{\text{auto}} = \frac{w_{\text{auto}}}{A_{\text{shaft}}}$$

Express the pressure the force applied to the piston produces:

$$P = \frac{F}{A_{\text{piston}}}$$

Because the pressures are the same, we can equate them to obtain:

$$\frac{w_{\text{auto}}}{A_{\text{shaft}}} = \frac{F}{A_{\text{piston}}}$$

Solving for F yields:

$$F = w_{\text{auto}}\frac{A_{\text{piston}}}{A_{\text{shaft}}} = m_{\text{auto}}g\frac{A_{\text{piston}}}{A_{\text{shaft}}}$$

Substitute numerical values and evaluate F:

$$F = (1500\,\text{kg})(9.81\,\text{m/s}^2)\left(\frac{1.00\,\text{cm}}{8.00\,\text{cm}}\right)^2 = \boxed{230\ \text{N}}$$

Buoyancy

43 • **[SSM]** A block of an unknown material weighs 5.00 N in air and 4.55 N when submerged in water. (*a*) What is the density of the material? (*b*) From what material is the block likely to have been made?

Picture the Problem We can use the definition of density and Archimedes' principle to find the density of the unknown object. The difference between the weight of the object in air and in water is the buoyant force acting on the object.

(*a*) Using its definition, express the density of the object:

$$\rho_{\text{object}} = \frac{m_{\text{object}}}{V_{\text{object}}} \qquad (1)$$

Apply Archimedes' principle to obtain:

$$B = w_{\substack{\text{displaced}\\\text{fluid}}} = m_{\substack{\text{displaced}\\\text{fluid}}}g = \rho_{\substack{\text{displaced}\\\text{fluid}}}V_{\substack{\text{displaced}\\\text{fluid}}}g$$

Solve for $V_{\substack{\text{displaced}\\\text{fluid}}}$:

$$V_{\substack{\text{displaced}\\\text{fluid}}} = \frac{B}{\rho_{\substack{\text{displaced}\\\text{fluid}}}g}$$

Because $V_{\substack{\text{displaced}\\\text{fluid}}} = V_{\text{object}}$ and $\rho_{\substack{\text{displaced}\\\text{fluid}}} = \rho_{\text{water}}$:

$$V_{\text{object}} = \frac{B}{\rho_{\text{water}}g}$$

Substitute in equation (1) and simplify to obtain:

$$\rho_{\text{object}} = \frac{m_{\text{object}}g}{B}\rho_{\text{water}} = \frac{w_{\text{object}}}{B}\rho_{\text{water}}$$

Substitute numerical values and evaluate ρ_{object}:

$$\rho_{\text{object}} = \frac{5.00\text{N}}{5.00\,\text{N} - 4.55\,\text{N}}\left(1.00\times10^3\ \text{kg/m}^3\right) = \boxed{11\times10^3\ \text{kg/m}^3}$$

(*b*) From Table 13-1, we see that the density of the unknown material is close to that of lead.

49 •• **[SSM]** An object has "neutral buoyancy" when its density equals that of the liquid in which it is submerged, which means that it neither floats nor sinks. If the average density of an 85-kg diver is 0.96 kg/L, what mass of lead should, as dive master, suggest be added to give him neutral buoyancy?

Picture the Problem Let V = volume of diver, ρ_D the density of the diver, V_{Pb} the volume of added lead, and m_{Pb} the mass of lead. The diver is in equilibrium under the influence of his weight, the weight of the lead, and the buoyant force of the water.

Apply $\sum F_y = 0$ to the diver:

$$B - w_D - w_{Pb} = 0$$

Substitute to obtain:

$$\rho_w V_{D+Pb} g - \rho_D V_D g - m_{Pb} g = 0$$

or

$$\rho_w V_D + \rho_w V_{Pb} - \rho_D V_D - m_{Pb} = 0$$

Rewrite this expression in terms of masses and densities:

$$\rho_w \frac{m_D}{\rho_D} + \rho_w \frac{m_{Pb}}{\rho_{Pb}} - \rho_D \frac{m_D}{\rho_D} - m_{Pb} = 0$$

Solving for m_{Pb} yields:

$$m_{Pb} = \frac{\rho_{Pb}(\rho_w - \rho_D)m_D}{\rho_D(\rho_{Pb} - \rho_w)}$$

Substitute numerical values and evaluate m_{Pb}:

$$m_{Pb} = \frac{(11.3\times10^3\text{ kg/m}^3)(1.00\times10^3\text{ kg/m}^3 - 0.96\times10^3\text{ kg/m}^3)(85\text{ kg})}{(0.96\times10^3\text{ kg/m}^3)(11.3\times10^3\text{ kg/m}^3 - 1.00\times10^3\text{ kg/m}^3)} = \boxed{3.9\text{ kg}}$$

53 ••• **[SSM]** A ship sails from seawater (specific gravity 1.025) into freshwater, and therefore sinks slightly. When its 600,000-kg load is removed, it returns to its original level. Assuming that the sides of the ship are vertical at the water line, find the mass of the ship before it was unloaded.

Picture the Problem Let V = displacement of ship in the two cases, m be the mass of ship without load, and Δm be the load. The ship is in equilibrium under the influence of the buoyant force exerted by the water and its weight. We'll apply the condition for floating in the two cases and solve the equations simultaneously to determine the loaded mass of the ship.

Apply $\sum F_y = 0$ to the ship in fresh water:

$$\rho_w Vg - mg = 0 \qquad (1)$$

Apply $\sum F_y = 0$ to the ship in salt water:

$$\rho_{sw} Vg - (m + \Delta m)g = 0 \qquad (2)$$

Solve equation (1) for Vg:

$$Vg = \frac{mg}{\rho_w}$$

Substitute in equation (2) to obtain:

$$\rho_{sw}\frac{mg}{\rho_w} - (m + \Delta m)g = 0$$

Solving for m yields:

$$m = \frac{\rho_w \Delta m}{\rho_{sw} - \rho_w}$$

Add Δm to both sides of the equation and simplify to obtain:

$$m + \Delta m = \frac{\rho_w \Delta m}{\rho_{sw} - \rho_w} + \Delta m$$
$$= \Delta m\left(\frac{\rho_w}{\rho_{sw} - \rho_w} + 1\right)$$
$$= \frac{\Delta m \rho_{sw}}{\rho_{sw} - \rho_w}$$

Substitute numerical values and evaluate $m + \Delta m$:

$$m + \Delta m = \frac{(6.00\times10^5\ \text{kg})(1.025\rho_w)}{1.025\rho_w - \rho_w}$$
$$= \frac{(6.00\times10^5\ \text{kg})(1.025)}{1.025 - 1}$$
$$= \boxed{2.5\times10^7\ \text{kg}}$$

Continuity and Bernoulli's Equation

55 • **[SSM]** Water is flowing at 3.00 m/s in a horizontal pipe under a pressure of 200 kPa. The pipe narrows to half its original diameter. (*a*) What is the speed of flow in the narrow section? (*b*) What is the pressure in the narrow section? (*c*) How do the volume flow rates in the two sections compare?

Picture the Problem Let A_1 represent the cross-sectional area of the larger-diameter pipe, A_2 the cross-sectional area of the smaller-diameter pipe, v_1 the speed of the water in the larger-diameter pipe, and v_2 the velocity of the water in the smaller-diameter pipe. We can use the continuity equation to find v_2 and Bernoulli's equation for constant elevation to find the pressure in the smaller-diameter pipe.

(*a*) Using the continuity equation, relate the velocities of the water to the diameters of the pipe:

$$A_1v_1 = A_2v_2$$

or

$$\frac{\pi d_1^2}{4}v_1 = \frac{\pi d_2^2}{4}v_2 \Rightarrow v_2 = \frac{d_1^2}{d_2^2}v_1$$

Substitute numerical values and evaluate v_2:

$$v_2 = \left(\frac{d_1}{\frac{1}{2}d_1}\right)^2 (3.00\,\text{m/s}) = \boxed{12.0\,\text{m/s}}$$

(*b*) Using Bernoulli's equation for constant elevation, relate the pressures in the two segments of the pipe to the velocities of the water in these segments:

$$P_1 + \tfrac{1}{2}\rho_w v_1^2 = P_2 + \tfrac{1}{2}\rho_w v_2^2$$

Solving for P_2 yields:

$$P_2 = P_1 + \tfrac{1}{2}\rho_w v_1^2 - \tfrac{1}{2}\rho_w v_2^2 = P_1 + \tfrac{1}{2}\rho_w\left(v_1^2 - v_2^2\right)$$

Substitute numerical values and evaluate P_2:

$$P_2 = 200\,\text{kPa} + \tfrac{1}{2}\left(1.00\times10^3\,\text{kg/m}^3\right)\left[(3.00\,\text{m/s})^2 - (12.0\,\text{m/s})^2\right] = \boxed{133\,\text{kPa}}$$

(*c*) Using the continuity equation, evaluate I_{V1}:

$$I_{V1} = A_1 v_1 = \frac{\pi d_1^2}{4} v_1 = \frac{\pi d_1^2}{4}(3.00\,\text{m/s})$$

Using the continuity equation, express I_{V2}:

$$I_{V2} = A_2 v_2 = \frac{\pi d_2^2}{4} v_2$$

Substitute numerical values and evaluate I_{V2}:

$$I_{V2} = \frac{\pi\left(\frac{d_1}{2}\right)^2}{4}(12.0\,\text{m/s}) = \frac{\pi d_1^2}{4}(3.00\,\text{m/s})$$

Thus, as we expected would be the case:

$$\boxed{I_{V1} = I_{V2}}$$

57 •• **[SSM]** Blood flows at at 30 cm/s in an aorta of radius 9.0 mm. (*a*) Calculate the volume flow rate in liters per minute. (*b*) Although the cross-sectional area of a capillary is much smaller than that of the aorta, there are many capillaries, so their total cross-sectional area is much larger. If all the blood from the aorta flows into the capillaries and the speed of flow through the capillaries is 1.0 mm/s, calculate the total cross-sectional area of the capillaries. Assume laminar nonviscous, steady-state flow.

Picture the Problem We can use the definition of the volume flow rate to find the volume flow rate of blood in an aorta and to find the total cross-sectional area of the capillaries.

(*a*) Use the definition of the volume flow rate to find the volume flow rate through an aorta:

$$I_V = Av$$

Substitute numerical values and evaluate I_V:

$$I_V = \pi\left(9.0\times10^{-3}\,\text{m}^3\right)^2(0.30\,\text{m/s})$$

$$= 7.634\times10^{-5}\,\frac{\text{m}^3}{\text{s}}\times\frac{60\,\text{s}}{\text{min}}\times\frac{1\,\text{L}}{10^{-3}\,\text{m}^3}$$

$$= 4.58\,\text{L/min} = \boxed{4.6\,\text{L/min}}$$

(*b*) Use the definition of the volume flow rate to express the volume flow rate through the capillaries:

$$I_V = A_{cap}v_{cap} \Rightarrow A_{cap} = \frac{I_V}{v_{cap}}$$

Substitute numerical values and evaluate A_{cap}:

$$A_{cap} = \frac{7.63\times10^{-5}\,\text{m}^3/\text{s}}{0.0010\,\text{m/s}} = \boxed{7.6\times10^{-2}\,\text{m}^2}$$

61 •• **[SSM]** Horizontal flexible tubing for carrying cooling water flows through a large electromagnet used in your physics experiment at Fermi National Accelerator Laboratory. A minimum volume flow rate of 0.050 L/s through the tubing is necessary in order to keep your magnet cool. Within the magnet volume, the tubing has a circular cross section of radius 0.500 cm. In regions outside the magnet, the tubing widens to a radius of 1.25 cm. You have attached pressure sensors to measure differences in pressure between the 0.500 cm and 1.25 cm sections. The lab technicians tell you that if the flow rate in the system drops below 0.050 L/s, the magnet is in danger of overheating and that you should install an alarm to sound a warning when the flow rate drops below that level. What is the critical pressure difference at which you should program the sensors to send the alarm signal (and is this a minimum, or maximum, pressure difference)? Assume laminar nonviscous steady-state flow.

Picture the Problem The pictorial representation shows the narrowing of the cold-water supply tubes as they enter the magnet. We can apply Bernoulli's equation and the continuity equation to derive an expression for the pressure difference $P_1 - P_2$.

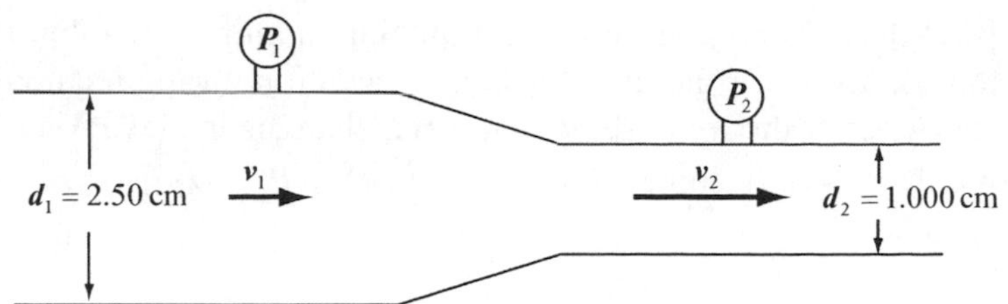

Apply Bernoulli's equation to the two sections of tubing to obtain:

$$P_1 + \tfrac{1}{2}\rho v_1^2 = P_2 + \tfrac{1}{2}\rho v_2^2$$

Solving for the pressure difference $P_1 - P_2$ yields:

$$\Delta P = P_1 - P_2 = \tfrac{1}{2}\rho v_2^2 - \tfrac{1}{2}\rho v_1^2 = \tfrac{1}{2}\rho\left(v_2^2 - v_1^2\right)$$

Factor v_1^2 from the parentheses to obtain:

$$\Delta P = \tfrac{1}{2}\rho v_1^2\left(\frac{v_2^2}{v_1^2} - 1\right) = \tfrac{1}{2}\rho v_1^2\left(\left(\frac{v_2}{v_1}\right)^2 - 1\right)$$

From the continuity equation we have:

$$A_1 v_1 = A_2 v_2$$

Solving for the ratio of v_2 to v_1, expressing the areas in terms of the diameters, and simplifying yields:

$$\frac{v_2}{v_1} = \frac{A_1}{A_2} = \frac{\frac{1}{4}\pi d_1^2}{\frac{1}{4}\pi d_2^2} = \left(\frac{d_1}{d_2}\right)^2$$

Use the expression for the volume flow rate to express v_1:

$$v_1 = \frac{I_V}{A_1} = \frac{I_V}{\frac{1}{4}\pi d_1^2} = \frac{4I_V}{\pi d_1^2}$$

Substituting for v_1 and v_2/v_1 in the expression for ΔP yields:

$$\Delta P = \tfrac{1}{2}\rho\left(\frac{4I_V}{\pi d_1^2}\right)^2\left(\left(\frac{d_1}{d_2}\right)^4 - 1\right)$$

Substitute numerical values and evaluate ΔP:

$$\Delta P = \tfrac{1}{2}\left(1.00\times10^3\ \text{kg/m}^3\right)\left(\frac{4\left(0.050\,\frac{\text{L}}{\text{s}}\times\frac{10^{-3}\ \text{m}^3}{\text{L}}\right)}{\pi(0.0250\ \text{m})^2}\right)^2\left(\left(\frac{2.50\ \text{cm}}{1.000\ \text{cm}}\right)^4 - 1\right) = \boxed{0.20\ \text{kPa}}$$

Because $\Delta P \propto I_V^2$ (ΔP as a function of I_V is a parabola that opens upward), this pressure difference is the minimum pressure difference.

63 ••• **[SSM]** Derive the Bernoulli Equation in more generality than done in the text, that is, allow for the fluid to change elevation during its movement. Using the work-energy theorem, show that when changes in elevation are allowed, Equation 13-16 becomes $P_1 + \rho g h_1 + \frac{1}{2}\rho v_1^2 = P_2 + \rho g h_2 + \frac{1}{2}\rho v_2^2$ (Equation 13-17).

Picture the Problem Consider a fluid flowing in a tube that varies in elevation as well as in cross-sectional area, as shown in the pictorial representation below. We can apply the work-energy theorem to a parcel of fluid that initially is contained between points 1 and 2. During time Δt this parcel moves along the tube to the region between point 1′ and 2′. Let ΔV be the volume of fluid passing point 1′ during time Δt. The same volume passes point 2 during the same time. Also, let $\Delta m = \rho\Delta V$ be the mass of the fluid with volume ΔV. The net effect on the parcel during time Δt is that mass Δm initially at height h_1 moving with speed v_1 is "transferred" to height h_2 with speed v_2.

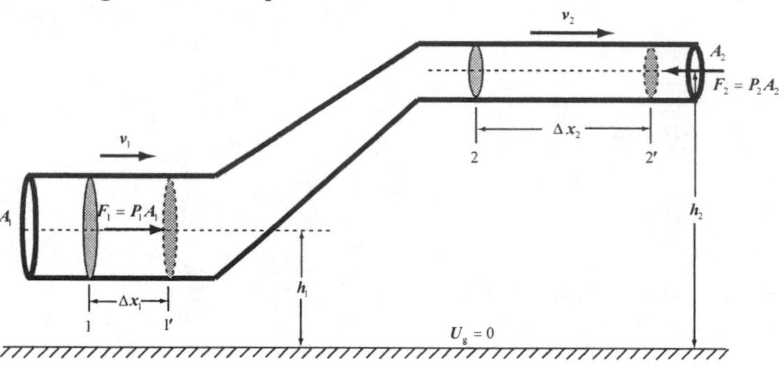

Express the work-energy theorem:

$$W_{\text{total}} = \Delta U + \Delta K \quad (1)$$

The change in the potential energy of the parcel is given by:

$$\begin{aligned}\Delta U &= (\Delta m)gh_2 - (\Delta m)gh_1 \\ &= (\Delta m)g(h_2 - h) \\ &= \rho\Delta Vg(h_2 - h_1)\end{aligned}$$

The change in the kinetic energy of the parcel is given by:

$$\begin{aligned}\Delta K &= \tfrac{1}{2}(\Delta m)v_2^2 - \tfrac{1}{2}(\Delta m)v_1^2 \\ &= \tfrac{1}{2}(\Delta m)(v_2^2 - v_1^2) \\ &= \tfrac{1}{2}\rho\Delta V(v_2^2 - v_1^2)\end{aligned}$$

Express the work done by the fluid behind the parcel (to the parcel's left in the diagram) as it pushes on the parcel with a force of magnitude $F_1 = P_1A_1$, where P_1 is the pressure at point 1:

$$W_1 = F_1\Delta x_1 = P_1A_1\Delta x_1 = P_1\Delta V$$

Express the work done by the fluid in front of the parcel (to the parcel's right in the diagram) as it pushes on the parcel with a force of magnitude $F_2 = P_2A_2$, where P_2 is the pressure at point 2:

$W_2 = -F_2\Delta x_2 = -P_2A_2\Delta x_2 = -P_2\Delta V$
where the work is negative because the applied force and the displacement are in opposite directions.

The total work done on the parcel is:

$W_{\text{total}} = P_1\Delta V - P_2\Delta V = (P_1 - P_2)\Delta V$

Substitute for ΔU, ΔK, and W_{total} in equation (1) to obtain:

$$(P_1 - P_2)\Delta V = \rho\Delta Vg(h_2 - h_1) + \tfrac{1}{2}\rho\Delta V(v_2^2 - v_1^2)$$

Simplifying this expression by dividing out ΔV yields:

$P_1 - P_2 = \rho g(h_2 - h_1) + \tfrac{1}{2}\rho(v_2^2 - v_1^2)$

Collect all the quantities having a subscript 1 on one side and those having a subscript 2 on the other to obtain:

$\boxed{P_1 + \rho gh_1 + \tfrac{1}{2}\rho v_1^2 = P_2 + \rho gh_2 + \tfrac{1}{2}\rho v_2^2}$

Remarks: This equation is known as Bernoulli's equation for the steady, nonviscous flow of an incompressible fluid.

*Viscous Flow

73 • [SSM] An abrupt transition occurs at Reynolds numbers of about 3×10^5, where the drag on a sphere moving through a fluid abruptly decreases. Estimate the speed at which this transition occurs for a baseball, and comment on whether it should play a role in the physics of the game.

Picture the Problem We can use the definition of Reynolds number to find the speed of a baseball at which the drag crisis occurs.

Using its definition, relate Reynolds number to the speed v of the baseball:

$$N_R = \frac{2r\rho v}{\eta} \Rightarrow v = \frac{\eta N_R}{2r\rho}$$

Substitute numerical values (see Figure 13-1 for the density of air and Table 13-3 for the coefficient of viscosity for air) and evaluate v:

$$v = \frac{(0.018\,\text{mPa}\cdot\text{s})(3\times10^5)}{2(0.05\,\text{m})(1.293\,\text{kg/m}^3)}$$

$$= 41.8\,\text{m/s} \times \frac{1\,\text{mi/h}}{0.447\,\text{m/s}} \approx \boxed{90\,\text{mi/h}}$$

Because most major league pitchers can throw a fastball in the low-to-mid-90s, this drag crisis may very well play a role in the game.

Remarks: This is a topic that has been fiercely debated by people who study the physics of baseball.

General Problems

77 • **[SSM]** Several teenagers are swimming toward a rectangular, wooden raft that is 3.00 m wide and 2.00 m long. If the raft is 9.00 cm thick, how many 75.0 kg teenage boys can stand on top of the raft without the raft becoming submerged? Assume the wood density is 650 kg/m^3.

Picture the Problem If the raft is to be just barely submerged, then the buoyant force on it will be equal in magnitude to the weight of the raft plus the weight of the boys. We can apply Archimedes' principle to find the buoyant force on the raft.

The buoyant force acting on the raft is the sum of the weights of the raft and the boys:

$$B = w_{\text{raft}} + w_{\text{boys}} \quad (1)$$

Express the buoyant force acting on the raft:

$$B = w_{\text{displaced water}} = m_{\text{displaced water}}g = \rho_{\text{water}}V_{\text{displaced water}}g = \rho_{\text{water}}V_{\text{raft}}g$$

Express the weight of the raft:

$$w_{\text{raft}} = m_{\text{raft}}g = \rho_{\text{raft}}V_{\text{raft}}g$$

Express the weight of the boys:

$$w_{\text{boys}} = m_{\text{boys}}g = Nm_{\text{1 boy}}g$$

Substituting for B, w_{raft}, and w_{boys} in equation (1) yields:

$$\rho_{\text{water}}V_{\text{raft}}g = \rho_{\text{raft}}V_{\text{raft}}g + Nm_{\text{1 boy}}g$$

Solve for N and simplify to obtain:

$$N = \frac{\rho_{\text{water}}V_{\text{raft}}g - \rho_{\text{raft}}V_{\text{raft}}g}{m_{\text{1 boy}}g} = \frac{(\rho_{\text{water}} - \rho_{\text{raft}})V_{\text{raft}}}{m_{\text{1 boy}}}$$

Substitute numerical values and evaluate N:

$$N = \frac{(1.00\times10^3\ \text{kg/m}^3 - 650\ \text{kg/m}^3)(3.00\ \text{m})(2.00\ \text{m})(0.0900\ \text{m})}{75.0\ \text{kg}} = 2.52$$

Hence, a maximum of $\boxed{2}$ boys can be on the raft under these circumstances.

81 •• **[SSM]** A 1.5-kg block of wood floats on water with 68 percent of its volume submerged. A lead block is placed on the wood, fully submerging the wood to a depth where the lead remains entirely out of the water. Find the mass of the lead block.

Picture the Problem Let m and V represent the mass and volume of the block of wood. Because the block is in equilibrium when it is floating, we can apply the condition for translational equilibrium and Archimedes' principle to express the dependence of the volume of water it displaces when it is fully submerged on its weight. We'll repeat this process for the situation in which the lead block is resting on the wood block with the latter fully submerged. Let the upward direction be the positive y direction.

Apply $\sum F_y = 0$ to floating block:

$$B - mg = 0 \qquad (1)$$

Use Archimedes' principle to relate the density of water to the volume of the block of wood:

$$B = w_{\substack{\text{displaced}\\ \text{water}}} = m_{\substack{\text{displaced}\\ \text{water}}} g = \rho_{\text{water}} V_{\substack{\text{displaced}\\ \text{water}}} g = \rho_{\text{water}}(0.68V)g$$

Using the definition of density, express the weight of the block in terms of its density:

$$mg = \rho_{\text{wood}} Vg$$

Substitute for B and mg in equation (1) to obtain:

$$\rho_{\text{water}}(0.68V)g - \rho_{\text{wood}} Vg = 0$$

Solving for ρ_{wood} yields:

$$\rho_{\text{wood}} = 0.68\rho_{\text{water}}$$

Use the definition of density to express the volume of the wood:

$$V = \frac{m}{\rho_{\text{wood}}}$$

Apply $\sum F_y = 0$ to the floating block when the lead block is placed on it:

$B' - m'g = 0$, where B' is the new buoyant force on the block and m' is the combined mass of the wood block and the lead block.

Use Archimedes' principle and the definition of density to obtain:

$$\rho_{\text{water}} Vg - (m_{\text{Pb}} + m)g = 0$$

Solve for the mass of the lead block to obtain:

$$m_{\text{Pb}} = \rho_{\text{water}} V - m$$

Substituting for V and ρ_{water} yields:

$$m_{\text{Pb}} = \frac{\rho_{\text{wood}}}{0.68} \frac{m}{\rho_{\text{wood}}} - m = \left(\frac{1}{0.68} - 1\right) m$$

Substitute numerical values and evaluate m_{Pb}:

$$m_{\text{Pb}} = \left(\frac{1}{0.68} - 1\right)(1.5\ \text{kg}) = \boxed{0.71\ \text{kg}}$$

85 •• **[SSM]** Crude oil has a viscosity of about 0.800 Pa·s at normal temperature. You are the chief design engineer in charge of constructing a 50.0-km horizontal pipeline that connects an oil field to a tanker terminal. The pipeline is to deliver oil at the terminal at a rate of 500 L/s and the flow through the pipeline is to be laminar. Assuming that the density of crude oil is 700 kg/m^3, estimate the diameter of the pipeline that should be used.

Picture the Problem We can use the definition of Reynolds number and assume a value for N_{R} of 1000 (well within the laminar-flow range) to obtain a trial value for the radius of the pipe. We'll then use Poiseuille's law to determine the pressure difference between the ends of the pipe that would be required to maintain a volume flow rate of 500 L/s.

Use the definition of Reynolds number to relate N_{R} to the radius of the pipe:

$$N_{\text{R}} = \frac{2r\rho v}{\eta}$$

Use the definition of I_{V} to relate the volume flow rate of the pipe to its radius:

$$I_{\text{V}} = Av = \pi r^2 v \Rightarrow v = \frac{I_{\text{V}}}{\pi r^2}$$

Substitute to obtain:

$$N_{\text{R}} = \frac{2\rho I_{\text{V}}}{\eta \pi r} \Rightarrow r = \frac{2\rho I_{\text{V}}}{\eta \pi N_{\text{R}}}$$

Substitute numerical values and evaluate r:

$$r = \frac{2(700\ \text{kg/m}^3)(0.500\ \text{m}^3/\text{s})}{\pi(0.800\ \text{Pa}\cdot\text{s})(1000)} = 27.9\ \text{cm}$$

Using Poiseuille's law, relate the pressure difference between the ends of the pipe to its radius:

$$\Delta P = \frac{8\eta L}{\pi r^4} I_{\text{V}}$$

Substitute numerical values and evaluate ΔP:

$$\Delta P = \frac{8(0.800\,\text{Pa}\cdot\text{s})(50\,\text{km})}{\pi(0.279\,\text{m})^4}(0.500\text{m}^3/\text{s})$$
$$= 8.41\times10^6\ \text{Pa} = 83.0\,\text{atm}$$

This pressure is too large to maintain in the pipe. Evaluate ΔP for a pipe of 50 cm radius:

$$\Delta P = \frac{8(0.800\,\text{Pa}\cdot\text{s})(50\,\text{km})}{\pi(0.50\,\text{m})^4}(0.500\,\text{m}^3/\text{s})$$
$$= 8.15\times10^5\ \text{Pa} = 8.04\,\text{atm}$$

1 m is a reasonable diameter for the pipeline.

87 •• **[SSM]** You are employed as a tanker truck driver for the summer. Heating oil is delivered to customers for winter usage by your large tanker truck. The delivery hose has a radius 1.00 cm. The specific gravity of the oil is 0.875, and its coefficient of viscosity is 200 mPa·s. What is the minimum time it will take you fill a customer's 55-gal oil drum if laminar flow through the hose must be maintained?

Picture the Problem We can use the volume of the drum and the volume flow rate equation to express the time to fill the customer's oil drum. The Reynolds number equation relates Reynolds number to the speed with which oil flows through the hose to the volume flow rate. Because the upper limit on the Reynolds number for laminar flow is approximately 2000, we'll use this value in our calculation of the fill time.

The time t_{fill} is related to the volume flow rate in the hose:

$$t_{\text{fill}} = \frac{V}{I_V} = \frac{V}{Av} \qquad (1)$$

where A is the cross-sectional area of the hose and V is the volume of the oil drum.

Reynolds number is defined by the equation:

$$N_R = \frac{2r\rho v}{\eta} \Rightarrow v = \frac{\eta N_R}{2r\rho}$$

Substitute for v and A in equation (1) to obtain:

$$t_{\text{fill}} = \frac{V}{\pi r^2\left(\dfrac{\eta N_R}{2r\rho}\right)} = \frac{2\rho V}{\pi r \eta N_R}$$

Substitute numerical values and evaluate t_{fill}:

$$t_{\text{fill}} = \frac{2(875\text{ kg/m}^3)\left(55\text{ gal}\times\dfrac{3.785\text{ L}}{\text{gal}}\right)}{\pi(0.010\text{ m})(200\text{ mPa}\cdot\text{s})(2000)}$$
$$= \boxed{29\text{ s}}$$

89 •• **[SSM]** A helium balloon can just lift a load that weighs 750 N and has a negligible volume. The skin of the balloon has a mass of 1.5 kg. (*a*) What is the volume of the balloon? (*b*) If the volume of the balloon were twice that found in Part (*a*), what would be the initial acceleration of the balloon when released at sea level carrying a load weighing 900 N?

Picture the Problem Because the balloon is in equilibrium under the influence of the buoyant force exerted by the air, the weight of its basket and load *w*, the weight of the skin of the balloon, and the weight of the helium. Choose upward to be the +*y* direction and apply the condition for translational equilibrium to relate these forces. Archimedes' principle relates the buoyant force on the balloon to the density of the air it displaces and the volume of the balloon.

(*a*) Apply $\sum F_y = 0$ to the balloon:

$$B - m_{\text{skin}}g - m_{\text{He}}g - w = 0$$

Letting V represent the volume of the balloon, use Archimedes' principle to express the buoyant force:

$$\rho_{\text{air}}Vg - m_{\text{skin}}g - m_{\text{He}}g - w = 0$$

Substituting for m_{He} yields:

$$\rho_{\text{air}}Vg - m_{\text{skin}}g - \rho_{\text{He}}Vg - w = 0$$

Solve for V to obtain:

$$V = \frac{m_{\text{skin}}g + w}{(\rho_{\text{air}} - \rho_{\text{He}})g}$$

Substitute numerical values and evaluate V:

$$V = \frac{(1.5\,\text{kg})(9.81\,\text{m/s}^2) + 750\,\text{N}}{(1.293\,\text{kg/m}^3 - 0.1786\,\text{kg/m}^3)(9.81\,\text{m/s}^2)} = \boxed{70\,\text{m}^3}$$

(*b*) Apply $\sum F_y = ma$ to the balloon:

$$B - m_{\text{tot}}g = m_{\text{tot}}a \Rightarrow a = \frac{B}{m_{\text{tot}}} - g \quad (1)$$

Assuming that the mass of the skin has not changed and letting V' represent the doubled volume of the balloon, express m_{tot}:

$$m_{\text{tot}} = m_{\text{load}} + m_{\text{He}} + m_{\text{skin}} = \frac{w_{\text{load}}}{g} + \rho_{\text{He}}V' + m_{\text{skin}}$$

Express the buoyant force acting on the balloon:

$$B = w_{\text{displaced fluid}} = \rho_{\text{air}}V'g$$

Substituting for m_{tot} and B in equation (1) yields:

$$a=\frac{\rho_{air}V'g}{\frac{w_{load}}{g}+\rho_{He}V'+m_{skin}}-g$$

Substitute numerical values and evaluate a:

$$\boldsymbol{a}=\frac{(1.293\,\text{kg/m}^3)(140\,\text{m}^3)(9.81\,\text{m/s}^2)}{\frac{900\text{ N}}{9.81\,\text{m/s}^2}+(0.1786\text{ kg/m}^3)(140\,\text{m}^3)+1.5\text{ kg}}-9.81\,\text{m/s}^2=\boxed{5.2\text{ m/s}^2}$$

Chapter 14
Oscillations

Conceptual Problems

3 •• [SSM] An object attached to a spring exhibits simple harmonic motion with an amplitude of 4.0 cm. When the object is 2.0 cm from the equilibrium position, what percentage of its total mechanical energy is in the form of potential energy? (*a*) One-quarter. (*b*) One-third. (*c*) One-half. (*d*) Two-thirds. (*e*) Three-quarters.

Picture the Problem The total energy of an object undergoing simple harmonic motion is given by $E_{\text{tot}} = \frac{1}{2}kA^2$, where k is the force constant and A is the amplitude of the motion. The potential energy of the oscillator when it is a distance x from its equilibrium position is $U(x) = \frac{1}{2}kx^2$.

Express the ratio of the potential energy of the object when it is 2.0 cm from the equilibrium position to its total energy:

$$\frac{U(x)}{E_{\text{tot}}} = \frac{\frac{1}{2}kx^2}{\frac{1}{2}kA^2} = \frac{x^2}{A^2}$$

Evaluate this ratio for $x = 2.0$ cm and $A = 4.0$ cm:

$$\frac{U(2\,\text{cm})}{E_{\text{tot}}} = \frac{(2.0\,\text{cm})^2}{(4.0\,\text{cm})^2} = \frac{1}{4}$$

and $\boxed{(a)}$ is correct.

7 •• [SSM] Two systems each consist of a spring with one end attached to a block and the other end attached to a wall. The identical springs are horizontal, and the blocks are supported from below by a frictionless horizontal table. The blocks are oscillating in simple harmonic motions with equal amplitudes. However, the mass of block A is four times as large as the mass of block B. How do their maximum speeds compare? (*a*) $v_{\text{A max}} = v_{\text{B max}}$, (*b*) $v_{\text{A max}} = 2v_{\text{B max}}$, (*c*) $v_{\text{A max}} = \frac{1}{2}v_{\text{B max}}$, (*d*) This comparison cannot be done by using the data given.

Determine the Concept The maximum speed of a simple harmonic oscillator is the product of its angular frequency and its amplitude. The angular frequency of a simple harmonic oscillator is the square root of the quotient of the force constant of the spring and the mass of the oscillator.

Relate the maximum speed of system A to its force constant:

$$v_{\text{A max}} = \omega_{\text{A}} A_{\text{A}} = \sqrt{\frac{k_{\text{A}}}{m_{\text{A}}}} A_{\text{A}}$$

Relate the maximum speed of system B to its force constant:

$$v_{\text{B max}} = \omega_{\text{B}} A_{\text{B}} = \sqrt{\frac{k_{\text{B}}}{m_{\text{B}}}} A_{\text{B}}$$

Divide the first of these equations by the second and simplify to obtain:

$$\frac{v_{\text{A max}}}{v_{\text{B max}}} = \frac{\sqrt{\frac{k_{\text{A}}}{m_{\text{A}}}} A_{\text{A}}}{\sqrt{\frac{k_{\text{B}}}{m_{\text{B}}}} A_{\text{B}}} = \sqrt{\frac{m_{\text{B}}}{m_{\text{A}}} \frac{k_{\text{A}}}{k_{\text{B}}}} \frac{A_{\text{A}}}{A_{\text{B}}}$$

Because the systems differ only in the masses attached to the springs:

$$\frac{v_{\text{A max}}}{v_{\text{B max}}} = \sqrt{\frac{m_{\text{B}}}{m_{\text{A}}}}$$

Substituting for m_{A} and simplifying yields:

$$\frac{v_{\text{A max}}}{v_{\text{B max}}} = \sqrt{\frac{m_{\text{B}}}{4m_{\text{B}}}} = \tfrac{1}{2} \Rightarrow v_{\text{A max}} = \tfrac{1}{2} v_{\text{B max}}$$

$\boxed{(c)}$ is correct.

9 •• **[SSM]** In general physics courses, the mass of the spring in simple harmonic motion is usually neglected because its mass is usually much smaller than the mass of the object attached to it. However, this is not always the case. If you neglect the mass of the spring when it is not negligible, how will your calculation of the system's period, frequency and total energy compare to the actual values of these parameters? Explain.

Determine the Concept Neglecting the mass of the spring, the period of a simple harmonic oscillator is given by $T = 2\pi/\omega = 2\pi\sqrt{m/k}$ where m is the mass of the oscillating system (spring plus object) and its total energy is given by $E_{\text{total}} = \frac{1}{2}kA^2$.

Neglecting the mass of the spring results in your using a value for the mass of the oscillating system that is smaller than its actual value. Hence your calculated value for the period will be smaller than the actual period of the system.

Because $\omega = \sqrt{k/m}$, neglecting the mass of the spring will result in your using a value for the mass of the oscillating system that is smaller than its actual value. Hence your calculated value for the frequency of the system will be larger than the actual frequency of the system.

Because the total energy of the oscillating system is the sum of its potential and kinetic energies, ignoring the mass of the spring will cause your calculation of the system's kinetic energy to be too small and, hence, your calculation of the total energy to be too small.

13 •• **[SSM]** Two mass–spring systems *A* and *B* oscillate so that their total mechanical energies are equal. If the force constant of spring A is two times the force constant of spring B, then which expression best relates their amplitudes? (*a*) $A_A = A_B/4$, (*b*) $A_A = A_B/\sqrt{2}$, (*c*) $A_A = A_B$, (*d*) Not enough information is given to determine the ratio of the amplitudes.

Picture the Problem We can express the energy of each system using $E = \frac{1}{2}kA^2$ and, because the energies are equal, equate them and solve for A_A.

Express the energy of mass-spring system A in terms of the amplitude of its motion:

$$E_A = \tfrac{1}{2}k_A A_A^2$$

Express the energy of mass-spring system B in terms of the amplitude of its motion:

$$E_B = \tfrac{1}{2}k_B A_B^2$$

Because the energies of the two systems are equal we can equate them to obtain:

$$\tfrac{1}{2}k_A A_A^2 = \tfrac{1}{2}k_B A_B^2 \Rightarrow A_A = \sqrt{\frac{k_B}{k_A}}A_B$$

Substitute for k_A and simplify to obtain:

$$A_A = \sqrt{\frac{k_B}{2k_B}}A_B = \frac{A_B}{\sqrt{2}}$$

$\boxed{(b)}$ is correct.

17 •• **[SSM]** Two simple pendulums are related as follows. Pendulum A has a length L_A and a bob of mass m_A; pendulum B has a length L_B and a bob of mass m_B. If the frequency of A is one-third that of B, then (*a*) $L_A = 3L_B$ and $m_A = 3m_B$, (*b*) $L_A = 9L_B$ and $m_A = m_B$, (*c*) $L_A = 9L_B$ regardless of the ratio m_A/m_B, (*d*) $L_A - \sqrt{3}L_B$ regardless of the ratio m_A/m_B.

Picture the Problem The frequency of a simple pendulum is independent of the mass of its bob and is given by $f = \frac{1}{2\pi}\sqrt{g/L}$.

Express the frequency of pendulum A:

$$f_A = \frac{1}{2\pi}\sqrt{\frac{g}{L_A}} \Rightarrow L_A = \frac{g}{4\pi^2 f_A^2}$$

Similarly, the length of pendulum B is given by:

$$L_B = \frac{g}{4\pi^2 f_B^2}$$

Divide the first of these equations by the second and simplify to obtain:

$$\frac{L_A}{L_B} = \frac{\dfrac{g}{4\pi^2 f_A^2}}{\dfrac{g}{4\pi^2 f_B^2}} = \frac{f_B^2}{f_A^2} = \left(\frac{f_B}{f_A}\right)^2$$

Substitute for f_A to obtain:

$$\frac{L_A}{L_B} = \left(\frac{f_B}{\frac{1}{3} f_B}\right)^2 = 9 \Rightarrow L_A = 9L_B$$

$\boxed{(c)}$ is correct.

23 •• **[SSM]** Two damped, driven spring-mass oscillating systems have identical masses, driving forces, and damping constants. However, system A's force constant k_A is four times system B's force constant k_B. Assume they are both very weakly damped. How do their resonant frequencies compare? (*a*) $\omega_A = \omega_B$, (*b*) $\omega_A = 2\omega_B$, (*c*) $\omega_A = \frac{1}{2}\omega_B$, (*d*) $\omega_A = \frac{1}{4}\omega_B$, (*e*) Their resonant frequencies cannot be compared, given the information provided.

Picture the Problem For very weak damping, the resonant frequency of a spring-mass oscillator is the same as its natural frequency and is given by $\omega_0 = \sqrt{k/m}$, where *m* is the oscillator's mass and *k* is the force constant of the spring.

Express the resonant frequency of System A:

$$\omega_A = \sqrt{\frac{k_A}{m_A}}$$

The resonant frequency of System B is given by:

$$\omega_B = \sqrt{\frac{k_B}{m_B}}$$

Dividing the first of these equations by the second and simplifying yields:

$$\frac{\omega_A}{\omega_B} = \frac{\sqrt{\dfrac{k_A}{m_A}}}{\sqrt{\dfrac{k_B}{m_B}}} = \sqrt{\frac{k_A}{k_B}\frac{m_B}{m_A}}$$

Because their masses are the same:

$$\frac{\omega_A}{\omega_B} = \sqrt{\frac{k_A}{k_B}}$$

Substituting for k_A yields:

$$\frac{\omega_A}{\omega_B} = \sqrt{\frac{4k_B}{k_B}} = 2 \Rightarrow \omega_A = 2\omega_B$$

$\boxed{(b)}$ is correct.

Estimation and Approximation

25 • **[SSM]** Estimate the width of a typical grandfather clocks' cabinet relative to the width of the pendulum bob, presuming the desired motion of the pendulum is simple harmonic.

Picture the Problem If the motion of the pendulum in a grandfather clock is to be simple harmonic motion, then its period must be independent of the angular amplitude of its oscillations. The period of the motion for large-amplitude oscillations is given by Equation 14-30 and we can use this expression to obtain a maximum value for the amplitude of swinging pendulum in the clock. We can then use this value and an assumed value for the length of the pendulum to estimate the width of the grandfather clocks' cabinet.

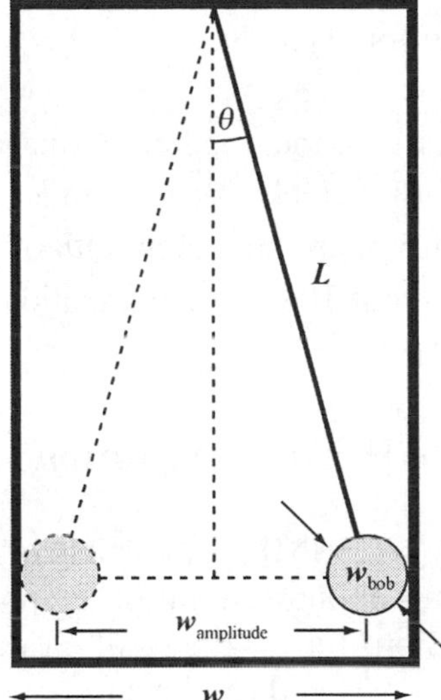

Referring to the diagram, we see that the minimum width of the cabinet is determined by the width of the bob and the width required to accommodate the swinging pendulum:

$$w = w_{\text{bob}} + w_{\text{amplitude}}$$

and

$$\frac{w}{w_{\text{bob}}} = 1 + \frac{w_{\text{amplitude}}}{w_{\text{bob}}} \qquad (1)$$

Express $w_{\text{amplitude}}$ in terms of the angular amplitude θ and the length of the pendulum L:

$$w_{\text{amplitude}} = 2L\sin\theta$$

Substituting for $w_{\text{amplitude}}$ in equation (1) yields:

$$\frac{w}{w_{\text{bob}}} = 1 + \frac{2L\sin\theta}{w_{\text{bob}}} \qquad (2)$$

Equation 14-30 gives us the period of a simple pendulum as a function of its angular amplitude:

$$T = T_0\left[1 + \frac{1}{2^2}\sin^2\frac{1}{2}\theta + \ldots\right]$$

If T is to be approximately equal to T_0, the second term in the brackets must be small compared to the first term. Suppose that:

$$\frac{1}{4}\sin^2\frac{1}{2}\theta \le 0.001$$

Solving for θ yields:

$$\theta \le 2\sin^{-1}(0.0632) \approx 7.25°$$

If we assume that the length of a grandfather clock's pendulum is about 1.5 m and that the width of the bob is about 10 cm, then equation (2) yields:

$$\frac{w}{w_{\text{bob}}} = 1 + \frac{2(1.5\text{ m})\sin 7.25°}{0.10\text{ m}} \approx \boxed{5}$$

Simple Harmonic Motion

31 • **[SSM]** A particle of mass m begins at rest from $x = +25$ cm and oscillates about its equilibrium position at $x = 0$ with a period of 1.5 s. Write expressions for (*a*) the position x as a function of t, (*b*) the velocity v_x as a function of t, and (*c*) the acceleration a_x as a function of t.

Picture the Problem The position of the particle as a function of time is given by $x = A\cos(\omega t + \delta)$. Its velocity as a function of time is $v_x = -A\omega\sin(\omega t + \delta)$ and its acceleration is $a_x = -A\omega^2\cos(\omega t + \delta)$. The initial position and velocity give us two equations from which to determine the amplitude A and phase constant δ.

(*a*) Express the position, velocity, and acceleration of the particle as a function of t:

$$x = A\cos(\omega t + \delta) \quad (1)$$
$$v_x = -A\omega\sin(\omega t + \delta) \quad (2)$$
$$a_x = -A\omega^2\cos(\omega t + \delta) \quad (3)$$

Find the angular frequency of the particle's motion:

$$\omega = \frac{2\pi}{T} = \frac{4\pi}{3}\text{s}^{-1} = 4.19\text{s}^{-1}$$

Relate the initial position and velocity to the amplitude and phase constant:

$$x_0 = A\cos\delta$$
and
$$v_0 = -\omega A\sin\delta$$

Divide the equation for v_0 by the equation for x_0 to eliminate A:

$$\frac{v_0}{x_0} = \frac{-\omega A\sin\delta}{A\cos\delta} = -\omega\tan\delta$$

Solving for δ yields:

$$\delta = \tan^{-1}\left(-\frac{v_0}{x_0\omega}\right) = \tan^{-1}\left(-\frac{0}{x_0\omega}\right) = 0$$

Substitute in equation (1) to obtain:

$$x = (25\,\text{cm})\cos\left[\left(\frac{4\pi}{3}\,\text{s}^{-1}\right)t\right] = \boxed{(0.25\,\text{m})\cos\left[\left(4.2\,\text{s}^{-1}\right)t\right]}$$

(*b*) Substitute in equation (2) to obtain:

$$v_x = -(25\,\text{cm})\left(\frac{4\pi}{3}\,\text{s}^{-1}\right)\sin\left[\left(\frac{4\pi}{3}\,\text{s}^{-1}\right)t\right] = \boxed{-(1.0\,\text{m/s})\sin\left[\left(4.2\,\text{s}^{-1}\right)t\right]}$$

(*c*) Substitute in equation (3) to obtain:

$$a_x = -(25\,\text{cm})\left(\frac{4\pi}{3}\,\text{s}^{-1}\right)^2\cos\left[\left(\frac{4\pi}{3}\,\text{s}^{-1}\right)t\right] = \boxed{-\left(4.4\,\text{m/s}^2\right)\cos\left[\left(4.2\,\text{s}^{-1}\right)t\right]}$$

37 •• **[SSM]** The position of a particle is given by $x = 2.5\cos\pi t$, where x is in meters and t is in seconds. (*a*) Find the maximum speed and maximum acceleration of the particle. (*b*) Find the speed and acceleration of the particle when $x = 1.5$ m.

Picture the Problem The position of the particle is given by $x = A\cos\omega t$, where $A = 2.5$ m and $\omega = \pi$ rad/s. The velocity is the time derivative of the position and the acceleration is the time derivative of the velocity.

(*a*) The velocity is the time derivative of the position and the acceleration is the time derivative of the acceleration:

$$x = A\cos\omega t \Rightarrow v = \frac{dx}{dt} = -\omega A\sin\omega t$$

and $a = \frac{dv}{dt} = -\omega^2 A\cos\omega t$

The maximum value of $\sin\omega t$ is +1 and the minimum value of $\sin\omega t$ is −1. A and ω are positive constants:

$$v_{\text{max}} = A\omega = (2.5\,\text{m})\left(\pi\,\text{s}^{-1}\right) = \boxed{7.9\,\text{m/s}}$$

The maximum value of $\cos\omega t$ is +1 and the minimum value of $\cos\omega t$ is −1:

$$a_{\text{max}} = A\omega^2 = (2.5\,\text{m})\left(\pi\,\text{s}^{-1}\right)^2 = \boxed{25\,\text{m/s}^2}$$

(*b*) Use the Pythagorean identity $\sin^2\omega t + \cos^2\omega t = 1$ to eliminate t from the equations for x and v:

$$\frac{v^2}{\omega^2 A^2} + \frac{x^2}{A^2} = 1 \Rightarrow |v| = \omega\sqrt{A^2 - x^2}$$

Substitute numerical values and evaluate $|v(1.5\text{ m})|$:

$$|v(1.5\text{ m})| = (\pi\text{ rad/s})\sqrt{(2.5\text{ m})^2 - (1.5\text{ m})^2} = \boxed{6.3\text{ m/s}}$$

Substitute x for $A\cos\omega t$ in the equation for a to obtain:

$$a = -\omega^2 x$$

Substitute numerical values and evaluate a:

$$a = -(\pi\text{ rad/s})^2(1.5\text{ m}) = \boxed{-15\text{ m/s}^2}$$

Simple Harmonic Motion as Related to Circular Motion

39 • **[SSM]** A particle moves at a constant speed of 80 cm/s in a circle of radius 40 cm centered at the origin. (*a*) Find the frequency and period of the x component of its position. (*b*) Write an expression for the x component of its position as a function of time t, assuming that the particle is located on the $+y$-axis at time $t = 0$.

Picture the Problem We can find the period of the motion from the time required for the particle to travel completely around the circle. The frequency of the motion is the reciprocal of its period and the x-component of the particle's position is given by $x = A\cos(\omega t + \delta)$. We can use the initial position of the particle to determine the phase constant δ.

(*a*) Use the definition of speed to find the period of the motion:

$$T = \frac{2\pi r}{v} = \frac{2\pi(0.40\text{ m})}{0.80\text{ m/s}} = 3.14 = \boxed{3.1\text{ s}}$$

Because the frequency and the period are reciprocals of each other:

$$f = \frac{1}{T} = \frac{1}{3.14\text{ s}} = \boxed{0.32\text{ Hz}}$$

(*b*) Express the x component of the position of the particle:

$$x = A\cos(\omega t + \delta) = A\cos(2\pi f t + \delta) \quad (1)$$

The initial condition on the particle's position is:

$$x(0) = 0$$

Substitute in the expression for x to obtain:

$$0 = A\cos\delta \Rightarrow \delta = \cos^{-1}(0) = \frac{\pi}{2}$$

Substitute for A, ω, and δ in equation (1) to obtain:

$$x = \boxed{(40\text{ cm})\cos\left[(2.0\text{ s}^{-1})t + \frac{\pi}{2}\right]}$$

Energy in Simple Harmonic Motion

43 • **[SSM]** A 1.50-kg object on a frictionless horizontal surface oscillates at the end of a spring of force constant $k = 500$ N/m. The object's maximum speed is 70.0 cm/s. (*a*) What is the system's total mechanical energy? (*b*) What is the amplitude of the motion?

Picture the Problem The total mechanical energy of the oscillating object can be expressed in terms of its kinetic energy as it passes through its equilibrium position: $E_{tot} = \frac{1}{2}mv_{max}^2$. Its total energy is also given by $E_{tot} = \frac{1}{2}kA^2$. We can equate these expressions to obtain an expression for A.

(*a*) Express the total mechanical energy of the object in terms of its maximum kinetic energy:

$$E = \frac{1}{2}mv_{max}^2$$

Substitute numerical values and evaluate E:

$$E = \frac{1}{2}(1.50\,\text{kg})(0.700\,\text{m/s})^2 = 0.3675\,\text{J} = \boxed{0.368\,\text{J}}$$

(*b*) Express the total mechanical energy of the object in terms of the amplitude of its motion:

$$E_{tot} = \frac{1}{2}kA^2 \Rightarrow A = \sqrt{\frac{2E_{tot}}{k}}$$

Substitute numerical values and evaluate A:

$$A = \sqrt{\frac{2(0.3675\,\text{J})}{500\,\text{N/m}}} = \boxed{3.83\,\text{cm}}$$

Simple Harmonic Motion and Springs

49 • **[SSM]** A 3.0-kg object on a frictionless horizontal surface is attached to one end of a horizontal spring, oscillates with an amplitude of 10 cm and a frequency of 2.4 Hz. (*a*) What is the force constant of the spring? (*b*) What is the period of the motion? (*c*) What is the maximum speed of the object? (*d*) What is the maximum acceleration of the object?

Picture the Problem (*a*) The angular frequency of the motion is related to the force constant of the spring through $\omega^2 = k/m$. (*b*) The period of the motion is the reciprocal of its frequency. (*c*) and (*d*) The maximum speed and acceleration of an object executing simple harmonic motion are $v_{max} = A\omega$ and $a_{max} = A\omega^2$, respectively.

(*a*) Relate the angular frequency of the motion to the force constant of the spring:

$$\omega^2 = \frac{k}{m} \Rightarrow k = m\omega^2 = 4\pi^2 f^2 m$$

Substitute numerical values to obtain:

$$k = 4\pi^2 \left(2.4\,\text{s}^{-1}\right)^2 (3.0\,\text{kg}) = 682\,\text{N/m} = \boxed{0.68\,\text{kN/m}}$$

(*b*) Relate the period of the motion to its frequency:

$$T = \frac{1}{f} = \frac{1}{2.4\,\text{s}^{-1}} = 0.417\,\text{s} = \boxed{0.42\,\text{s}}$$

(*c*) The maximum speed of the object is given by:

$$v_{\text{max}} = A\omega = 2\pi f A$$

Substitute numerical values and evaluate v_{max}:

$$v_{\text{max}} = 2\pi\left(2.4\,\text{s}^{-1}\right)(0.10\,\text{m}) = 1.51\,\text{m/s} = \boxed{1.5\,\text{m/s}}$$

(*d*) The maximum acceleration of the object is given by:

$$a_{\text{max}} = A\omega^2 = 4\pi^2 f^2 A$$

Substitute numerical values and evaluate a_{max}:

$$a_{\text{max}} = 4\pi^2 \left(2.4\,\text{s}^{-1}\right)^2 (0.10\,\text{m}) = \boxed{23\,\text{m/s}^2}$$

Simple Pendulum Systems

59 • **[SSM]** Find the length of a simple pendulum if its frequency for small amplitudes is 0.75 Hz.

Picture the Problem The frequency of a simple pendulum depends on its length and on the local gravitational field and is given by $f = \frac{1}{2\pi}\sqrt{\frac{g}{L}}$.

The frequency of a simple pendulum oscillating with small amplitude is given by:

$$f = \frac{1}{2\pi}\sqrt{\frac{g}{L}} \Rightarrow L = \frac{g}{4\pi^2 f^2}$$

Substitute numerical values and evaluate L:

$$L = \frac{9.81\,\text{m/s}^2}{4\pi^2\left(0.75\,\text{s}^{-1}\right)^2} = \boxed{44\text{ cm}}$$

65 ••• **[SSM]** A simple pendulum of length L is attached to a massive cart that slides without friction down a plane inclined at angle θ with the horizontal, as shown in Figure 14-28. Find the period of oscillation for small oscillations of this pendulum.

Picture the Problem The cart accelerates down the ramp with a constant acceleration of $g\sin\theta$. This happens because the cart is much more massive than the bob, so the motion of the cart is unaffected by the motion of the bob oscillating back and forth. The path of the bob is quite complex in the reference frame of the ramp, but in the reference frame moving with the cart the path of the bob is much simpler—in this frame the bob moves back and forth along a circular arc. To solve this problem we first apply Newton's second law (to the bob) in the inertial reference frame of the ramp. Then we transform to the reference frame moving with the cart in order to exploit the simplicity of the motion in that frame.

Draw the free-body diagram for the bob. Let ϕ denote the angle that the string makes with the normal to the ramp. The forces on the bob are the tension force and the force of gravity:

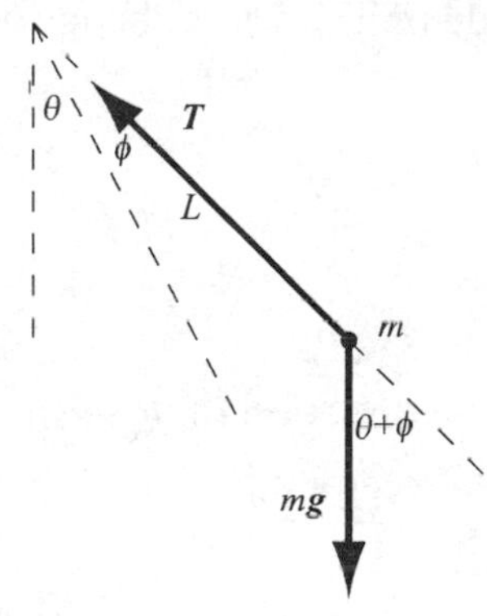

Apply Newton's 2nd law to the bob, labeling the acceleration of the bob relative to the ramp $\vec{a}_{\text{BR}}$:

$$\vec{T} + m\vec{g} = m\vec{a}_{\text{BR}}$$

The acceleration of the bob relative to the ramp is equal to the acceleration of the bob relative to the cart plus the acceleration of the cart relative to the ramp:

$$\vec{a}_{\text{BR}} = \vec{a}_{\text{BC}} + \vec{a}_{\text{CR}}$$

Substitute for $\vec{a}_{\text{BR}}$ in $\vec{T} + m\vec{g} = m\vec{a}_{\text{BR}}$:

$$\vec{T} + m\vec{g} = m\left(\vec{a}_{\text{BC}} + \vec{a}_{\text{CR}}\right)$$

Rearrange terms and label $\vec{g} - \vec{a}_{\text{CR}}$ as $\vec{g}_{\text{eff}}$, where $\vec{g}_{\text{eff}}$ is the acceleration, relative to the cart, of an object in free fall. (If the tension force is set to zero the bob is in free fall.):

$$\vec{T} + m\left(\vec{g} - \vec{a}_{\text{CR}}\right) = m\vec{a}_{\text{BC}}$$

Label $\vec{g} - \vec{a}_{\text{CR}}$ as $\vec{g}_{\text{eff}}$ to obtain

$$\vec{T} + m\vec{g}_{\text{eff}} = m\vec{a}_{\text{BC}} \qquad (1)$$

To find the magnitude of $\vec{g}_{\text{eff}}$, first draw the vector addition diagram representing the equation $\vec{g}_{\text{eff}} = \vec{g} - \vec{a}_{\text{CR}}$. Recall that $a_{\text{CR}} = g \sin \theta$.

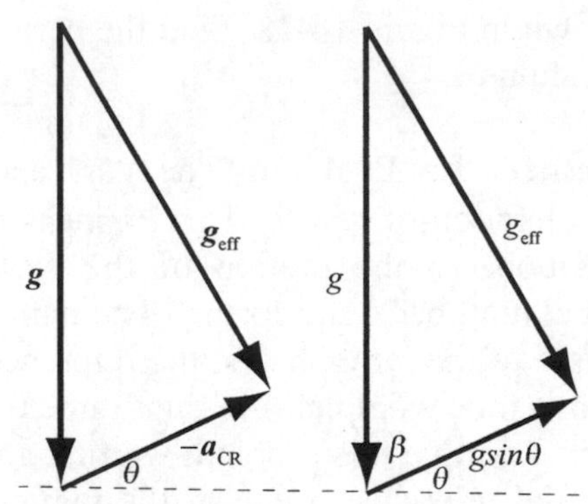

From the diagram, find the magnitude of $\vec{g}_{\text{eff}}$. Use the law of cosines:

$$g_{\text{eff}}^2 = g^2 + g^2 \sin^2 \theta - 2g(g \sin \theta) \cos \beta$$

But $\cos \beta = \sin \theta$, so

$$g_{\text{eff}}^2 = g^2 + g^2 \sin^2 \theta - 2g^2 \sin^2 \theta$$
$$= g^2 (1 - \sin^2 \theta) = g^2 \cos^2 \theta$$

Thus $g_{\text{eff}} = g \cos \theta$

To find the direction of $\vec{g}_{\text{eff}}$, first redraw the vector addition diagram as shown:

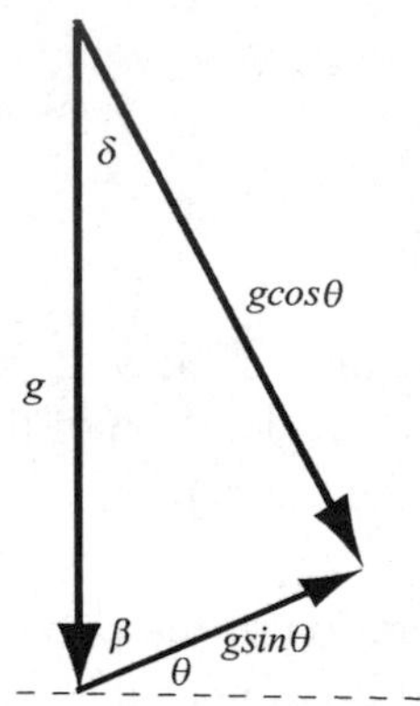

From the diagram find the direction of $\vec{g}_{\text{eff}}$. Use the law of cosines again and solve for δ:

$$g^2 \sin^2 \theta = g^2 + g^2 \cos^2 \theta - 2g^2 \cos \theta \cos \delta$$

and so $\delta = \theta$

To find an equation for the motion of the bob draw the "free-body diagram" for the "forces" that appear in equation (1). Draw the path of the bob in the reference frame moving with the cart:

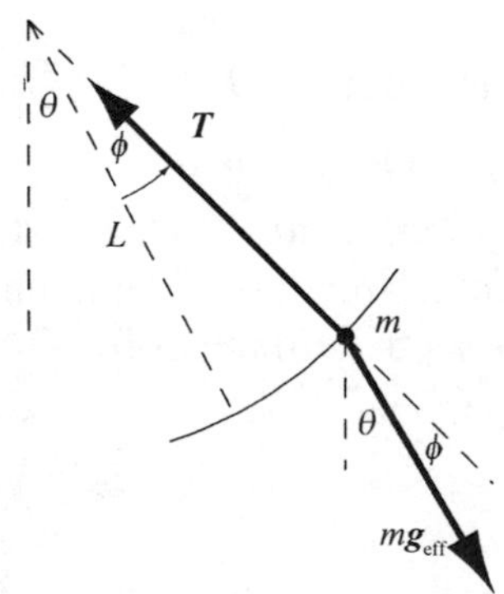

Take the tangential components of each vector in equation (1) in the frame of the cart yields. The tangential component of the acceleration is equal to the radius of the circle times the angular acceleration $(a_t = r\alpha)$:

$$0 - mg_{\text{eff}} \sin\phi = mL\frac{d^2\phi}{dt^2}$$

where L is the length of the string and $\frac{d^2\phi}{dt^2}$ is the angular acceleration of the bob. The positive tangential "direction" is counterclockwise.

Rearranging this equation yields:

$$mL\frac{d^2\phi}{dt^2} + mg_{\text{eff}} \sin\phi = 0 \qquad (2)$$

For small oscillations of the pendulum:

$$|\phi| << 1 \text{ and } \sin\phi \approx \phi$$

Substituting for $\sin\phi$ in equation (2) yields:

$$mL\frac{d^2\phi}{dt^2} + mg_{\text{eff}}\phi = 0$$

or

$$\frac{d^2\phi}{dt^2} + \frac{g_{\text{eff}}}{L}\phi = 0 \qquad (3)$$

Equation (3) is the equation of motion for simple harmonic motion with angular frequency:

$$\omega = \sqrt{\frac{g_{\text{eff}}}{L}}$$

where ω is the angular frequency of the oscillations (and not the angular speed of the bob).

The period of this motion is:

$$T = \frac{2\pi}{\omega} = 2\pi\sqrt{\frac{L}{g_{\text{eff}}}} \qquad (4)$$

Substitute $g\cos\theta$ for g_{eff} in equation (4) to obtain:

$$\boldsymbol{T} = 2\pi\sqrt{\frac{L}{g_{\text{eff}}}} = \boxed{2\pi\sqrt{\frac{L}{g\cos\theta}}}$$

Remarks: Note that, in the limiting case $\theta = 0$, $T = 2\pi\sqrt{L/g}$ and $T \to 0$. As $\theta \to 90°$, $T \to \infty$.

*Physical Pendulums

67 • **[SSM]** A thin 5.0-kg disk with a 20-cm radius is free to rotate about a fixed horizontal axis perpendicular to the disk and passing through its rim. The disk is displaced slightly from equilibrium and released. Find the period of the subsequent simple harmonic motion.

Picture the Problem The period of this physical pendulum is given by $T = 2\pi\sqrt{I/MgD}$ where I is the moment of inertia of the thin disk about the fixed horizontal axis passing through its rim. We can use the parallel-axis theorem to express I in terms of the moment of inertia of the disk about an axis through its center of mass and the distance from its center of mass to its pivot point.

Express the period of a physical pendulum:

$$T = 2\pi\sqrt{\frac{I}{MgD}}$$

Using the parallel-axis theorem, find the moment of inertia of the thin disk about an axis through the pivot point:

$$I = I_{cm} + MR^2 = \tfrac{1}{2}MR^2 + MR^2 = \tfrac{3}{2}MR^2$$

Substituting for I and simplifying yields:

$$T = 2\pi\sqrt{\frac{\tfrac{3}{2}MR^2}{MgR}} = 2\pi\sqrt{\frac{3R}{2g}}$$

Substitute numerical values and evaluate T:

$$T = 2\pi\sqrt{\frac{3(0.20\,\text{m})}{2(9.81\,\text{m/s}^2)}} = \boxed{1.1\,\text{s}}$$

73 ••• **[SSM]** Points P_1 and P_2 on a plane object (Figure 14-30) are distances h_1 and h_2, respectively, from the center of mass. The object oscillates with the same period T when it is free to rotate about an axis through P_1 and when it is free to rotate about an axis through P_2. Both of these axes are perpendicular to the plane of the object. Show that $h_1 + h_2 = gT^2/(4\pi)^2$, where $h_1 \neq h_2$.

Picture the Problem We can use the equation for the period of a physical pendulum and the parallel-axis theorem to show that $h_1 + h_2 = gT^2/4\pi^2$.

Express the period of the physical pendulum:

$$T = 2\pi\sqrt{\frac{I}{mgd}}$$

Using the parallel-axis theorem, relate the moment of inertia about an axis through P_1 to the moment of inertia about an axis through the plane's center of mass:

$$I = I_{cm} + mh_1^2$$

Substitute for I to obtain:

$$T = 2\pi\sqrt{\frac{I_{\text{cm}} + mh_1^2}{mgh_1}}$$

Square both sides of this equation and rearrange terms to obtain:

$$\frac{mgT^2}{4\pi^2} = \frac{I_{\text{cm}}}{h_1} + mh_1 \qquad (1)$$

Because the period of oscillation is the same for point P_2:

$$\frac{I_{\text{cm}}}{h_1} + mh_1 = \frac{I_{\text{cm}}}{h_2} + mh_2$$

Combining like terms yields:

$$\left(\frac{1}{h_1} - \frac{1}{h_2}\right) I_{\text{cm}} = m(h_2 - h_1)$$

Provided $h_1 \neq h_2$:

$$I_{\text{cm}} = mh_1h_2$$

Substitute in equation (1) and simplify to obtain:

$$\frac{mgT^2}{4\pi^2} = \frac{mh_1h_2}{h_1} + mh_1 \Rightarrow h_1 + h_2 = \boxed{\frac{gT^2}{4\pi^2}}$$

Damped Oscillations

77 •• **[SSM]** Show that the ratio of the amplitudes for two successive oscillations is constant for a linearly damped oscillator.

Picture the Problem The amplitude of the oscillation at time t is $A(t) = A_0e^{-t/2\tau}$ where $\tau = m/b$ is the decay constant. We can express the amplitudes one period apart and then show that their ratio is constant.

Relate the amplitude of a given oscillation peak to the time at which the peak occurs:

$$A(t) = A_0e^{-t/2\tau}$$

Express the amplitude of the oscillation peak at $t' = t + T$:

$$A(t+T) = A_0e^{-(t+T)/2\tau}$$

Express the ratio of these consecutive peaks:

$$\frac{A(t)}{A(t+T)} = \frac{A_0e^{-t/2\tau}}{A_0e^{-(t+T)/2\tau}} = e^{-T/2\tau} = \boxed{\text{constant}}$$

81 •• **[SSM]** Seismologists and geophysicists have determined that the vibrating Earth has a resonance period of 54 min and a Q factor of about 400. After a large earthquake, Earth will "ring" (continue to vibrate) for up to 2 months. (*a*) Find the percentage of the energy of vibration lost to damping forces during each cycle. (*b*) Show that after n periods the vibrational energy is given by $E_n = (0.984)^n E_0$, where E_0 is the original energy. (*c*) If the original energy of vibration of an earthquake is E_0, what is the energy after 2.0 d?

Picture the Problem (*a*) We can find the fractional loss of energy per cycle from the physical interpretation of Q for small damping. (*b*) We will also find a general expression for the earth's vibrational energy as a function of the number of cycles it has completed. (*c*) We can then solve this equation for the earth's vibrational energy after any number of days.

(*a*) Express the fractional change in energy as a function of Q:

$$\frac{\Delta E}{E} = \frac{2\pi}{Q} = \frac{2\pi}{400} = \boxed{1.57\%}$$

(*b*) Express the energy of the damped oscillator after one cycle:

$$E_1 = E_0\left(1 - \frac{\Delta E}{E}\right)$$

Express the energy after two cycles:

$$E_2 = E_1\left(1 - \frac{\Delta E}{E}\right) = E_0\left(1 - \frac{\Delta E}{E}\right)^2$$

Generalizing to n cycles:

$$E_n = E_0\left(1 - \frac{\Delta E}{E}\right)^n = E_0(1 - 0.0157)^n$$
$$= \boxed{E_0(0.984)^n}$$

(*c*) Express 2.0 d in terms of the number of cycles; that is, the number of vibrations the earth will have experienced:

$$2.0\,\text{d} = 2.0\,\text{d} \times \frac{24\,\text{h}}{\text{d}} \times \frac{60\,\text{m}}{\text{h}}$$
$$= 2880\,\text{min} \times \frac{1T}{54\,\text{min}}$$
$$= 53.3T$$

Evaluate E(2 d):

$$E(2\text{d}) = E_0(0.9843)^{53.3} = \boxed{0.43E_0}$$

83 ••• **[SSM]** You are in charge of monitoring the viscosity of oils at a manufacturing plant and you determine the viscosity of an oil by using the following method: The viscosity of a fluid can be measured by determining the decay time of oscillations for an oscillator that has known properties and operates while immersed in the fluid. As long as the speed of the oscillator through the fluid is relatively small, so that turbulence is not a factor, the drag force of the fluid on a sphere is proportional to the sphere's speed relative to the fluid: $F_d = 6\pi a\eta v$, where η is the viscosity of the fluid and a is the sphere's radius. Thus, the constant b is given by $6\pi a\eta$. Suppose your apparatus consists of a stiff spring that has a force constant equal to 350 N/cm and a gold sphere (radius 6.00 cm) hanging on the spring. (*a*) What is the viscosity of an oil do you measure if the decay time for this system is 2.80 s? (*b*) What is the Q factor for your system?

Picture the Problem (*a*) The decay time for a damped oscillator (with speed-dependent damping) system is defined as the ratio of the mass of the oscillator to the coefficient of v in the damping force expression. (*b*) The Q factor is the product of the resonance frequency and the damping time.

(*a*) From $F_d = 6\pi a\eta v$ and $F_d = -bv$, it follows that:

$$b = 6\pi a\eta \Rightarrow \eta = \frac{b}{6\pi a}$$

Because $\tau = m/b$, we can substitute for b to obtain:

$$\eta = \frac{m}{6\pi a\tau}$$

Substituting $m = \rho V$ and simplifying yields:

$$\eta = \frac{\rho V}{6\pi a\tau} = \frac{\frac{4}{3}\pi a^3\rho}{6\pi a\tau} = \frac{2a^2\rho}{9\tau}$$

Substitute numerical values and evaluate η (see Table 13-1 for the density of gold):

$$\eta = \frac{2(0.0600\text{ m})^2(19.3\times10^3\text{ kg/m}^3)}{9(2.8\text{ s})} = \boxed{5.51\text{ Pa}\cdot\text{s}}$$

(*b*) The Q factor is the product of the resonance frequency and the damping time:

$$Q = \omega_0\tau = \sqrt{\frac{k}{m}}\tau = \sqrt{\frac{k}{\rho V}}\tau = \sqrt{\frac{k}{\frac{4}{3}\pi a^3\rho}}\tau$$

Substitute numerical values and evaluate Q:

$$Q = \sqrt{\frac{3\left(350\dfrac{\text{N}}{\text{cm}}\times\dfrac{100\text{ cm}}{\text{m}}\right)}{4\pi(0.0600\text{ m})^3(19.3\times10^3\text{ kg/m}^3)}}(2.80\text{ s}) \approx \boxed{125}$$

Driven Oscillations and Resonance

87 •• **[SSM]** A 2.00-kg object oscillates on a spring of force constant 400 N/m. The linear damping constant has a value of 2.00 kg/s. The system is driven by a sinusoidal force of maximum value 10.0 N and angular frequency 10.0 rad/s. (*a*) What is the amplitude of the oscillations? (*b*) If the driving frequency is varied, at what frequency will resonance occur? (*c*) What is the amplitude of oscillation at resonance? (*d*) What is the width of the resonance curve $\Delta\omega$?

Picture the Problem (*a*) The amplitude of the damped oscillations is related to the damping constant, mass of the system, the amplitude of the driving force, and the natural and driving frequencies through $A=\dfrac{F_0}{\sqrt{m^2\left(\omega_0^2-\omega^2\right)^2+b^2\omega^2}}$.

(*b*) Resonance occurs when $\omega=\omega_0$. (*c*) At resonance, the amplitude of the oscillations is $A=F_0\big/\sqrt{b^2\omega^2}$. (*d*) The width of the resonance curve is related to the damping constant and the mass of the system according to $\Delta\omega=b/m$.

(*a*) Express the amplitude of the oscillations as a function of the driving frequency:

$$A=\frac{F_0}{\sqrt{m^2\left(\omega_0^2-\omega^2\right)^2+b^2\omega^2}}$$

Because $\omega_0=\sqrt{\dfrac{k}{m}}$:

$$A=\frac{F_0}{\sqrt{m^2\left(\dfrac{k}{m}-\omega^2\right)^2+b^2\omega^2}}$$

Substitute numerical values and evaluate A:

$$A=\frac{10.0\text{ N}}{\sqrt{(2.00\text{ kg})^2\left(\dfrac{400\text{ N/m}}{2.00\text{ kg}}-(10.0\text{ rad/s})^2\right)^2+(2.00\text{ kg/s})^2(10.0\text{ rad/s})^2}}=\boxed{4.98\text{ cm}}$$

(*b*) Resonance occurs when:

$$\omega=\omega_0=\sqrt{\frac{k}{m}}$$

Substitute numerical values and evaluate ω:

$$\omega=\sqrt{\frac{400\text{ N/m}}{2.00\text{ kg}}}=14.14\text{ rad/s}$$
$$=\boxed{14.1\text{ rad/s}}$$

(*c*) The amplitude of the motion at resonance is given by:

$$A=\frac{F_0}{\sqrt{b^2\omega_0^2}}$$

Substitute numerical values and evaluate A:

$$A=\frac{10.0\,\text{N}}{\sqrt{(2.00\,\text{kg/s})^2(14.14\,\text{rad/s})^2}}=\boxed{35.4\,\text{cm}}$$

(*d*) The width of the resonance curve is:

$$\Delta\omega=\frac{b}{m}=\frac{2.00\,\text{kg/s}}{2.00\,\text{kg}}=\boxed{1.00\,\text{rad/s}}$$

General Problems

93 •• **[SSM]** A block that has a mass equal to m_1 is supported from below by a frictionless horizontal surface. The block, which is attached to the end of a horizontal spring with a force constant k, oscillates with an amplitude A. When the spring is at its greatest extension and the block is instantaneously at rest, a second block of mass m_2 is placed on top of it. (*a*) What is the smallest value for the coefficient of static friction μ_s such that the second object does not slip on the first? (*b*) Explain how the total mechanical energy E, the amplitude A, the angular frequency ω, and the period T of the system are affected by the placing of m_2 on m_1, assuming that the coefficient of friction is great enough to prevent slippage.

Picture the Problem Applying Newton's 2nd law to the first object as it is about to slip will allow us to express μ_s in terms of the maximum acceleration of the system which, in turn, depends on the amplitude and angular frequency of the oscillatory motion.

(*a*) Apply $\sum F_x = ma_x$ to the second object as it is about to slip:

$$f_{\text{s,max}}=m_2a_{\text{max}}$$

Apply $\sum F_y = 0$ to the second object:

$$F_\text{n}-m_2g=0$$

Use $f_{\text{s,max}}=\mu_\text{s}F_\text{n}$ to eliminate $f_{\text{s,max}}$ and F_n between the two equations and solve for μ_s:

$$\mu_\text{s}m_2g=m_2a_{\text{max}}\Rightarrow\mu_\text{s}=\frac{a_{\text{max}}}{g}$$

Relate the maximum acceleration of the oscillator to its amplitude and angular frequency and substitute for ω^2:

$$a_{\text{max}}=A\omega^2=A\frac{k}{m_1+m_2}$$

Finally, substitute for a_{max} to obtain:

$$\boxed{\mu_s = \frac{Ak}{(m_1 + m_2)g}}$$

(*b*) *A* is unchanged. *E* is unchanged because $E = \frac{1}{2}kA^2$. ω is reduced and *T* is increased by increasing the total mass of the system.

97 •• **[SSM]** Show that for the situations in Figure 14-35*a* and Figure 14-35*b* the object oscillates with a frequency $f = (1/2\pi)\sqrt{k_{eff}/m}$, where k_{eff} is given by (*a*) $k_{eff} = k_1 + k_2$, and (*b*) $1/k_{eff} = 1/k_1 + 1/k_2$. *Hint: Find the magnitude of the net force F on the object for a small displacement x and write* $F = -k_{eff}x$. *Note that in Part* (*b*) *the springs stretch by different amounts, the sum of which is x.*

Picture the Problem Choose a coordinate system in which the +*x* direction is to the right and assume that the object is displaced to the right. In case (*a*), note that the two springs undergo the same displacement whereas in (*b*) they experience the same force.

(*a*) Express the net force acting on the object:

$$F_{net} = -k_1x - k_2x = -(k_1 + k_2)x = -k_{eff}x$$

where $k_{eff} = \boxed{k_1 + k_2}$

(*b*) Express the force acting on each spring and solve for x_2:

$$F = -k_1x_1 = -k_2x_2 \Rightarrow x_2 = \frac{k_1}{k_2}x_1$$

Express the total extension of the springs:

$$x_1 + x_2 = -\frac{F}{k_{eff}}$$

Solving for k_{eff} yields:

$$k_{eff} = -\frac{F}{x_1 + x_2} = -\frac{-k_1x_1}{x_1 + x_2} = \frac{k_1x_1}{x_1 + \frac{k_1}{k_2}x_1} = \frac{1}{\frac{1}{k_1} + \frac{1}{k_2}}$$

Take the reciprocal of both sides of the equation to obtain:

$$\frac{1}{k_{eff}} = \boxed{\frac{1}{k_1} + \frac{1}{k_2}}$$

105 ••• **[SSM]** In this problem, derive the expression for the average power delivered by a driving force to a driven oscillator (Figure 14-39).

(*a*) Show that the instantaneous power input of the driving force is given by

$$P = Fv = -A\,\omega F_0 \cos \omega t \sin(\omega t - \delta).$$

(*b*) Use the identity $\sin(\theta_1 - \theta_2) = \sin\theta_1\cos\theta_2 - \cos\theta_1\sin\theta_2$ to show that the equation in (*a*) can be written as

$$P = A\omega F_0 \sin\delta\cos^2\omega t - A\omega F_0\cos\delta\cos\omega t\sin\omega t$$

(*c*) Show that the average value of the second term in your result for (*b*) over one or more periods is zero, and that therefore $P_{av} = \frac{1}{2}A\omega F_0\sin\delta$.

(*d*) From Equation 14-56 for tan δ, construct a right triangle in which the side opposite the angle δ is $b\omega$ and the side adjacent is $m(\omega_0^2 - \omega^2)$, and use this triangle to show that

$$\sin\delta = \frac{b\omega}{\sqrt{m^2(\omega_0^2-\omega^2)^2 + b^2\omega^2}} = \frac{b\omega A}{F_0}.$$

(*e*) Use your result for Part (*d*) to eliminate ωA from your result for Part (*c*) so that the average power input can be written as

$$P_{av} = \frac{1}{2}\frac{F_0^2}{b}\sin^2\delta = \frac{1}{2}\left[\frac{b\omega^2F_0^2}{m^2(\omega_0^2-\omega^2)^2 + b^2\omega^2}\right].$$

Picture the Problem We can follow the step-by-step instructions provided in the problem statement to obtain the desired results.

(*a*) Express the average power delivered by a driving force to a driven oscillator:

$$P = \vec{F}\cdot\vec{v} = Fv\cos\theta$$

or, because θ is 0°,

$$P = Fv$$

Express F as a function of time:

$$F = F_0\cos\omega t$$

Express the position of the driven oscillator as a function of time:

$$x = A\cos(\omega t - \delta)$$

Differentiate this expression with respect to time to express the velocity of the oscillator as a function of time:

$$v = -A\omega\sin(\omega t - \delta)$$

Substitute to express the average power delivered to the driven oscillator:

$$P = (F_0\cos\omega t)[-A\omega\sin(\omega t - \delta)]$$
$$= \boxed{-A\omega F_0\cos\omega t\sin(\omega t - \delta)}$$

(*b*) Expand $\sin(\omega t - \delta)$ to obtain:

$$\sin(\omega t - \delta) = \sin\omega t\cos\delta - \cos\omega t\sin\delta$$

Substitute in your result from (*a*) and simplify to obtain:

$$P = -A\omega F_0 \cos\omega t(\sin\omega t\cos\delta - \cos\omega t\sin\delta)$$
$$= \boxed{A\omega F_0 \sin\delta\cos^2\omega t - A\omega F_0\cos\delta\cos\omega t\sin\omega t}$$

(*c*) Integrate $\sin\theta\cos\theta$ over one period to determine $\langle\sin\theta\cos\theta\rangle$:

$$\langle\sin\theta\cos\theta\rangle = \frac{1}{2\pi}\left[\int_0^{2\pi}\sin\theta\cos\theta d\theta\right] = \frac{1}{2\pi}\left[\frac{1}{2}\sin^2\theta\Big|_0^{2\pi}\right] = 0$$

Integrate $\cos^2\theta$ over one period to determine $\langle\cos^2\theta\rangle$:

$$\langle\cos^2\theta\rangle = \frac{1}{2\pi}\int_0^{2\pi}\cos^2\theta d\theta = \frac{1}{2\pi}\left[\frac{1}{2}\int_0^{2\pi}(1+\cos 2\theta)d\theta\right] = \frac{1}{2\pi}\left[\tfrac{1}{2}\int_0^{2\pi}d\theta + \tfrac{1}{2}\int_0^{2\pi}\cos 2\theta d\theta\right] = \frac{1}{2\pi}(\pi + 0) = \frac{1}{2}$$

Substitute and simplify to express P_{av}:

$$P_{av} = A\omega F_0\sin\delta\langle\cos^2\omega t\rangle - A\omega F_0\cos\delta\langle\cos\omega t\sin\omega t\rangle = \tfrac{1}{2}A\omega F_0\sin\delta - A\omega F_0\cos\delta(0) = \boxed{\tfrac{1}{2}A\omega F_0\sin\delta}$$

(*d*) Construct a triangle that is consistent with

$$\tan\delta = \frac{b\omega}{m(\omega_0^2 - \omega^2)}:$$

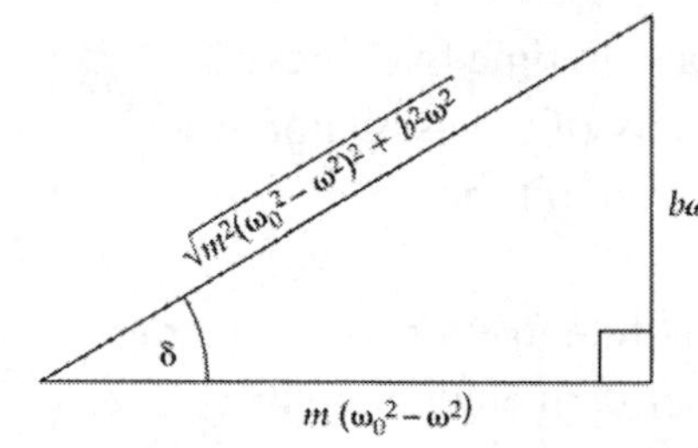

Using the triangle, express $\sin\delta$:

$$\sin\delta = \boxed{\frac{b\omega}{\sqrt{m^2(\omega_0^2 - \omega^2)^2 + b^2\omega^2}}}$$

Using Equation 14-56, reduce this expression to the simpler form:

$$\boxed{\sin\delta = \frac{b\omega A}{F_0}}$$

(*e*) Solve $\sin\delta = \dfrac{b\omega A}{F_0}$ for ω:

$$\omega = \frac{F_0}{bA}\sin\delta$$

Substitute in the expression for P_{av} to eliminate ω:

$$P_{av} = \boxed{\frac{F_0^2}{2b}\sin^2\delta}$$

Substitute for $\sin\delta$ from (*d*) to obtain:

$$P_{av} = \boxed{\frac{1}{2}\left[\frac{b\omega^2 F_0^2}{m^2\left(\omega_0^2 - \omega^2\right)^2 + b^2\omega^2}\right]}$$

Chapter 15
Traveling Waves

Conceptual Problems

1 • **[SSM]** A rope hangs vertically from the ceiling. A pulse is sent up the rope. Does the pulse travel faster, slower, or at a constant speed as it moves toward the ceiling? Explain your answer.

Determine the Concept The speed of a transverse wave on a rope is given by $v=\sqrt{F_T/\mu}$ where F_T is the tension in the rope and μ is its linear density. The waves on the rope move faster as they move toward the ceiling because the tension increases due to the weight of the rope below the pulse.

5 • **[SSM]** To keep all of the lengths of the treble strings (unwrapped steel wires) in a piano all about the same order of magnitude, wires of different linear mass densities are employed. Explain how this allows a piano manufacturer to use wires with lengths that are the same order of magnitude.

Determine the Concept The resonant (standing wave) frequencies on a string are inversely proportional to the square root of the linear density of the string $\left(f=\sqrt{T_T/\mu}/\lambda\right)$. Thus extremely high frequencies (which might otherwise require very long strings) can be accommodated on relatively short strings if the strings are linearly denser that the high frequency strings. High frequencies are not a problem as they utilize short strings anyway.

11 • **[SSM]** At a given location, two harmonic sound waves have the same amplitude, but the frequency of sound A is twice the frequency of sound B. How do their average energy densities compare? (*a*) The average energy density of A is twice the average energy density of B. (*b*) The average energy density of A is four times the average energy density of B. (*c*) The average energy density of A is 16 times the average energy density of B. (*d*) You cannot compare the average energy densities from the data given.

Determine the Concept The average energy density of a sound wave is given by $\eta_{av}=\frac{1}{2}\rho\omega^2 s_0^2$ where ρ is the average density of the medium, s_0 is the displacement amplitude of the molecules making up the medium, and ω is the angular frequency of the sound waves.

Express the average energy density of sound A:

$$\eta_{av,A}=\frac{1}{2}\rho_A\omega_A^2 s_{0,A}^2$$

The average energy density of sound B is given by:

$$\eta_{av,B}=\frac{1}{2}\rho_B\omega_B^2 s_{0,B}^2$$

Dividing the first of these equation by the second yields:

$$\frac{\eta_{\text{av, A}}}{\eta_{\text{av, B}}} = \frac{\frac{1}{2}\rho_A \omega_A^2 s_{0,A}^2}{\frac{1}{2}\rho_B \omega_B^2 s_{0,B}^2}$$

Because the sound waves are identical except for their frequencies:

$$\frac{\eta_{\text{av, A}}}{\eta_{\text{av, B}}} = \frac{\omega_A^2}{\omega_B^2} = \left(\frac{2\pi f_A}{2\pi f_B}\right)^2 = \left(\frac{f_A}{f_B}\right)^2$$

Because $f_A = 2f_B$:

$$\frac{\eta_{\text{av, A}}}{\eta_{\text{av, B}}} = \left(\frac{2f_B}{f_B}\right)^2 = 4 \Rightarrow \eta_{\text{av, A}} = 4\eta_{\text{av, B}}$$

$\boxed{(b)}$ is correct.

19 • **[SSM]** Sound waves in air encounter a 1.0-m wide door into a classroom. Due to the effects of refraction, the sound of which frequency is least likely to be heard by all the students in the room, assuming the room is full? (*a*) 600 Hz, (*b*) 300 Hz, (*c*) 100 Hz, (*d*) All the sounds are equally likely to be heard in the room. (*e*) Diffraction depends on wavelength not frequency, so you cannot tell from the data given.

Determine the Concept If the wavelength is large relative to the door, the diffraction effects are large and the waves spread out as they pass through the door. Because we're interested in sounds that are least likely to be heard everywhere in the room, we want the wavelength to be short and the frequency to be high. Hence $\boxed{(a)}$ is correct.

21 •• **[SSM]** Stars often occur in pairs revolving around their common center of mass. If one of the stars is a black hole, it is invisible. Explain how the existence of such a black hole might be inferred by measuring the Doppler frequency shift of the light observed from the other, visible star.

Determine the Concept The light from the companion star will be shifted about its mean frequency periodically due to the relative approach to and recession from the earth of the companion star as it revolves about the black hole.

Speed of Waves

33 •• **[SSM]** (*a*) Compute the derivative of the speed of a wave on a string with respect to the tension dv/dF_T, and show that the differentials dv and dF_T obey $dv/v = \frac{1}{2}dF_T/F_T$. (*b*) A wave moves with a speed of 300 m/s on a string that is under a tension of 500 N. Using to the differential approximation, estimate how much the tension must be changed to increase the speed to 312 m/s. (*c*) Calculate

ΔF_T exactly and compare it to the differential approximation result in Part (*b*). Assume that the string does not stretch with the increase in tension.

Picture the Problem (*a*) The speed of a transverse wave on a string is given by $v=\sqrt{F_T/\mu}$ where F_T is the tension in the wire and μ is its linear density. We can differentiate this expression with respect to F_T and then separate the variables to show that the differentials satisfy $dv/v=\frac{1}{2}dF_T/F_T$. (*b*) We'll approximate the differential quantities to determine by how much the tension must be changed to increase the speed of the wave to 312 m/s. (*c*) We can use $v=\sqrt{F_T/\mu}$ to obtain an exact expression for ΔF_T,

(*a*) Evaluate dv/dF_T:

$$\frac{dv}{dF}=\frac{d}{dF_T}\left[\sqrt{\frac{F_T}{\mu}}\right]=\frac{1}{2}\sqrt{\frac{1}{F_T\mu}}=\frac{1}{2}\cdot\frac{v}{F_T}$$

Separate the variables to obtain:

$$\frac{dv}{v}=\boxed{\frac{1}{2}\frac{dF_T}{F_T}}$$

(*b*) Solve the equation derived in Part (*a*) for dF_T:

$$dF_T=2F_T\frac{dv}{v}$$

Approximate dF_T with ΔF_T and dv with Δv to obtain:

$$\Delta F_T=2F_T\frac{\Delta v}{v}$$

Substitute numerical values and evaluate ΔF_T:

$$\Delta F_T=2(500\text{ N})\left(\frac{312\text{ m/s}-300\text{ m/s}}{300\text{ m/s}}\right)=\boxed{40\text{ N}}$$

(*c*) The exact value for $(\Delta F)_{\text{exact}}$ is given by:

$$(\Delta F)_{\text{exact}}=F_{T,2}-F_{T,1} \qquad (1)$$

Express the wave speeds for the two tensions:

$$v_1=\sqrt{\frac{F_{T,1}}{\mu}}\text{ and }v_2=\sqrt{\frac{F_{T,2}}{\mu}}$$

Dividing the second equation by the first and simplifying yields:

$$\frac{v_2}{v_1}=\frac{\sqrt{\frac{F_{T,2}}{\mu}}}{\sqrt{\frac{F_{T,1}}{\mu}}}=\sqrt{\frac{F_{T,2}}{F_{T,1}}}\Rightarrow F_{T,2}=F_{F,1}\left(\frac{v_2}{v_1}\right)^2$$

Substituting for $F_{T,2}$ in equation (1) yields:

$$(\Delta F_T)_{exact} = F_{F,1}\left(\frac{v_2}{v_1}\right)^2 - F_{T,1}$$

$$= F_{F,1}\left[\left(\frac{v_2}{v_1}\right)^2 - 1\right]$$

Substitute numerical values and evaluate $(\Delta F_T)_{exact}$:

$$(\Delta F_T)_{exact} = (500\text{ N})\left[\left(\frac{312\text{ m/s}}{300\text{ m/s}}\right)^2 - 1\right]$$

$$= \boxed{40.8\text{ N}}$$

The percent error between the exact and approximate values for ΔF_T is:

$$\frac{(\Delta F_T)_{exact} - \Delta F_T}{(\Delta F_T)_{exact}} = \frac{40.8\text{ N} - 40.0\text{ N}}{40.8\text{ N}}$$

$$\approx \boxed{2\%}$$

Harmonic Waves on a String

39 • **[SSM]** A harmonic wave on a string with a mass per unit length of 0.050 kg/m and a tension of 80 N, has an amplitude of 5.0 cm. Each point on the string moves with simple harmonic motion at a frequency of 10 Hz. What is the power carried by the wave propagating along the string?

Picture the Problem The average power propagated along the string by a harmonic wave is $P_{av} = \frac{1}{2}\mu\omega^2 A^2 v$, where v is the speed of the wave, and μ, ω, and A are the linear density of the string, the angular frequency of the wave, and the amplitude of the wave, respectively.

Express and evaluate the power propagated along the string:

$$P_{av} = \tfrac{1}{2}\mu\omega^2 A^2 v$$

The speed of the wave on the string is given by:

$$v = \sqrt{\frac{F_T}{\mu}}$$

Substitute for v to obtain:

$$P_{av} = \tfrac{1}{2}\mu\omega^2 A^2\sqrt{\frac{F_T}{\mu}}$$

Substitute numerical values and evaluate P_{av}:

$$P_{av} = \tfrac{1}{2}(4\pi^2)(0.050\,\text{kg/m})(10\,\text{s}^{-1})^2(0.050\,\text{m})^2\sqrt{\frac{80\,\text{N}}{0.05\,\text{kg/m}}} = \boxed{9.9\,\text{W}}$$

45 •• **[SSM]** Power is to be transmitted along a taut string by means of transverse harmonic waves. The wave speed is 10 m/s and the linear mass density of the string is 0.010 kg/m. The power source oscillates with an amplitude of 0.50 mm. (*a*) What average power is transmitted along the string if the frequency is 400 Hz? (*b*) The power transmitted can be increased by increasing the tension in the string, the frequency of the source, or the amplitude of the waves. By how much would each of these quantities have to increase to cause an increase in power by a factor of 100 if it is the only quantity changed?

Picture the Problem The average power propagated along a string by a harmonic wave is $P_{av} = \frac{1}{2}\mu\omega^2A^2v$ where v is the speed of the wave, and μ, ω, and A are the linear density of the string, the angular frequency of the wave, and the amplitude of the wave, respectively.

(*a*) Express the average power transmitted along the string:

$$P_{av} = \tfrac{1}{2}\mu\omega^2A^2v = 2\pi^2\mu f^2A^2v$$

Substitute numerical values and evaluate P_{av}:

$$P_{av} = 2\pi^2(0.010\,\text{kg/m})(400\,\text{s}^{-1})^2 \times(0.50\times10^{-3}\,\text{m})^2(10\,\text{m/s}) = \boxed{79\,\text{mW}}$$

(*b*) Because $P_{av} \propto f^2$, increasing f by a factor of 10 would increase P_{av} by a factor of 100.

Because $P_{av} \propto A^2$, increasing A by a factor of 10 would increase P_{av} by a factor of 100.

Because $P_{av} \propto v$ and $v \propto \sqrt{F}$, increasing F by a factor of 100 would increase v by a factor of 100 and P_{av} by a factor of 100.

Harmonic Sound Waves

49 • **[SSM]** (*a*) What is the displacement amplitude for a sound wave with a frequency of 100 Hz and a pressure amplitude of 1.00×10^{-4} atm? (*b*) The displacement amplitude of a sound wave of frequency 300 Hz is 1.00×10^{-7} m. Assuming the density of air is 1.29 kg/m^3, what is the pressure amplitude of this wave?

Picture the Problem The pressure amplitude depends on the density of the medium ρ, the angular frequency of the sound wave μ, the speed of the wave v, and the displacement amplitude s_0 according to $p_0 = \rho\omega v s_0$.

(*a*) Solve $p_0 = \rho\omega v s_0$ for s_0:

$$s_0 = \frac{p_0}{\rho\omega v}$$

Substitute numerical values and evaluate s_0:

$$s_0 = \frac{(1.00\times10^{-4}\text{ atm})(1.01325\times10^{5}\text{ Pa/atm})}{2\pi(1.29\text{ kg/m}^3)(100\text{ s}^{-1})(343\text{ m/s})} = 3.64\times10^{-5}\text{ m} = \boxed{36.4\ \mu\text{m}}$$

(*b*) Use $p_0 = \rho\omega v s_0$ to find p_0:

$$p_0 = 2\pi(1.29\,\text{kg/m}^3)(300\,\text{s}^{-1})(343\,\text{m/s})(1.00\times10^{-7}\text{ m}) = \boxed{83.4\,\text{mPa}}$$

Waves in Three Dimensions: Intensity

55 • **[SSM]** A loudspeaker at a rock concert generates a sound that has an intensity level equal to 1.00×10^{-2} W/m^2 at 20.0 m and has a frequency of 1.00 kHz. Assume that the speaker spreads its energy uniformly in three dimensions. (*a*) What is the total acoustic power output of the speaker? (*b*) At what distance will the sound intensity be at the pain threshold of 1.00 W/m^2? (*c*) What is the sound intensity at 30.0 m?

Picture the Problem Because the power radiated by the loudspeaker is the product of the intensity of the sound and the area over which it is distributed, we can use this relationship to find the average power, the intensity of the radiation, or the distance to the speaker for a given intensity or average power.

(*a*) Use $P_{\text{av}} = 4\pi r^2 I$ to find the total acoustic power output of the speaker:

$$P_{\text{av}} = 4\pi(20.0\,\text{m})^2(1.00\times10^{-2}\text{ W/m}^2) = 50.27\text{ W} = \boxed{50.3\text{ W}}$$

(*b*) Relate the intensity of the sound at 20 m to the distance from the speaker:

$$1.00\times10^{-2}\ \text{W/m}^2=\frac{P_{\text{av}}}{4\pi(20.0\ \text{m})^2}$$

Relate the threshold-of-pain intensity to the distance from the speaker:

$$1.00\ \text{W/m}^2=\frac{P_{\text{av}}}{4\pi r^2}$$

Divide the first of these equations by the second and solve for r:

$$r=\sqrt{(1.00\times10^{-2})(20.0\,\text{m})^2}=\boxed{2.00\,\text{m}}$$

(*c*) Use $I=\dfrac{P_{\text{av}}}{4\pi r^2}$ to find the intensity at 30.0 m:

$$I(30.0\,\text{m})=\frac{50.3\ \text{W}}{4\pi(30.0\,\text{m})^2}$$

$$=\boxed{4.45\times10^{-3}\ \text{W/m}^2}$$

*Intensity Level

57 • **[SSM]** What is the intensity level in decibels of a sound wave of intensity (*a*) 1.00×10^{-10} W/m² and (*b*) 1.00×10^{-2} W/m²?

Picture the Problem The intensity level β of a sound wave, measured in decibels, is given by $\beta=(10\,\text{dB})\log(I/I_0)$ where $I_0 = 10^{-12}$ W/m² is defined to be the threshold of hearing.

(*a*) Using its definition, calculate the intensity level of a sound wave whose intensity is 1.00×10^{-10} W/m²:

$$\beta=(10\,\text{dB})\log\left(\frac{1.00\times10^{-10}\ \text{W/m}^2}{10^{-12}\ \text{W/m}^2}\right)$$

$$=10\log10^2=\boxed{20.0\,\text{dB}}$$

(*b*) Proceed as in (*a*) with $I = 1.00 \times 10^{-2}$ W/m²:

$$\beta=(10\,\text{dB})\log\left(\frac{1.00\times10^{-2}\ \text{W/m}^2}{10^{-12}\ \text{W/m}^2}\right)$$

$$=10\log10^{10}=\boxed{100\,\text{dB}}$$

67 ••• **[SSM]** The noise intensity level at some location in an empty examination hall is 40 dB. When 100 students are writing an exam, the noise level at that location increases to 60 dB. Assuming that the noise produced by each student contributes an equal amount of acoustic power, find the noise intensity level at that location when 50 students have left.

Picture the Problem Because the sound intensities are additive, we'll find the noise intensity level due to one student by subtracting the background noise intensity from the intensity due to the students and dividing by 100. Then, we'll use this result to calculate the intensity level due to 50 students.

Express the intensity level due to 50 students:

$$\beta_{50} = (10\,\text{dB})\log\left(\frac{50I_1}{I_0}\right)$$

Find the sound intensity when 100 students are writing the exam:

$$60\,\text{dB} = (10\,\text{dB})\log\left(\frac{I_{100}}{I_0}\right)$$

and

$$I_{100} = 10^6 I_0 = 10^{-6}\ \text{W/m}^2$$

Find the sound intensity due to the background noise:

$$40\,\text{dB} = (10\,\text{dB})\log\left(\frac{I_{\text{background}}}{I_0}\right)$$

and

$$I_{\text{background}} = 10^4 I_0 = 10^{-8}\ \text{W/m}^2$$

Express the sound intensity due to the 100 students:

$$I_{100} - I_{\text{background}} = 10^{-6}\ \text{W/m}^2 - 10^{-8}\ \text{W/m}^2 \approx 10^{-6}\ \text{W/m}^2$$

Find the sound intensity due to 1 student:

$$\frac{I_{100} - I_{\text{background}}}{100} = 10^{-8}\ \text{W/m}^2$$

Substitute numerical values and evaluate the noise intensity level due to 50 students:

$$\beta_{50} = (10\,\text{dB})\log\frac{50(1.00\times10^{-8}\ \text{W/m}^2)}{10^{-12}\ \text{W/m}^2} = \boxed{57\,\text{dB}}$$

String Waves Experiencing Speed Changes

69 • **[SSM]** Consider a taut string with a mass per unit length μ_1, carrying transverse wave pulses that are incident upon a point where the string connects to a second string with a mass per unit length μ_2. (*a*) Show that if $\mu_2 = \mu_1$, then the reflection coefficient r equals zero and the transmission coefficient τ equals +1. (*b*) Show that if $\mu_2 >> \mu_1$, then $r \approx -1$ and $\tau \approx 0$; and (*c*) if $\mu_2 << \mu_1$ then $r \approx +1$ and $\tau \approx +2$.

Picture the Problem We can use the definitions of the reflection and transmission coefficients and the expression for the speed of waves on a string (Equation 15-3) to *r* and *t* in terms of the linear densities of the strings.

(*a*) Use their definitions to express the reflection and transmission coefficients:

$$r = \frac{v_2 - v_1}{v_2 + v_1} = \frac{1 - \frac{v_1}{v_2}}{1 + \frac{v_1}{v_2}} \quad (1)$$

and

$$\tau = \frac{2v_2}{v_2 + v_1} = \frac{2}{1 + \frac{v_1}{v_2}} \quad (2)$$

Use Equation 15-3 to express v_2 and v_1:

$$v_2 = \sqrt{\frac{F_T}{\mu_2}} \text{ and } v_1 = \sqrt{\frac{F_T}{\mu_1}}$$

Dividing the expression for v_1 by the expression for v_2 and simplifying yields:

$$\frac{v_1}{v_2} = \frac{\sqrt{\frac{F_T}{\mu_1}}}{\sqrt{\frac{F_T}{\mu_2}}} = \sqrt{\frac{\mu_2}{\mu_1}}$$

Substitute for $\frac{v_1}{v_2}$ in equation (1) to obtain:

$$r = \frac{1 - \sqrt{\frac{\mu_2}{\mu_1}}}{1 + \sqrt{\frac{\mu_2}{\mu_1}}} \quad (3)$$

Substitute for $\frac{v_1}{v_2}$ in equation (2) to obtain:

$$\tau = \frac{2}{1 + \sqrt{\frac{\mu_2}{\mu_1}}} \quad (4)$$

If $\mu_2 = \mu_1$:

$$r = \frac{1 - \sqrt{1}}{1 + \sqrt{1}} = \boxed{0}$$

and

$$\tau = \frac{2}{1 + \sqrt{\frac{\mu_2}{\mu_1}}} = \frac{2}{1 + \sqrt{1}} = \boxed{1}$$

(*b*) From equations (1) and (2), if $\mu_2 >> \mu_1$ then $v_1 >> v_2$:

$$r = \frac{v_2 - v_1}{v_2 + v_1} \approx \frac{-v_1}{v_1} = \boxed{-1}$$

and

$$\tau = \frac{2v_2}{v_2 + v_1} = \frac{2}{1 + \frac{v_1}{v_2}} \approx \boxed{0}$$

(*c*) If $\mu_2 << \mu_1$, then $v_2 >> v_1$:

$$r = \frac{v_2 - v_1}{v_2 + v_1} = \frac{1 - \frac{v_1}{v_2}}{1 + \frac{v_1}{v_2}} \approx \boxed{1}$$

and

$$\tau = \frac{2v_2}{v_2 + v_1} = \frac{2}{1 + \frac{v_1}{v_2}} \approx \boxed{2}$$

The Doppler Effect

79 •• **[SSM]** The Doppler effect is routinely used to measure the speed of winds in storm systems. As the manager of a weather monitoring station in the Midwest, you are using a Doppler radar system that has a frequency of 625 MHz to bounce a radar pulse off of the raindrops in a swirling thunderstorm system 50 km away. You measure the reflected radar pulse to be up-shifted in frequency by 325 Hz. Assuming the wind is headed directly toward you, how fast are the winds in the storm system moving? *Hint: The radar system can only measure the component of the wind velocity along its "line of sight."*

Picture the Problem Because the wind is moving toward the weather station (radar device), the frequency f_r the raindrops receive will be greater than the frequency emitted by the radar device. The radar waves reflected from the raindrops, moving toward the stationary detector at the weather station, will be of a still higher frequency f_r'. We can use the Doppler shift equations to derive an expression for the radial speed u of the wind in terms of difference of these frequencies.

Use Equation 15-41*a* to express the frequency f_r received by the raindrops in terms of f_s, u_r, and c:

$$f_r = \frac{c \pm u_r}{c \pm u_s} f_s = \left(\frac{c + u_r}{c \pm 0}\right) f_s \quad (1)$$

Express the frequency f_r' received by the stationary source at the weather station in terms of f_r, u_r, and c:

$$f_r' = \frac{c \pm u_r}{c \pm u_s} f_r = \left(\frac{c + u_r}{c \pm 0}\right) f_r \quad (2)$$

Substitute equation (1) in equation (2) to eliminate f_r:

$$f_r' = \left(\frac{c+u_r}{c}\right)^2 f_s = \left(1+\frac{u_r}{c}\right)^2 f_s$$

$$\approx \left(1+\frac{2u_r}{c}\right) f_s \text{ provided } u_r << c.$$

The frequency difference detected at the source is:

$$\Delta f = f_r' - f_s = \left(1+\frac{2u_r}{c}\right) f_s - f_s$$

$$= \frac{2u_r}{c} f_s$$

Solving for u_r yields:

$$u_r = \frac{c}{2f_s}\Delta f$$

Substitute numerical values and evaluate u_r:

$$u_r = \frac{2.998\times10^8 \text{ m/s}}{2(625\,\text{MHz})}(325\,\text{Hz})$$

$$= 77.95 \text{ m/s} \times \frac{1\,\text{mi/h}}{0.4470\,\text{m/s}}$$

$$= \boxed{174\,\text{mi/h}}$$

83 •• **[SSM]** A sound source of frequency f_0 moves with speed u_s relative to still air toward a receiver who is moving with speed u_r relative to still air away from the source. (*a*) Write an expression for the received frequency f'. (*b*) Use the result that $(1-x)^{-1} \approx 1+x$ to show that if both u_s and u_r are small compared to v, then the received frequency is approximately $f' = \left(1+\frac{u_{rel}}{v}\right) f_s$, where $u_{rel} = u_s - u_r$ is the velocity of the source relative to the receiver.

Picture the Problem The received and transmitted frequencies are related through $f_r = \frac{v \pm u_r}{v \pm u_s} f_s$ (Equation 15-41a), where the variables have the meanings given in the problem statement. Because the source and receiver are moving in the same direction, we use the minus signs in both the numerator and denominator.

(*a*) Relate the received frequency f_r to the frequency f_s of the source:

$$f_r = \frac{v \pm u_r}{v \pm u_s} f_s = \frac{1-\frac{u_r}{v}}{1-\frac{u_s}{v}} f_s$$

$$= \boxed{\left(1-\frac{u_r}{v}\right)\left(1-\frac{u_s}{v}\right)^{-1} f_s}$$

(*b*) Expand $(1-u_s/v)^{-1}$ binomially and discard the higher-order terms:

$$\left(1-\frac{u_s}{v}\right)^{-1} \approx 1+\frac{u_s}{v}$$

Substitute to obtain:

$$f_r = \left(1-\frac{u_r}{v}\right)\left(1+\frac{u_s}{v}\right)f_s$$

$$= \left[1+\frac{u_s}{v}-\frac{u_r}{v}-\left(\frac{u_r}{v}\right)\left(\frac{u_s}{v}\right)\right]f_s$$

$$\approx \left(1+\frac{u_s-u_r}{v}\right)f_s$$

because both u_s and u_r are small compared to v.

Because $u_{rel} = u_s - u_r$

$$\boxed{f_r' \approx \left(1+\frac{u_{rel}}{v}\right)f_s}$$

General Problems

91 • **[SSM]** A whistle that has a frequency of 500 Hz moves in a circle of radius 1.00 m at 3.00 rev/s. What are the maximum and minimum frequencies heard by a stationary listener in the plane of the circle and 5.00 m away from its center?

Picture the Problem The diagram depicts the whistle traveling in a circular path of radius r = 1.00 m. The stationary listener will hear the maximum frequency when the whistle is at point 1 and the minimum frequency when it is at point 2. These maximum and minimum frequencies are determined by f_0 and the tangential speed $u_s = 2\pi\, r/T$. We can relate the frequencies heard at point P to the speed of the approaching whistle at point 1 and the speed of the receding whistle at point 2.

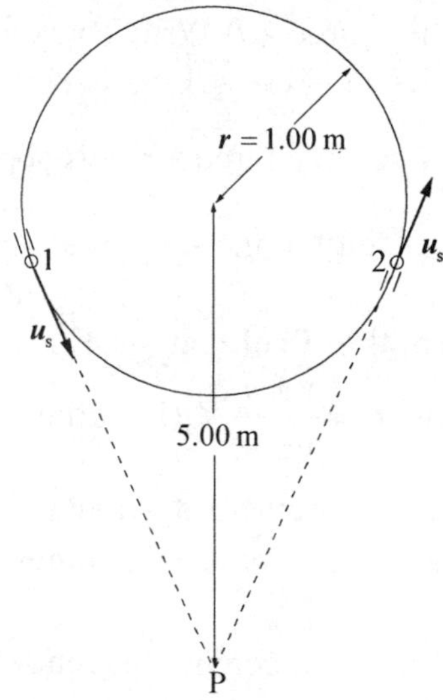

Relate the frequency heard at point P to the speed of the approaching whistle at point 1:

$$f_{max} = \frac{1}{1-\frac{u_s}{v}}f_s$$

Because $u_s = r\omega$:

$$f_{max} = \frac{1}{1-\frac{r\omega}{v}} f_s$$

Substitute numerical values and evaluate f_{max}:

$$f_{max} = \frac{1}{1-\frac{(1.00\,\text{m})\left(3.00\frac{\text{rev}}{\text{s}}\times\frac{2\pi\,\text{rad}}{\text{rev}}\right)}{343\,\text{m/s}}}(500\,\text{Hz}) = \boxed{529\,\text{Hz}}$$

Relate the frequency heard at point P to the speed of the receding whistle at point 2:

$$f_{min} = \frac{1}{1+\frac{u_s}{v}} f_s$$

Substitute numerical values and evaluate f_{min}:

$$f_{min} = \frac{1}{1+\frac{(1.00\,\text{m})\left(3.00\frac{\text{rev}}{\text{s}}\times\frac{2\pi\,\text{rad}}{\text{rev}}\right)}{343\,\text{m/s}}}(500\,\text{Hz}) = \boxed{474\,\text{Hz}}$$

95 •• **[SSM]** A loudspeaker driver 20.0 cm in diameter is vibrating at 800 Hz with an amplitude of 0.0250 mm. Assuming that the air molecules in the vicinity have the same amplitude of vibration, find (*a*) the pressure amplitude immediately in front of the driver, (*b*) the sound intensity, and (*c*) the acoustic power being radiated by the front surface of the driver.

Picture the Problem (*a*) and (*b*) The pressure amplitude can be calculated directly from $p_0 = \rho\omega v s_0$, and the intensity from $I = \frac{1}{2}\rho\omega^2 s_0^2 v$. (*c*) The power radiated is the intensity times the area of the driver.

(*a*) Relate the pressure amplitude to the displacement amplitude, angular frequency, wave velocity, and air density:

$$p_0 = \rho\omega v s_0$$

Substitute numerical values and evaluate p_0:

$$p_0 = \left(1.29\,\text{kg/m}^3\right)\left[2\pi\left(800\,\text{s}^{-1}\right)\right]\times\left(343\,\text{m/s}\right)\left(0.0250\times10^{-3}\,\text{m}\right) = \boxed{55.6\,\text{N/m}^2}$$

(*b*) Relate the intensity to these same quantities:

$$I = \tfrac{1}{2}\rho\omega^2 s_0^2 v$$

Substitute numerical values and evaluate *I*:

$$I = \tfrac{1}{2}\left(1.29\,\text{kg/m}^3\right)\left[2\pi\left(800\,\text{s}^{-1}\right)\right]^2 \times \left(0.0250\times10^{-3}\,\text{m}\right)^2\left(343\,\text{m/s}\right) = 3.494\,\text{W/m}^2 = \boxed{3.49\,\text{W/m}^2}$$

(*c*) Express the power in terms of the intensity and the area of the driver:

$$P = IA = \pi r^2 I$$

Substitute numerical values and evaluate *P*:

$$P = \pi\left(0.100\,\text{m}\right)^2\left(3.494\,\text{W/m}^2\right) = \boxed{0.110\,\text{W}}$$

Chapter 16
Superposition and Standing Waves

Conceptual Problems

1 • **[SSM]** Two rectangular wave pulses are traveling in opposite directions along a string. At $t = 0$, the two pulses are as shown in Figure 16-29. Sketch the wave functions for $t = 1.0$, 2.0, and 3.0 s.

Picture the Problem We can use the speeds of the pulses to determine their positions at the given times.

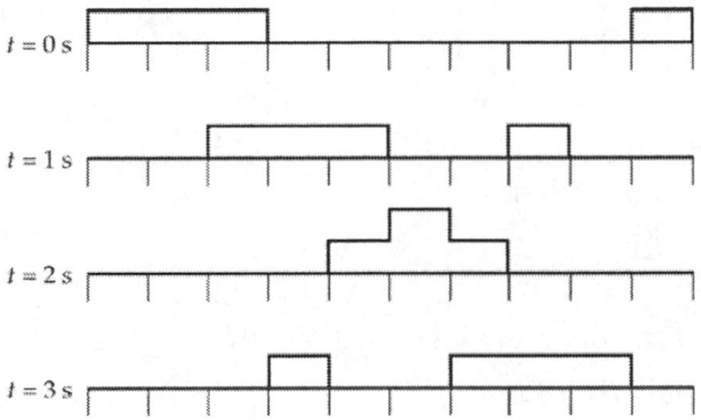

11 • **[SSM]** An organ pipe that is open at both ends has a fundamental frequency of 400 Hz. If one end of this pipe is now stopped, the fundamental frequency is (*a*) 200 Hz, (*b*) 400 Hz, (*c*) 546 Hz, (*d*) 800 Hz.

Picture the Problem The first harmonic displacement-wave pattern in an organ pipe open at both ends (open-open) and vibrating in its fundamental mode is represented in part (*a*) of the diagram. Part (*b*) of the diagram shows the wave pattern corresponding to the fundamental frequency for a pipe of the same length L that is stopped. Letting unprimed quantities refer to the open pipe and primed quantities refer to the stopped pipe, we can relate the wavelength and, hence, the frequency of the fundamental modes using $v = f\lambda$.

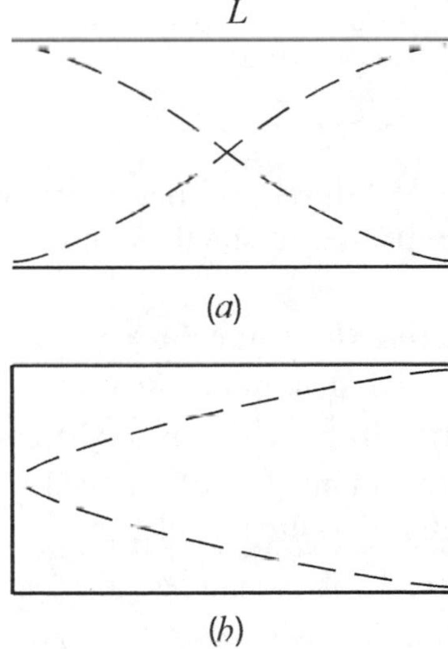

Express the frequency of the first harmonic in the open-open pipe in terms of the speed and wavelength of the waves:

$$f_1 = \frac{v}{\lambda_1}$$

Relate the length of the open pipe to the wavelength of the fundamental mode:

$$\lambda_1 = 2L$$

Substitute for λ_1 to obtain:

$$f_1 = \frac{v}{2L}$$

Express the frequency of the first harmonic in the closed pipe in terms of the speed and wavelength of the waves:

$$f_1' = \frac{v}{\lambda_1'}$$

Relate the length of the open-closed pipe to the wavelength of it's fundamental mode:

$$\lambda_1' = 4L$$

Substitute for λ_1' to obtain:

$$f_1' = \frac{v}{4L} = \frac{1}{2}\left(\frac{v}{2L}\right) = \frac{1}{2}f_1$$

Substitute numerical values and evaluate f_1':

$$f_1' = \frac{1}{2}(400\text{ Hz}) = 200\text{ Hz}$$

and $\boxed{(a)}$ is correct.

13 •• **[SSM]** Explain how you might use the resonance frequencies of an organ pipe to estimate the temperature of the air in the pipe.

Determine the Concept You could measure the frequencies at which resonance occurs and determine their mode from the standing-wave pattern (open-open supports all the harmonics whereas the resonance frequencies of an open-closed pipe are odd integers of the fundamental frequency). Using one of the modes, you can determine the wavelength from the tube length, and then use $v = f\lambda$ to find the speed of sound at the ambient temperature. Finally you could use $v = \sqrt{\gamma RT/M}$,where γ= 1.4 for a diatomic gas such as air, M is the molar mass of air, R is the universal gas constant, and T is the absolute temperature, to find the temperature of the air in the pipe.

15 •• **[SSM]** (*a*) When a guitar string is vibrating in its fundamental mode, is the wavelength of the sound it produces in air typically the same as the wavelength of the standing wave on the string? Explain. (*b*) When an organ pipe is in any one of its standing wave modes, is the wavelength of the traveling sound wave it produces in air typically the same as the wavelength of the standing sound wave in the pipe? Explain.

Determine the Concept
(*a*) No; the wavelength of a wave is related to its frequency and speed of propagation ($\lambda = v/f$). The frequency of the plucked string will be the same as the frequency of the wave it produces in air, but the speeds of the waves depend on the media in which they are propagating. Because the velocities of propagation differ, the wavelengths will not be the same.

(*b*) Yes. Because both the standing waves in the pipe and the traveling waves have the same speed and frequency, they must have the same wavelength.

Superposition and Interference

25 • **[SSM]** Two harmonic waves having the same frequency, wave speed and amplitude are traveling in the same direction and in the same propagating medium. In addition, they overlap each other. If they differ in phase by $\pi/2$ and each has an amplitude of 0.050 m, what is the amplitude of the resultant wave?

Picture the Problem We can use $A = 2y_0 \cos\frac{1}{2}\delta$ to find the amplitude of the resultant wave.

Evaluate the amplitude of the resultant wave when $\delta = \pi/2$:

$$A = 2y_0 \cos\tfrac{1}{2}\delta = 2(0.050\,\text{m})\cos\frac{1}{2}\left(\frac{\pi}{2}\right)$$

$$= \boxed{7.1\,\text{cm}}$$

29 • **[SSM]** Two speakers separated by some distance emit sound waves of the same frequency. At some point *P*, the intensity due to each speaker separately is I_0. The distance from *P* to one of the speakers is $\frac{1}{2}\lambda$ longer than that from *P* to the other speaker. What is the intensity at *P* if (*a*) the speakers are coherent and in phase, (*b*) the speakers are incoherent, and (*c*) the speakers are coherent 180° out of phase?

Picture the Problem The intensity at the point of interest is dependent on whether the speakers are coherent and on the total phase difference in the waves arriving at the given point. We can use $\delta = 2\pi\dfrac{\Delta x}{\lambda}$ to determine the phase difference δ, $A=\left|2p_0\cos\frac{1}{2}\delta\right|$ to find the amplitude of the resultant wave, and the fact that the intensity I is proportional to the square of the amplitude to find the intensity at P for the given conditions.

(*a*) Find the phase difference δ:

$$\delta=2\pi\frac{\frac{1}{2}\lambda}{\lambda}=\pi$$

Find the amplitude of the resultant wave:

$$A=\left|2p_0\cos\tfrac{1}{2}\pi\right|=0$$

Because the intensity is proportional to A^2:

$$I=\boxed{0}$$

(*b*) The sources are incoherent and the intensities add:

$$I=\boxed{2I_0}$$

(*c*) The total phase difference is the sum of the phase difference of the sources and the phase difference due to the path difference:

$$\begin{aligned}\delta_{\text{tot}}&=\delta_{\text{sources}}+\delta_{\text{path difference}}\\&=\pi+2\pi\frac{\Delta x}{\lambda}=\pi+2\pi\left(\frac{1}{2}\right)\\&=2\pi\end{aligned}$$

Find the amplitude of the resultant wave:

$$A=\left|2p_0\cos\tfrac{1}{2}(2\pi)\right|=2p_0$$

Because the intensity is proportional to A^2:

$$I=\frac{A^2}{p_0^2}I_0=\frac{(2p_0)^2}{p_0^2}I_0=\boxed{4I_0}$$

33 •• **[SSM]** Sound source A is located at $x = 0$, $y = 0$, and sound source B is located at $x = 0$, $y = 2.4$ m. The two sources radiate coherently and in phase. An observer at $x = 40$ m, $y = 0$ notes that as he takes a few steps from $y = 0$ in either the $+y$ or $-y$ direction, the sound intensity diminishes. What is the lowest frequency and the next to lowest frequency of the sources that can account for that observation?

Picture the Problem Because the sound intensity diminishes as the observer moves, parallel to a line through the sources, away from his initial position, we can conclude that his initial position is one at which there is constructive interference of the sound coming from the two sources. We can apply the condition for constructive interference to relate the wavelength of the sound to the path difference at his initial position and the relationship between the velocity, frequency, and wavelength of the waves to express this path difference in terms of the frequency of the sources.

Express the condition for constructive interference at (15.0 m, 0):

$$\Delta r = n\lambda,\ n = 1,2,3,... \qquad (1)$$

Express the path difference Δr:

$$\Delta r = r_B - r_A$$

Using the Pythagorean theorem, express r_B:

$$r_B = \sqrt{(15.0\,\text{m})^2 + (2.4\,\text{m})^2}$$

Substitute for r_B to obtain:

$$\Delta r = \sqrt{(15.0\,\text{m})^2 + (2.4\,\text{m})^2} - 15.0\,\text{m}$$

Using $v = f\lambda$ and equation (1), express f_n in terms of Δr and n:

$$f_n = n\frac{v}{\Delta r}$$

Substituting numerical values yields:

$$f_n = n\frac{343\,\text{m/s}}{\sqrt{(15.0\,\text{m})^2 + (2.4\,\text{m})^2} - 15.0\,\text{m}} = (1.798\,\text{kHz})n$$

Evaluate f_n for $n = 1$ and 2:

$$f_1 = \boxed{2\,\text{kHz}} \text{ and } f_2 = \boxed{5\,\text{kHz}}$$

35 •• **[SSM]** Two harmonic water waves of equal amplitudes but different frequencies, wave numbers, and speeds are traveling in the same direction. In addition, they are superposed on each other. The total displacement of the wave can be written as $y(x,t) = A[\cos(k_1x - \omega_1t) + \cos(k_2x - \omega_2t)]$, where $\omega_1/k_1 = v_1$ (the speed of the first wave) and $\omega_2/k_2 = v_2$ (the speed of the second wave). (*a*) Show that $y(x,t)$ can be written in the form $y(x,t) = Y(x, t)\cos(k_{av}x - \omega_{av}t)$, where $\omega_{av} = (\omega_1 + \omega_2)/2$, $k_{av} = (k_1 + k_2)/2$, $Y(x, t) = 2A\cos[(\Delta k/2)x - (\Delta\omega/2)t]$, $\Delta\omega = \omega_1 - \omega_2$, and $\Delta k = k_1 - k_2$. The factor $Y(x, t)$ is called the *envelope* of the wave. (*b*) Let $A = 1.00$ cm, $\omega_1 = 1.00$ rad/s, $k_1 = 1.00\ \text{m}^{-1}$, $\omega_2 = 0.900$ rad/s, and $k_2 = 0.800\text{m}^{-1}$. Using a **spreadsheet program** or **graphing calculator**, make a plot of $y(x,t)$ versus x at $t = 0.00$ s for $0 < x < 5.00$ m. (*c*) Using a **spreadsheet program** or **graphing calculator**, make three plots of $Y(x,t)$ versus x for $-5.00\ \text{m} < x < 5.00$ m on the same graph. Make one plot for $t = 0.00$ s, the second for $t = 5.00$ s, and the third for $t = 10.00$ s. Estimate the speed at which the envelope moves from the three plots, and compare this estimate with the speed obtained using $v_{envelope} = \Delta\omega/\Delta k$.

Picture the Problem We can use the trigonometric identity

$$\cos A + \cos B = 2\cos\left(\frac{A+B}{2}\right)\cos\left(\frac{A-B}{2}\right)$$

to derive the expression given in (a) and the speed of the envelope can be found from the second factor in this expression; i.e., from $\cos[(\Delta k/2)x - (\Delta\omega/2)t]$.

(a) Express the amplitude of the resultant wave function $y(x,t)$:

$$y(x,t) = A(\cos(k_1x - \omega_1t) + \cos(k_2x - \omega_2t))$$

Use the trigonometric identity $\cos A + \cos B = 2\cos\left(\frac{A+B}{2}\right)\cos\left(\frac{A-B}{2}\right)$ to obtain:

$$y(x,t) = 2A\left[\cos\frac{k_1x - \omega_1t + k_2x - \omega_2t}{2}\cos\frac{k_1x - \omega_1t - k_2x + \omega_2t}{2}\right]$$

$$= 2A\left[\cos\left(\frac{k_1 + k_2}{2}x - \frac{\omega_1 + \omega_2}{2}t\right)\cos\left(\frac{k_1 - k_2}{2}x + \frac{\omega_2 - \omega_1}{2}t\right)\right]$$

Substitute $\omega_{ave} = (\omega_1 + \omega_2)/2$, $k_{ave} = (k_1 + k_2)/2$, $\Delta\omega = \omega_1 - \omega_2$ and $\Delta k = k_1 - k_2$ to obtain:

$$y(x,t) = 2A\left[\cos(k_{ave}x - \omega_{ave}t)\cos\left(\frac{\Delta k}{2}x - \frac{\Delta\omega}{2}t\right)\right]$$

$$= \boxed{Y(x,t)[\cos(k_{ave}x - \omega_{ave}t)]}$$

where

$$Y(x,t) = \boxed{2A\cos\left(\frac{\Delta k}{2}x - \frac{\Delta\omega}{2}t\right)}$$

(b) A spreadsheet program to calculate $y(x,t)$ between 0 m and 5.00 m at t = 0.00 s, follows. The constants and cell formulas used are shown in the table.

Cell	Content/Formula	Algebraic Form
A15	0	
A16	A15+0.25	$x + \Delta x$
B15	2*$B$2*COS(0.5*($B$3 –$B$4)*A15 –0.5*($B$5–$B$6)*$B$8)	$Y(x, 0.00\text{ s})$
C15	B15*COS(0.5*($B$3+$B$4)*A15 –0.5*($B$5+$B$6)*$B$7)	$y(x, 0.00\text{ s})$

	A	B	C
1			
2	$A=$	1	cm
3	$k_1=$	1	m^{-1}
4	$k_2=$	0.8	m^{-1}
5	$\omega_1=$	1	rad/s
6	$\omega_2=$	0.9	rad/s
7	$t=$	0.00	s
8			
9			
10			
11			
12	x	$Y(x,0)$	$y(x,0)$
13	(m)	(cm)	(cm)
14			
15	0.00	2.000	2.000
16	0.25	1.999	1.949
17	0.50	1.998	1.799
18	0.75	1.994	1.557
31	4.00	1.842	−1.652
32	4.25	1.822	−1.413
33	4.50	1.801	−1.108
34	4.75	1.779	−0.753
35	5.00	1.755	−0.370

A graph of $y(x,0)$ follows:

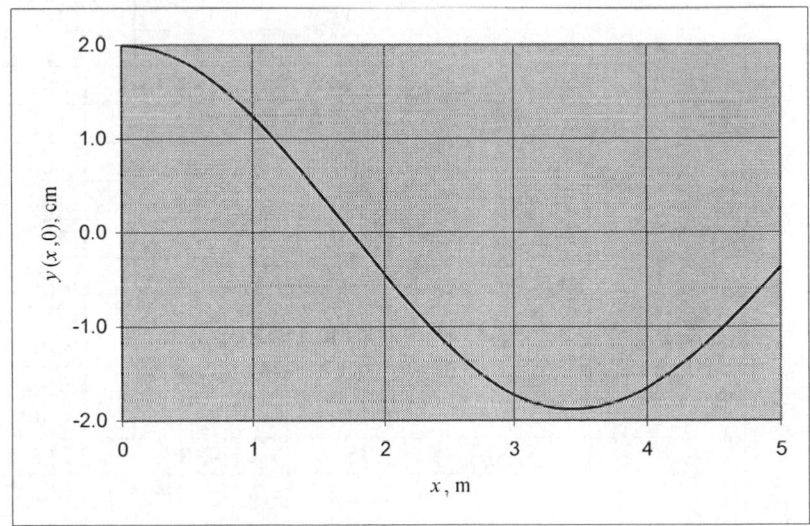

(*c*) A spreadsheet program to calculate $Y(x,t)$ for -5.00 m $< x <$ 5.00 m and $t = 0.00$ s, $t = 5.00$ s and $t = 10.00$ s follow: The constants and cell formulas used are shown in the table.

Cell	Content/Formula	Algebraic Form
A15	0	
A16	A15+0.25	$x + \Delta x$
B15	2*$B$2*COS(0.5*($B$3 –$B$4)*A15 –0.5*($B$5–$B$6)*$B$7)	$Y(x,0.00\text{ s})$
C15	2*$B$2*COS(0.5*($B$3 –$B$4)*A15 –0.5*($B$5–$B$6)*$B$8)	$Y(x,5.00\text{ s})$
D15	2*$B$2*COS(0.5*($B$3 –$B$4)*A15 –0.5*($B$5–$B$6)*$B$9)	$Y(x,10.00\text{ s})$

	A	B	C	D
1				
2	$A=$	1	cm	
3	$k_1=$	1	m^{-1}	
4	$k_2=$	0.8	m^{-1}	
5	$\omega_1=$	1	rad/s	
6	$\omega_2=$	0.9	rad/s	
7	$t=$	0	s	
8	$t=$	5	s	
9	$t=$	10	s	
10				
11				
12	x	$Y(x,0)$	$Y(x,5\text{ s})$	$Y(x,10\text{ s})$
13	(m)	(cm)	(cm)	(cm)
14				
15	–5.00	1.755	1.463	1.081
16	–4.75	1.779	1.497	1.122
17	–4.50	1.801	1.530	1.163
18	–4.25	1.822	1.561	1.204
19	–4.00	1.842	1.592	1.243
51	4.00	1.842	1.978	1.990
52	4.25	1.822	1.969	1.994
53	4.50	1.801	1.960	1.998
54	4.75	1.779	1.950	1.999
55	5.00	1.755	1.938	2.000

(*c*) Graphs of $Y(x,t)$ versus x for $-5.00\text{ m} < x < 5.00\text{ m}$ and $t = 0.00$ s, $t = 5.00$ s and $t = 10.00$ s follow:

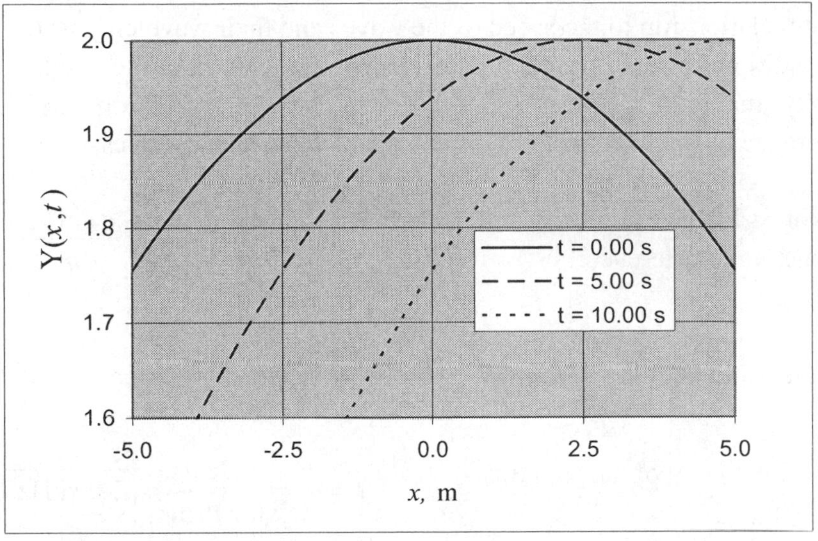

To estimate the speed of the envelope, we can use its horizontal displacement between $t = 0.00$ s and $t = 5.00$ s:

$$v_{\text{est}} = \frac{\Delta x}{\Delta t} \qquad (1)$$

From the graph we note that the wave traveled 2.5 m in 5.00 s:

$$v_{\text{est}} = \frac{2.50\text{ m}}{5.00\text{ s}} = \boxed{50\text{ cm/s}}$$

The speed of the envelope is given by:

$$v_{\text{envelope}} = \frac{\Delta\omega}{\Delta k} = \frac{\omega_1 - \omega_2}{k_1 - k_2}$$

Substitute numerical values and evaluate v_{envelope}:

$$v_{\text{envelope}} = \frac{1.00\text{ rad/s} - 0.900\text{ rad/s}}{1.00\text{ m}^{-1} - 0.800\text{ m}^{-1}} = 50\text{ cm/s}$$

in agreement with our graphical estimate.

39 ••• **[SSM]** Two sound sources driven in phase by the same amplifier are 2.00 m apart on the y axis, one at $y = +1.00$ m and the other at $y = -1.00$ m. At points large distances from the y axis, constructive interference is heard at at angles with the x axis of $\theta_0 = 0.000$ rad, $\theta_1 = 0.140$ rad and $\theta_2 = 0.283$ rad, and at no angles in between (see Figure 16-31). (*a*) What is the wavelength of the sound waves from the sources? (*b*) What is the frequency of the sources? (*c*) At what other angles is constructive interference heard? (*d*) What is the smallest angle for which the sound waves cancel?

Picture the Problem (*a*) Let d be the separation of the two sound sources. We can express the wavelength of the sound in terms of the d and either of the angles at which intensity maxima are heard. (*b*) We can find the frequency of the sources from its relationship to the speed of the waves and their wavelengths. (*c*) Using the condition for constructive interference, we can find the angles at which intensity maxima are heard. (*d*) We can use the condition for destructive interference to find the smallest angle for which the sound waves cancel.

(*a*) Express the condition for constructive interference:

$$d\sin\theta_m = m\lambda \Rightarrow \lambda = \frac{d\sin\theta_m}{m} \quad (1)$$

where $m = 0, 1, 2, 3,\ldots$

Evaluate λ for $m = 1$:

$$\lambda = (2.00\,\text{m})\sin(0.140\,\text{rad}) = \boxed{0.279\,\text{m}}$$

(*b*) The frequency of the sound is given by:

$$f = \frac{v}{\lambda} = \frac{343\,\text{m/s}}{0.279\,\text{m}} = \boxed{1.23\,\text{kHz}}$$

(*c*) Solve equation (1) for θ_m:

$$\theta_m = \sin^{-1}\left(\frac{m\lambda}{d}\right) = \sin^{-1}\left[\frac{m(0.279\,\text{m})}{2.00\,\text{m}}\right]$$
$$= \sin^{-1}[(0.1395)m]$$

The table shows the values for θ as a function of m:

m	θ_m
	(rad)
3	0.432
4	0.592
5	0.772
6	0.992
7	1.35
8	undefined

(*d*) Express the condition for destructive interference:

$$d\sin\theta_m = m\frac{\lambda}{2}$$

where $m = 1, 3, 5,\ldots$

Solving for θ_m yields:

$$\theta_m = \sin^{-1}\left(m\frac{\lambda}{2d}\right)$$

Evaluate this expression for $m = 1$:

$$\theta_1 = \sin^{-1}\left[\frac{0.279\,\text{m}}{2(2.00\,\text{m})}\right] = \boxed{0.0698\,\text{rad}}$$

Beats

43 ••• **[SSM]** A stationary police radar gun emits microwaves at 5.00 GHz. When the gun is aimed at a car, it superimposes the transmitted and reflected waves. Because the frequencies of these two waves differ, beats are generated, with the speed of the car proportional to the beat frequency. The speed of the car, 83 mi/h, appears on the display of the radar gun. Assuming the car is moving along the line-of-sight of the police officer, and using the Doppler shift equations, (*a*) show that, for a fixed radar gun frequency, the beat frequency is proportional to the speed of the car. *HINT: Car speeds are tiny compared to the speed of light.* (*b*) What is the beat frequency in this case? (*c*) What is the calibration factor for this radar gun? That is, what is the beat frequency generated per mi/h of speed?

Picture the Problem The microwaves strike the speeding car at frequency f_r .This frequency will be less than f_s if the car is moving away from the radar gun and greater than f_s if the car is moving toward the radar gun. The frequency shift is given by Equation 15-42 (the low-speed, relative to light, approximation). The car then acts as a moving source emitting waves of frequency f_r. The radar gun detects waves of frequency f_r' that are either greater than or less than f_r depending on the direction the car is moving. The total frequency shift is the sum of the two frequency shifts.

(*a*) Express the frequency difference Δf as the sum of the frequency difference $\Delta f_1 = f_r - f_s$ and the frequency difference $\Delta f_2 = f_r' - f_r$:

$$\Delta f = \Delta f_1 + \Delta f_2 \quad (1)$$

Using Equation 15-42, substitute for the frequency differences in equation (1):

$$\Delta f = -\frac{u}{c} f_s - \frac{u}{c} f_r = -\frac{u}{c}(f_s + f_r) \quad (2)$$

where $u = u_s \pm u_r = u_r$ is the speed of the source relative to the receiver.

Apply Equation 15-42 to Δf_1 to obtain:

$$\frac{\Delta f_1}{f_s} = \frac{f_r - f_s}{f_s} = -\frac{u_r}{c}$$

where we've used the minus sign because we know the frequency difference is a downshift.

Solving for f_r yields:

$$f_r = \left(1 - \frac{u_r}{c}\right) f_s$$

Substitute for f_r in equation (2) and simplify to obtain:

$$\Delta f = -\frac{u_r}{c}\left(f_s + \left(1 - \frac{u_r}{c}\right) f_s\right)$$
$$= -\frac{u_r}{c}\left(2 - \frac{u_r}{c}\right) f_s$$
$$= -2\frac{u_r}{c} + \left(\frac{u_r}{c}\right)^2 f_s$$

Because $\frac{u_r}{c} << 1$:

$$\Delta f \approx -2 f_s \frac{u_r}{c} \Rightarrow \Delta f \propto \boxed{u_r}$$

(*b*) Substitute numerical values and evaluate $|\Delta f|$:

$$|\Delta f| \approx 2(5.00 \times 10^9\ \text{Hz}) \frac{\left(83\,\text{mi/h} \times \frac{0.4470\ \text{m/s}}{1\ \text{mi/h}}\right)}{2.998 \times 10^8\ \text{m/s}} = \boxed{1.2\ \text{kHz}}$$

(*c*) The calibration factor is $\frac{1.24\ \text{kHz}}{83\ \text{mi/h}} = \boxed{15\ \text{Hz/mi/h}}$.

Standing Waves

47 • **[SSM]** A 5.00-g, 1.40-m long flexible wire has a tension of 968 N and is fixed at both ends. (*a*) Find the speed of transverse waves on the wire. (*b*) Find the wavelength and frequency of the fundamental. (*c*) Find the frequencies of the second and third harmonics.

Picture the Problem We can find the speed of transverse waves on the wire using $v = \sqrt{F_T/\mu}$ and the wavelengths of any harmonic from $L = n\frac{\lambda_n}{2}$, where $n = 1, 2, 3, \ldots$ We can use $v = f\lambda$ to find the frequency of the fundamental. For a wire fixed at both ends, the higher harmonics are integer multiples of the first harmonic (fundamental).

(*a*) Relate the speed of transverse waves on the wire to the tension in the wire and its linear density:

$$v = \sqrt{\frac{F_T}{\mu}} = \sqrt{\frac{F_T L}{m}}$$

Substitute numerical values and evaluate v:

$$v = \sqrt{\frac{968\,\text{N}}{(0.00500\,\text{kg})/(1.40\,\text{m})}} = \boxed{521\,\text{m/s}}$$

(*b*) Using the standing-wave condition for a wire fixed at both ends, relate the length of the wire to the wavelength of the harmonic mode in which it is vibrating:

$$L = n\frac{\lambda_n}{2},\ n = 1, 2, 3, \ldots$$

Solve for λ_1:

$$\lambda_1 = 2L = 2(1.40\,\text{m}) = \boxed{2.80\,\text{m}}$$

Express the frequency of the first harmonic in terms of the speed and wavelength of the waves:

$$f_1 = \frac{v}{\lambda_1} = \frac{521\,\text{m/s}}{2.80\,\text{m}} = \boxed{186\,\text{Hz}}$$

(*c*) Because, for a wire fixed at both ends, the higher harmonics are integer multiples of the first harmonic:

$$f_2 = 2f_1 = 2(186\,\text{Hz}) = \boxed{372\,\text{Hz}}$$

and

$$f_3 = 3f_1 = 3(186\,\text{Hz}) = \boxed{558\,\text{Hz}}$$

51 •• **[SSM]** The wave function $y(x,t)$ for a certain standing wave on a string fixed at both ends is given by $y(x,t) = 4.20\sin(0.200x)\cos(300t)$, where y and x are in centimeters and t is in seconds. (*a*) A standing wave can be considered as the superposition of two traveling waves. What are the wavelength and frequency of the two traveling waves that make up the specified standing wave? (*b*) What is the speed of these waves on this string? (*c*) If the string is vibrating in its fourth harmonic, how long is it?

Picture the Problem We can find λ and f by comparing the given wave function to the general wave function for a string fixed at both ends. The speed of the waves can then be found from $v = f\lambda$. We can find the length of the string from its fourth harmonic wavelength.

(*a*) Using the wave function, relate k and λ:

$$k = \frac{2\pi}{\lambda} \Rightarrow \lambda = \frac{2\pi}{k}$$

Substitute numerical values and evaluate λ:

$$\lambda = \frac{2\pi}{0.200\,\text{cm}^{-1}} = 10\pi\,\text{cm} = \boxed{31.4\,\text{cm}}$$

Using the wave function, relate f and ω:

$$\omega = 2\pi f \Rightarrow f = \frac{\omega}{2\pi}$$

Substitute numerical values and evaluate *f*:

$$f=\frac{300\,\text{s}^{-1}}{2\pi}=\boxed{47.7\,\text{Hz}}$$

(*b*) The speed of the traveling waves is the ratio of their angular frequency and wave number:

$$v=\frac{\omega}{k}=\frac{300\,\text{s}^{-1}}{0.200\,\text{cm}^{-1}}=\boxed{15.0\,\text{m/s}}$$

(*c*) Relate the length of the string to the wavelengths of its standing-wave patterns:

$$L=n\frac{\lambda_n}{2},\,n=1,2,3,\ldots$$

Solve for L when $n = 4$:

$$L=2\lambda_4=2(31.4\,\text{cm})=\boxed{62.8\,\text{cm}}$$

55 •• **[SSM]** An organ pipe has a fundamental frequency of 440.0 Hz at 16.00°C. What will be the fundamental frequency of the pipe if the temperature increases to 32.00°C (assuming the length of the pipe remains constant)? Would it be better to construct organ pipes from a material that expands substantially as the temperature increases or, should the pipes be made of material that maintains the same length at all normal temperatures?

Picture the Problem We can use $v = f\lambda$ to express the fundamental frequency of the organ pipe in terms of the speed of sound and $v=\sqrt{\frac{\gamma RT}{M}}$ to relate the speed of sound and the fundamental frequency to the absolute temperature.

Express the fundamental frequency of the organ pipe in terms of the speed of sound:

$$f=\frac{v}{\lambda}$$

Relate the speed of sound to the temperature:

$$v=\sqrt{\frac{\gamma RT}{M}}$$

where γ and R are constants, M is the molar mass, and T is the absolute temperature.

Substitute for v to obtain:

$$f=\frac{1}{\lambda}\sqrt{\frac{\gamma RT}{M}}$$

Using primed quantities to represent the higher temperature, express the new frequency as a function of T:

$$f' = \frac{1}{\lambda'}\sqrt{\frac{\gamma RT'}{M}}$$

As we have seen, λ is proportional to the length of the pipe. For the first question, we assume the length of the pipe does not change, so $\lambda = \lambda'$. Then the ratio of f' to f is:

$$\frac{f'}{f} = \sqrt{\frac{T'}{T}} \Rightarrow f' = f\sqrt{\frac{T'}{T}}$$

Evaluate f' for $T' = 305$ K and $T = 289$ K:

$$f' = f_{305\,\text{K}} = f_{289\,\text{K}}\sqrt{\frac{305\,\text{K}}{289\,\text{K}}}$$

$$= (440.0\,\text{Hz})\sqrt{\frac{305\,\text{K}}{289\,\text{K}}} = \boxed{452\,\text{Hz}}$$

It would be better to have the pipe expand so that v/L, where L is the length of the pipe, is independent of temperature.

61 •• **[SSM]** The strings of a violin are tuned to the tones G, D, A, and E, which are separated by a fifth from one another. That is, $f(\text{D}) = 1.5f(\text{G})$, $f(\text{A}) = 1.5f(\text{D}) = 440$ Hz, and $f(\text{E}) = 1.5f(\text{A})$. The distance between the bridge at the scroll and the bridge over the body, the two fixed points on each string, is 30.0 cm. The tension on the E string is 90.0 N. (*a*) What is the linear mass density of the E string? (*b*) To prevent distortion of the instrument over time, it is important that the tension on all strings be the same. Find the linear mass densities of the other strings.

Picture the Problem (*a*) The mass densities of the strings are related to the transverse wave speed and tension through $v = \sqrt{F_{\text{T}}/\mu}$. (*b*) We can use $v = f\lambda = 2fL$ to relate the frequencies of the violin strings to their lengths and linear densities.

(*a*) Relate the speed of transverse waves on a string to the tension in the string and solve for the string's linear density:

$$v = \sqrt{\frac{F_{\text{T}}}{\mu}} \Rightarrow \mu = \frac{F_{\text{T}}}{v^2}$$

Express the dependence of the speed of the transverse waves on their frequency and wavelength:

$$v = f_{\text{E}}\lambda = 2f_{\text{E}}L$$

Substituting for v yields:

$$\mu_E = \frac{F_{T,E}}{4 f_E^2 L^2}$$

Substitute numerical values and evaluate μ_E:

$$\mu_E = \frac{90.0\,\text{N}}{4\left[1.5\left(440\,\text{s}^{-1}\right)\right]^2 (0.300\,\text{m})^2} = 5.74\times10^{-4}\ \text{kg/m} = \boxed{0.574\,\text{g/m}}$$

(*b*) Evaluate μ_A:

$$\mu_A = \frac{90.0\,\text{N}}{4\left(440\,\text{s}^{-1}\right)^2 (0.300\,\text{m})^2} = 1.29\times10^{-3}\ \text{kg/m} = \boxed{1.29\,\text{g/m}}$$

Evaluate μ_D:

$$\mu_D = \frac{90.0\,\text{N}}{4\left(293\,\text{s}^{-1}\right)^2 (0.300\,\text{m})^2} = 2.91\times10^{-3}\ \text{kg/m} = \boxed{2.91\,\text{g/m}}$$

Evaluate μ_G:

$$\mu_G = \frac{90.0\,\text{N}}{4\left(195\,\text{s}^{-1}\right)^2 (0.300\,\text{m})^2} = \boxed{6.57\,\text{g/m}}$$

65 •• **[SSM]** A commonly used physics experiment that examines resonances of transverse waves on a string is shown in Figure 16-34. A weight is attached to the end of a string draped over a pulley; the other end of the string is attached to a mechanical oscillator that moves up and down at a frequency f that is to remain fixed throughout the demonstration. The length L between the oscillator and the pulley is fixed, and the tension is equal to the gravitational force on the weight. For certain values of the tension the string resonates. Assume the string does not stretch or shrink as the tension is varied. You are in charge of setting up this apparatus for a lecture demonstration. (*a*) Explain why only certain discrete values of the tension result in standing waves on the string. (*b*) To produce a standing wave with an additional antinode, do you need to increase or decrease the tension? Explain. (*c*) Prove your reasoning in Part (*b*) by showing that the values for the tension F_{Tn} for the nth standing-wave mode are given by $F_{Tn} = 4L^2 f^2 \mu / n^2$, and thus the F_{Tn} is inversely proportional to n^2. (*d*) For your particular setup to fit onto the lecture table, you chose L = 1.00 m, f = 80.0 Hz, and μ = 0.750 g/m. Calculate how much tension is needed to produce each of the first three modes (standing waves) of the string.

Picture the Problem (*c*) and (*d*) We can equate the expression for the velocity of a wave on a string and the expression for the velocity of a wave in terms of its frequency and wavelength to obtain an expression for the weight that must be suspended from the end of the string in order to produce a given standing wave pattern. By using the condition on the wavelength that must be satisfied at resonance, we can express the weight on the end of the string in terms of μ, f, L, and an integer n and then evaluate this expression for $n = 1$, 2, and 3 for the first three standing wave patterns.

(*a*) Because the frequency is fixed, the wavelength depends only on the tension on the string. This is true because the only parameter that can affect the wave speed on the string is the tension on the string. The tension on the string is provided by the weight hanging from its end. Given that the length of the string is fixed, only certain wavelengths can resonate on the string. Thus, since only certain wavelengths are allowed, only certain wave speeds will work. This, in turn, means that only certain tensions, and therefore weights, will work.

(*b*) Higher frequency modes on the same length of string results in shorter wavelengths. To accomplish this without changing frequency, you need to reduce the wave speed. This is accomplished by reducing the tension in the string. Because the tension is provided by the weight on the end of the string, you must reduce the weight.

(*c*) Express the velocity of a wave on the string in terms of the tension F_T in the string and its linear density μ:

$$v = \sqrt{\frac{F_T}{\mu}} = \sqrt{\frac{w}{\mu}}$$

where w is the weight of the object suspended from the end of the string.

Express the wave speed in terms of its wavelength λ and frequency f:

$$v = f\lambda$$

Equate these expressions for v to obtain:

$$f\lambda = \sqrt{\frac{w}{\mu}} \rightarrow w = \mu f^2 \lambda^2$$

Express the condition on λ that corresponds to resonance:

$$\lambda = \frac{2L}{n}, n = 1, 2, 3, \ldots$$

Substitute to obtain:

$$w_n = \mu f^2 \left(\frac{2L}{n}\right)^2, n = 1, 2, 3, \ldots$$

or

$$w_n = \boxed{\frac{4\mu f^2 L^2}{n^2}}, n = 1, 2, 3, \ldots$$

(*d*) Substitute numerical values for L, f, and μ to obtain:

$$w_n = \frac{4(0.750\,\text{g/m})(80.0\,\text{s}^{-1})^2(1.00\,\text{m})^2}{n^2}$$

$$= \frac{19.20\,\text{N}}{n^2}$$

Evaluate w_n for $n = 1$:

$$w_1 = \frac{19.20\text{ N}}{(1)^2} = \boxed{19.2\text{ N}}$$

Evaluate w_n for $n = 2$:

$$w_2 = \frac{19.20\text{ N}}{(2)^2} = \boxed{4.80\text{ N}}$$

Evaluate w_n for $n = 3$:

$$w_3 = \frac{19.20\text{ N}}{(3)^2} = \boxed{2.13\text{ N}}$$

*Wave Packets

67 • **[SSM]** A tuning fork with natural frequency f_0 begins vibrating at time $t = 0$ and is stopped after a time interval Δt. The waveform of the sound at some later time is shown (Figure 16-35) as a function of x. Let N be an estimate of the number of cycles in this waveform. (*a*) If Δx is the length in space of this wave packet, what is the range in wave numbers Δk of the packet? (*b*) Estimate the average value of the wavelength λ in terms of N and Δx. (*c*) Estimate the average wave number k in terms of N and Δx. (*d*) If Δt is the time it takes the wave packet to pass a point in space, what is the range in angular frequencies $\Delta\omega$ of the packet? (*e*) Express f_0 in terms of N and Δt? (*f*) The number N is uncertain by about ±1 cycle. Use Figure 16-35 to explain why. (*g*) Show that the uncertainty in the wave number due to the uncertainty in N is $2\pi/\Delta x$.

Picture the Problem We can approximate the duration of the pulse from the product of the number of cycles in the interval and the period of each cycle and the wavelength from the number of complete wavelengths in Δx. We can use its definition to find the average wave number from the average wavelength.

(*a*) Relate the duration of the pulse to the number of cycles in the interval and the period of each cycle:

$$\Delta t \approx NT = \boxed{\frac{N}{f_0}}$$

(*b*) There are about N complete wavelengths in Δx; hence:

$$\lambda \approx \boxed{\frac{\Delta x}{N}}$$

(*c*) Use its definition to express the wave number k:

$$k = \frac{2\pi}{\lambda} = \boxed{\frac{2\pi N}{\Delta x}}$$

(*d*) N is uncertain because the waveform dies out gradually rather than stopping abruptly at some time; hence, where the pulse starts and stops is not well defined.

(*e*) Using our result in part (*c*), express the uncertainty in *k*:

$$\Delta k = \frac{2\pi\Delta N}{\Delta x} = \boxed{\frac{2\pi}{\Delta x}}$$

because $\Delta N = \pm 1$.

General Problems

71 •• **[SSM]** The wave function for a standing wave on a string is described by $y(x,t) = (0.020\ ) \sin (4\pi x\ ) \cos (60\pi t)$, where y and x are in meters and t is in seconds. Determine the maximum displacement and maximum speed of a point on the string at (*a*) $x = 0.10$ m, (*b*) $x = 0.25$ m, (*c*) $x = 0.30$ m, and (*d*) $x = 0.50$ m.

Picture the Problem The coefficient of the factor containing the time dependence in the wave function is the maximum displacement of any point on the string. The time derivative of the wave function is the instantaneous speed of any point on the string and the coefficient of the factor containing the time dependence is the maximum speed of any point on the string.

Differentiate the wave function with respect to t to find the speed of any point on the string:

$$v_y = \frac{\partial}{\partial t}\left[(0.020)\sin 4\pi x \cos 60\pi t\right] = -(0.020)(60\pi)\sin 4\pi x \sin 60\pi t = -1.2\pi \sin 4\pi x \sin 60\pi t$$

(*a*) Referring to the wave function, express the maximum displacement of the standing wave:

$$y_{\text{max}}(x) = (0.020\,\text{m})\sin\left[\left(4\pi\,\text{m}^{-1}\right)x\right] \quad (1)$$

Evaluate equation (1) at $x = 0.10$ m:

$$y_{\text{max}}(0.10\,\text{m}) = (0.020\,\text{m}) \times \sin\left[\left(4\pi\,\text{m}^{-1}\right)(0.10\,\text{m})\right] = \boxed{1.9\,\text{cm}}$$

Referring to the derivative of the wave function with respect to t, express the maximum speed of the standing wave:

$$v_{y,\text{max}}(x) = (1.2\pi\,\text{m/s})\sin\left[\left(4\pi\,\text{m}^{-1}\right)x\right] \quad (2)$$

Evaluate equation (2) at $x = 0.10$ m:

$$v_{y,\text{max}}(0.10\,\text{m}) = (1.2\pi\,\text{m/s}) \times \sin\left[\left(4\pi\,\text{m}^{-1}\right)(0.10\,\text{m})\right] = \boxed{3.6\,\text{m/s}}$$

(*b*) Evaluate equation (1) at $x = 0.25$ m:

$$y_{max}(0.25\,\text{m}) = (0.020\,\text{m})\times\sin\left[\left(4\pi\,\text{m}^{-1}\right)(0.25\,\text{m})\right] = \boxed{0}$$

Evaluate equation (2) at $x = 0.25$ m:

$$v_{y,max}(0.25\,\text{m}) = (1.2\pi\,\text{m/s})\times\sin\left[\left(4\pi\,\text{m}^{-1}\right)(0.25\,\text{m})\right] = \boxed{0}$$

(*c*) Evaluate equation (1) at $x = 0.30$ m:

$$y_{max}(0.30\,\text{m}) = (0.020\,\text{m})\times\sin\left[\left(4\pi\,\text{m}^{-1}\right)(0.30\,\text{m})\right] = \boxed{-1.2\,\text{cm}}$$

Evaluate equation (2) at $x = 0.30$ m:

$$v_{y,max}(0.30\,\text{m}) = (1.2\pi\,\text{m/s})\times\sin\left[\left(4\pi\,\text{m}^{-1}\right)(0.30\,\text{m})\right] = \boxed{-2.2\,\text{m/s}}$$

(*d*) Evaluate equation (1) at $x = 0.50$ m:

$$y_{max}(0.50\,\text{m}) = (0.020\,\text{m})\times\sin\left[\left(4\pi\,\text{m}^{-1}\right)(0.50\,\text{m})\right] = \boxed{0}$$

Evaluate equation (2) at $x = 0.50$ m:

$$v_{y,max}(0.50\,\text{m}) = (1.2\pi\,\text{m/s})\times\sin\left[\left(4\pi\,\text{m}^{-1}\right)(0.50\,\text{m})\right] = \boxed{0}$$

75 •• **[SSM]** Three successive resonance frequencies in an organ pipe are 1310, 1834, and 2358 Hz. (*a*) Is the pipe closed at one end or open at both ends? (*b*) What is the fundamental frequency? (*c*) What is the effective length of the pipe?

Picture the Problem (*a*) We can use the conditions $\Delta f = f_1$ and $f_n = nf_1$, where n is an integer, which must be satisfied if the pipe is open at both ends to decide whether the pipe is closed at one end or open at both ends. (*b*) Once we have decided this question, we can use the condition relating Δf and the fundamental frequency to determine the latter. In Part (*c*) we can use the standing-wave condition for the appropriate pipe to relate its length to its resonance wavelengths.

(*a*) Express the conditions on the frequencies for a pipe that is open at both ends:

$$\Delta f = f_1$$
and
$$f_n = nf_1$$

Evaluate $\Delta f = f_1$:

$$\Delta f = 1834\,\text{Hz} - 1310\,\text{Hz} = 524\,\text{Hz}$$

Using the 2nd condition, find n:

$$n = \frac{f_n}{f_1} = \frac{1310\,\text{Hz}}{524\,\text{Hz}} = 2.5$$

$$\boxed{\text{The pipe is closed at one end.}}$$

(*b*) Express the condition on the frequencies for a pipe that is open at both ends:

$$\Delta f = 2f_1 \Rightarrow f_1 = \tfrac{1}{2}\Delta f$$

Substitute numerical values and evaluate f_1:

$$f_1 = \tfrac{1}{2}(524\,\text{Hz}) = \boxed{262\,\text{Hz}}$$

(*c*) Using the standing-wave condition for a pipe open at one end, relate the effective length of the pipe to its resonance wavelengths:

$$L = n\frac{\lambda_n}{4},\ n = 1, 3, 5, \ldots$$

For $n = 1$ we have:

$$\lambda_1 = \frac{v}{f_1} \text{ and } L = \frac{\lambda_1}{4} = \frac{v}{4f_1}$$

Substitute numerical values and evaluate L:

$$L = \frac{343\,\text{m/s}}{4\left(262\,\text{s}^{-1}\right)} = \boxed{32.7\,\text{cm}}$$

81 •• **[SSM]** The speed of sound in air is proportional to the square root of the absolute temperature T (Equation 15-5). (*a*) Show that if the air temperature changes by a small amount, the fractional change in the fundamental frequency of an organ pipe is approximately equal to half the fractional change in the absolute temperature. That is, show that $\Delta f/f \approx \frac{1}{2}\Delta T/T$, where f is the frequency at absolute temperature T and Δf is the change in frequency when the temperature change by ΔT. (Ignore any change in the length of the pipe due to thermal expansion of the organ pipe.) (*b*) Suppose that an organ pipe that is stopped at one end has a fundamental frequency of 200.0 Hz when the temperature is 20.00°C. Use the approximate result from Part (*a*) to determine its fundamental frequency when the temperature is 30.00°C. (*c*) Compare your Part (*b*) result to what you would get using exact calculations. (Ignore any change in the length of the pipe due to thermal expansion.)

Picture the Problem We can express the fundamental frequency of the organ pipe as a function of the air temperature and differentiate this expression with respect to the temperature to express the rate at which the frequency changes with respect to temperature. For changes in temperature that are small compared to the temperature, we can approximate the differential changes in frequency and temperature with finite changes to complete the derivation of $\Delta f/f = \frac{1}{2}\Delta T/T$. In Part (*b*) we'll use this relationship and the data for the frequency at 20.00°C to find the frequency of the fundamental at 30.00°C.

(*a*) Express the fundamental frequency of an organ pipe in terms of its wavelength and the speed of sound:

$$f = \frac{v}{\lambda}$$

Relate the speed of sound in air to the absolute temperature:

$$v = \sqrt{\frac{\gamma RT}{M}} = C\sqrt{T}$$

where

$$C = \sqrt{\frac{\gamma R}{M}} = \text{constant}$$

Defining a new constant C', substitute to obtain:

$$f = \frac{C}{\lambda}\sqrt{T} = C'\sqrt{T}$$

because λ is constant for the fundamental frequency we ignore any change in the length of the pipe.

Differentiate this expression with respect to T:

$$\frac{df}{dT} = \frac{1}{2}C'T^{-1/2} = \frac{f}{2T}$$

Separate the variables to obtain:

$$\frac{df}{f} = \frac{1}{2}\frac{dT}{T}$$

For $\Delta T << T$, we can approximate df by Δf and dT by ΔT to obtain:

$$\frac{\Delta f}{f} = \boxed{\frac{1}{2}\frac{\Delta T}{T}}$$

(*b*) Express the fundamental frequency at 30.00°C in terms of its frequency at 20.00°C:

$$f_{30} = f_{20} + \Delta f$$

Solve the result in (*a*) for Δf:

$$\Delta f = \tfrac{1}{2} f \frac{\Delta T}{T}$$

Substitute for Δf to obtain:

$$f_{30}=f_{20}+\tfrac{1}{2}f_{20}\frac{\Delta T}{T}$$

Substitute numerical values and evaluate f_{30}:

$$f_{30}=200.0\,\text{Hz}+\tfrac{1}{2}(200.0\,\text{Hz})\frac{303.15\,\text{K}-293.15\,\text{K}}{293.15\,\text{K}}=\boxed{203.4\,\text{Hz}}$$

(*c*) The exact expression for f_{30} is:

$$f_{30}=\frac{v_{30}}{\lambda}=\frac{\sqrt{\dfrac{\gamma RT_{30}}{M}}}{\lambda}=\sqrt{\frac{\gamma RT_{30}}{\lambda^2 M}}$$

The exact expression for f_{20} is:

$$f_{20}=\frac{v_{20}}{\lambda}=\frac{\sqrt{\dfrac{\gamma RT_{20}}{M}}}{\lambda}=\sqrt{\frac{\gamma RT_{20}}{\lambda^2 M}}$$

Dividing the first of these equations by the second and simplifying yields:

$$\frac{f_{30}}{f_{20}}=\frac{\sqrt{\dfrac{\gamma RT_{30}}{\lambda^2 M}}}{\sqrt{\dfrac{\gamma RT_{20}}{\lambda^2 M}}}=\sqrt{\frac{T_{30}}{T_{20}}}$$

Solve for f_{30} to obtain:

$$f_{30}=\sqrt{\frac{T_{30}}{T_{20}}}f_{20}$$

Substitute numerical values and evaluate f_{30}:

$$f_{30}=\sqrt{\frac{303.15\,\text{K}}{293.15\,\text{K}}}(200.0\,\text{Hz})=\boxed{203.4\,\text{Hz}}$$

85 ••• **[SSM]** In principle, a wave with almost any arbitrary shape can be expressed as a sum of harmonic waves of different frequencies. (*a*) Consider the function defined by

$$f(x)=\frac{4}{\pi}\left(\frac{\cos x}{1}-\frac{\cos 3x}{3}+\frac{\cos 5x}{5}-\ldots\right)=\frac{4}{\pi}\sum_{n=0}(-1)^n\frac{\cos[(2n+1)x]}{2n+1}$$

Write a **spreadsheet program** to calculate this series using a finite number of terms, and make three graphs of the function in the range $x = 0$ to $x = 4\pi$.. To create the first graph, for each value of x that you plot, approximate the sum from $n = 0$ to $n = \infty$ with the first term of the sum. To create the second and third graphs, use only the first five term and the first ten terms, respectively. This function is sometimes called the *square wave*. (*b*) What is the relation between this function and Liebnitz' series for π, $\frac{\pi}{4} = 1 - \frac{1}{3} + \frac{1}{5} - \frac{1}{7} + \ldots$?

A spreadsheet program to evaluate $f(x)$ is shown below. Typical cell formulas used are shown in the table.

Cell	Content/Formula	Algebraic Form
A6	A5+0.1	$x + \Delta x$
B4	2*B3+1	$2n + 1$
B5	(−1)^B$3*COS(B$4*$A5) /B$4*4/PI()	$\frac{4}{\pi}\frac{(-1)^0 \cos((1)(0.0))}{1}$
C5	B5+(−1)^C$3*COS(C$4*$A5) /C$4*4/PI()	$1.2732 + \frac{4}{\pi}\frac{(-1)^1 \cos((3)(0.0))}{3}$

	A	B	C	D		K	L
1							
2							
3	$n =$	0	1	2		9	10
4	$2n$+1=	1	3	5		19	21
5	0.0	1.2732	0.8488	1.1035		0.9682	1.0289
6	0.1	1.2669	0.8614	1.0849		1.0134	0.9828
7	0.2	1.2479	0.8976	1.0352		1.0209	0.9912
8	0.3	1.2164	0.9526	0.9706		0.9680	1.0286
9	0.4	1.1727	1.0189	0.9130		1.0057	0.9742
10	0.5	1.1174	1.0874	0.8833		1.0298	1.0010
130	12.5	1.2704	0.8544	1.0952		0.9924	1.0031
131	12.6	1.2725	0.8503	1.1013		0.9752	1.0213
132	12.7	1.2619	0.8711	1.0710		1.0287	0.9714
133	12.8	1.2386	0.9143	1.0141		1.0009	1.0126
134	12.9	1.2030	0.9740	0.9493		0.9691	1.0146
135	13.0	1.1554	1.0422	0.8990		1.0261	0.9685

The graph of $f(x)$ versus x for $n = 1$ follows:

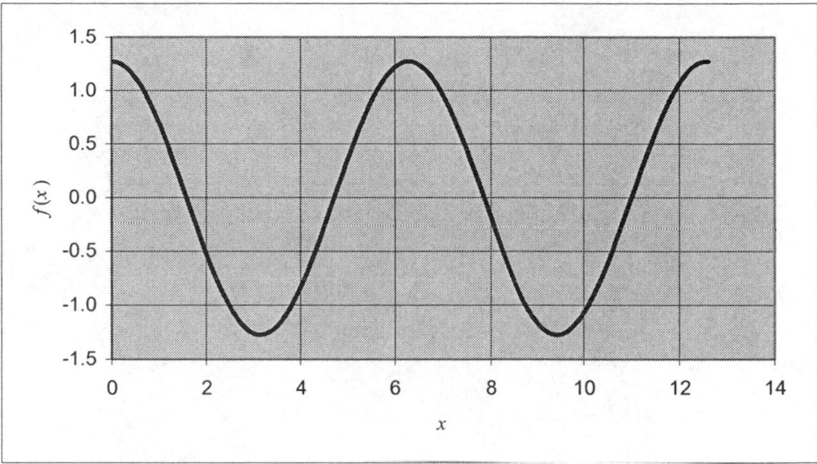

The graph of $f(x)$ versus x for $n = 5$ follows:

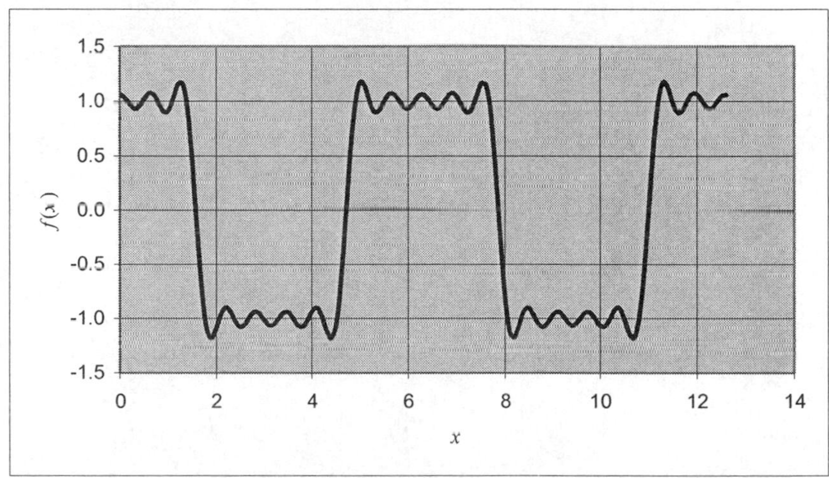

The graph of $f(x)$ versus x for $n = 10$ follows:

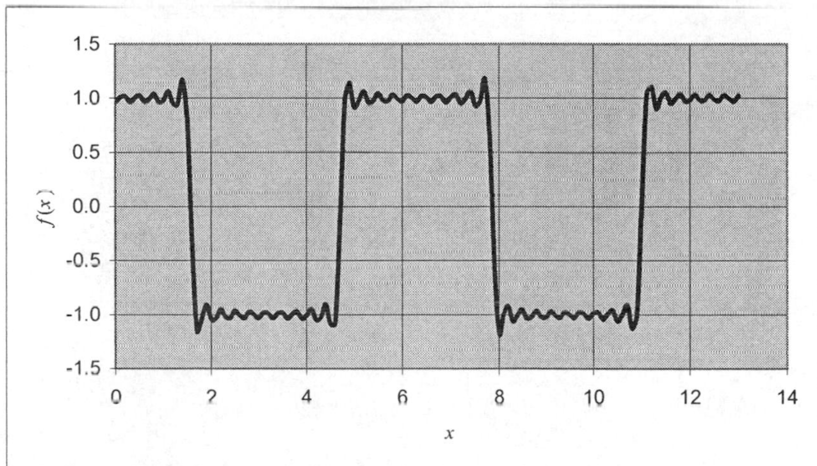

Evaluate $f(2\pi)$ to obtain:

$$f(2\pi) = \frac{4}{\pi}\left(\frac{\cos 2\pi}{1} - \frac{\cos 3(2\pi)}{3} + \frac{\cos 5(2\pi)}{5} - \ldots\right)$$

$$= \frac{4}{\pi}\left(1 - \frac{1}{3} + \frac{1}{5} - \frac{1}{7} + \ldots\right)$$

$$= 1$$

which is equivalent to the Liebnitz formula.

Chapter 17
Temperature and the Kinetic Theory of Gases

Conceptual Problems

3 • **[SSM]** "Yesterday I woke up and it was 20 °F in my bedroom," said Mert to his old friend Mort. "That's nothing," replied Mort. "My room was –5.0 °C." Who had the colder room, Mert or Mort?

Picture the Problem We can decide which room was colder by converting 20°F to the equivalent Celsius temperature.

Using the Fahrenheit-Celsius conversion, convert 20°F to the equivalent Celsius temperature:

$$t_C = \tfrac{5}{9}(t_F - 32°) = \tfrac{5}{9}(20° - 32°) = -6.7°C$$

so $\boxed{\text{Mert's room was colder.}}$

5 • **[SSM]** Figure 17-18 shows a plot of volume versus absolute temperature for a process that takes a fixed amount of an ideal gas from point A to point B. What happens to the pressure of the gas during this process?

Determine the Concept From the ideal-gas law, we have $P = nRT/V$. In the process depicted, both the temperature and the volume increase, but the temperature increases faster than does the volume. Hence, the pressure increases.

15 • **[SSM]** Suppose that you compress an ideal gas to half its original volume, while also halving its absolute temperature. During this process, the pressure of the gas (*a*) halves, (*b*) remains constant, (*c*) doubles, (*d*) quadruples.

Determine the Concept From the ideal-gas law, $PV = nRT$, halving both the temperature and volume of the gas leaves the pressure unchanged. $\boxed{(b)}$ is correct.

17 •• **[SSM]** Which speed is greater, the speed of sound in a gas or the rms speed of the molecules of the gas? Justify your answer, using the appropriate formulas, and explain why your answer is intuitively plausible.

Determine the Concept The rms speed of molecules of an ideal gas is given by $v_{rms} = \sqrt{\dfrac{3RT}{M}}$ and the speed of sound in a gas is given by $v_{sound} = \sqrt{\dfrac{\gamma RT}{M}}$.

The rms speed of the molecules of an ideal-gas is given by:

$$v_{rms} = \sqrt{\frac{3RT}{M}}$$

The speed of sound in a gas is given by:

$$v_{\text{sound}} = \sqrt{\frac{\gamma RT}{M}}$$

Divide the first of these equations by the second and simplify to obtain:

$$\frac{v_{\text{rms}}}{v_{\text{sound}}} = \frac{\sqrt{\dfrac{3RT}{M}}}{\sqrt{\dfrac{\gamma RT}{M}}} = \sqrt{\frac{3}{\gamma}}$$

For a monatomic gas, $\gamma = 1.67$ and:

$$\frac{v_{\text{rms, monatomic}}}{v_{\text{sound}}} = \sqrt{\frac{3}{1.67}} = 1.34$$

For a diatomic gas, $\gamma = 1.40$ and:

$$\frac{v_{\text{rms, diatomic}}}{v_{\text{sound}}} = \sqrt{\frac{3}{1.40}} = 1.46$$

In general, the rms speed is always somewhat greater than the speed of sound. However, it is only the component of the molecular velocities in the direction of propagation that is relevant to this issue. In addition, In a gas the mean free path is greater than the average intermolecular distance.

21 •• **[SSM]** Liquid nitrogen is relatively cheap, while liquid helium is relatively expensive. One reason for the difference in price is that while nitrogen is the most common constituent of the atmosphere, only small traces of helium can be found in the atmosphere. Use ideas from this chapter to explain why it is that only small traces of helium can be found in the atmosphere.

Determine the Concept The average molecular speed of He gas at 300 K is about 1.4 km/s, so a significant fraction of He molecules have speeds in excess of earth's escape velocity (11.2 km/s). Thus, they "leak" away into space. Over time, the He content of the atmosphere decreases to almost nothing.

Estimation and Approximation

25 •• **[SSM]** In Chapter 11, we found that the escape speed at the surface of a planet of radius R is $v_e = \sqrt{2gR}$, where g is the acceleration due to gravity at the surface of the planet. If the rms speed of a gas is greater than about 15 to 20 percent of the escape speed of a planet, virtually all of the molecules of that gas will escape the atmosphere of the planet.

(*a*) At what temperature is v_{rms} for O_2 equal to 15 percent of the escape speed for Earth?
(*b*) At what temperature is v_{rms} for H_2 equal to 15 percent of the escape speed for Earth?
(*c*) Temperatures in the upper atmosphere reach 1000 K. How does this help account for the low abundance of hydrogen in Earth's atmosphere?

(*d*) Compute the temperatures for which the rms speeds of O_2 and H_2 are equal to 15 percent of the escape speed at the surface of the moon, where g is about one-sixth of its value on Earth and $R = 1738$ km. How does this account for the absence of an atmosphere on the moon?

Picture the Problem We can find the escape temperatures for the earth and the moon by equating, in turn, $0.15v_e$ and v_{rms} of O_2 and H_2. We can compare these temperatures to explain the absence from Earth's upper atmosphere and from the surface of the moon. See Appendix C for the molar masses of O_2 and H_2.

(*a*) Express v_{rms} for O_2:

$$v_{rms,O_2} = \sqrt{\frac{3RT}{M_{O_2}}}$$

where R is the gas constant, T is the absolute temperature, and M_{O_2} is the molar mass of oxygen.

Equate $0.15v_e$ and v_{rms,O_2}:

$$0.15\sqrt{2gR_{earth}} = \sqrt{\frac{3RT}{M}}$$

Solve for T to obtain:

$$T = \frac{0.045gR_{earth}M}{3R} \qquad (1)$$

Substitute numerical values and evaluate T for O_2:

$$T = \frac{0.045(9.81\,\text{m/s}^2)(6.37\times10^6\,\text{m})(32.0\times10^{-3}\,\text{kg/mol})}{3(8.314\,\text{J/mol}\cdot\text{K})} = \boxed{3.60\times10^3\,\text{K}}$$

(*b*) Substitute numerical values and evaluate T for H_2:

$$T = \frac{0.045(9.81\,\text{m/s}^2)(6.37\times10^6\,\text{m})(2.02\times10^{-3}\,\text{kg/mol})}{3(8.314\,\text{J/mol}\cdot\text{K})} = \boxed{230\,\text{K}}$$

(*c*) Because hydrogen is lighter than air it rises to the top of the atmosphere. Because the temperature is high there, a greater fraction of the molecules reach escape speed.

(*d*) Express equation (1) at the surface of the moon:

$$T=\frac{0.045g_{\text{moon}}R_{\text{moon}}M}{3R}=\frac{0.045\left(\frac{1}{6}g_{\text{earth}}\right)R_{\text{moon}}M}{3R}=\frac{0.0025g_{\text{earth}}R_{\text{moon}}M}{R}$$

Substitute numerical values and evaluate T for O_2:

$$T=\frac{0.0025\left(9.81\,\text{m/s}^2\right)\left(1.738\times10^6\,\text{m}\right)\left(32.0\times10^{-3}\,\text{kg/mol}\right)}{8.314\,\text{J/mol}\cdot\text{K}}=\boxed{160\,\text{K}}$$

Substitute numerical values and evaluate T for H_2:

$$T=\frac{0.0025\left(9.81\,\text{m/s}^2\right)\left(1.738\times10^6\,\text{m}\right)\left(2.02\times10^{-3}\,\text{kg/mol}\right)}{8.314\,\text{J/mol}\cdot\text{K}}=\boxed{10\,\text{K}}$$

Because g is less on the moon, the escape speed is lower. Thus, a larger percentage of the molecules are moving at escape speed.

27 •• **[SSM]** The escape speed for gas molecules in the atmosphere of Jupiter is 60 km/s and the surface temperature of Jupiter is typically –150°C. Calculate the rms speeds for (*a*) H_2, (*b*) O_2, and (*c*) CO_2 at this temperature. (*d*) Are H_2, O_2, and CO_2 likely to be found in the atmosphere of Jupiter?

Picture the Problem We can use $v_{\text{rms}}=\sqrt{3RT/M}$ to calculate the rms speeds of H_2, O_2, and CO_2 at 123 K and then compare these speeds to 20% of the escape velocity on Jupiter to decide the likelihood of finding these gases in the atmosphere of Jupiter. See Appendix C for the molar masses of H_2, O_2, and CO_2.

Express the rms speed of an atom as a function of the temperature:

$$v_{\text{rms}}=\sqrt{\frac{3RT}{M}}$$

(*a*) Substitute numerical values and evaluate v_{rms} for H_2:

$$v_{\text{rms},H_2}=\sqrt{\frac{3\left(8.314\,\text{J/mol}\cdot\text{K}\right)\left(123\,\text{K}\right)}{2.02\times10^{-3}\,\text{kg/mol}}}=\boxed{1.23\text{ km/s}}$$

(*b*) Evaluate v_{rms} for O_2:

$$v_{\text{rms},O_2}=\sqrt{\frac{3\left(8.314\,\text{J/mol}\cdot\text{K}\right)\left(123\,\text{K}\right)}{32.0\times10^{-3}\,\text{kg/mol}}}=\boxed{310\text{ m/s}}$$

(*c*) Evaluate v_{rms} for CO_2:

$$v_{rms,CO_2} = \sqrt{\frac{3(8.314\ \text{J/mol}\cdot\text{K})(123\ \text{K})}{44.0\times10^{-3}\ \text{kg/mol}}} = \boxed{264\ \text{m/s}}$$

(*d*) Calculate 20% of v_{esc} for Jupiter:

$$v = \tfrac{1}{5}v_{esc} = \tfrac{1}{5}(60\ \text{km/s}) = 12\ \text{km/s}$$

Because v_e is greater than v_{rms} for O_2, CO_2 and H_2, O_2, all three gasses should be found on Jupiter.

31 ••• **[SSM]** In normal breathing conditions, approximately 5 percent of each exhaled breath is carbon dioxide. Given this information and neglecting any difference in water-vapor content, estimate the typical difference in mass between an inhaled breath and an exhaled breath.

Picture the Problem One breath (one's lung capacity) is about half a liter. The only thing that occurs in breathing is that oxygen is exchanged for carbon dioxide. Let's estimate that of the 20% of the air that is breathed in as oxygen, ¼ is exchanged for carbon dioxide. Then the mass difference between breaths will be 5% of a breath multiplied by the molar mass difference between oxygen and carbon dioxide and by the number of moles in a breath. Because this is an estimation problem, we'll use 32 g/mol as an approximation for the molar mass of oxygen and 44 g/mol as an approximation for the molar mass of carbon dioxide.

Express the difference in mass between an inhaled breath and an exhaled breath:

$$\Delta m = m_{O_2} - m_{CO_2} = f_{CO_2}\left(M_{O_2} - M_{CO_2}\right)n_{breath}$$

where f_{CO_2} is the fraction of the air breathed in that is exchanged for carbon dioxide.

The number of moles per breath is given by:

$$n_{breath} = \frac{V_{breath}}{22.4\ \text{L/mol}}$$

Substituting for n_{breath} yields:

$$\Delta m = f_{CO_2}\left(M_{O_2} - M_{CO_2}\right)\left(\frac{V_{breath}}{22.4\ \text{L/mol}}\right)$$

Substitute numerical values and evaluate Δm:

$$\Delta m = (0.05)(44\ \text{g/mol} - 32\ \text{g/mol})\frac{(0.5\ \text{L})}{22.4\ \text{L/mol}} \approx \boxed{1\times10^{-4}\ \text{g}}$$

Temperature Scales

33 • **[SSM]** The melting point of gold is 1945.4 °F. Express this temperature in degrees Celsius.

Picture the Problem We can use the Fahrenheit-Celsius conversion equation to find this temperature on the Celsius scale.

Convert 1945.4°F to the equivalent Celsius temperature:

$$t_C = \tfrac{5}{9}(t_F - 32°) = \tfrac{5}{9}(1945.4° - 32°) = \boxed{1063°C}$$

39 • **[SSM]** A constant-volume gas thermometer reads 50.0 torr at the triple point of water. (*a*) Sketch a graph of pressure vs. absolute temperature for this thermometer. (*b*) What will be the pressure when the thermometer measures a temperature of 300 K? (*c*) What ideal-gas temperature corresponds to a pressure of 678 torr?

Picture the Problem We can use the equation of the graph plotted in (*a*) to (*b*) find the pressure when the temperature is 300 K and (*c*) the ideal-gas temperature when the pressure is 678 torr.

(*a*) A graph of pressure (in torr) versus temperature (in kelvins) for this thermometer is shown to the right. The equation of this graph is:

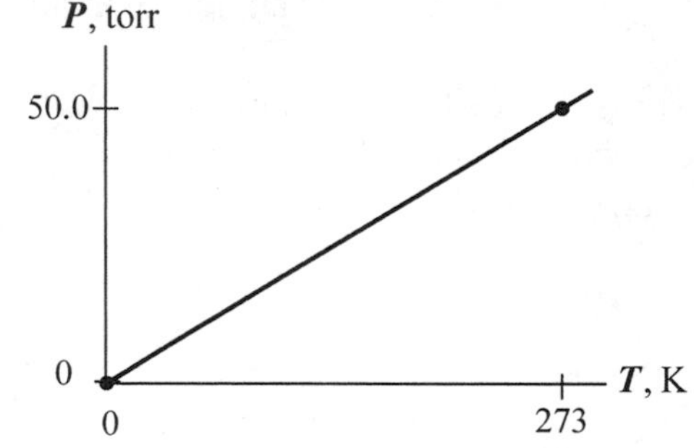

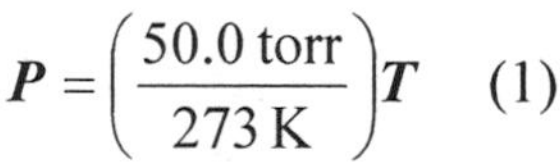

$$\boldsymbol{P} = \left(\frac{50.0\text{ torr}}{273\text{ K}}\right)\boldsymbol{T} \quad (1)$$

(*b*) Evaluate P when $T = 300$ K:

$$\boldsymbol{P}(300\text{ K}) = \left(\frac{50.0\text{ torr}}{273\text{ K}}\right)(300\text{ K}) = \boxed{54.9\text{ torr}}$$

(*c*) Solve equation (1) for T to obtain:

$$\boldsymbol{T} = \left(\frac{273\text{ K}}{50.0\text{ torr}}\right)\boldsymbol{P}$$

Evaluate $\boldsymbol{T}(678\text{ torr})$:

$$\boldsymbol{T}(678\text{ torr}) = \left(\frac{273\text{ K}}{50.0\text{ torr}}\right)(678\text{ torr}) = \boxed{3.70\times10^{3}\text{ K}}$$

45 ••• **[SSM]** A thermistor is a solid-state device widely used in a variety of engineering applications. Its primary characteristic is that its electrical resistance varies greatly with temperature. Its temperature dependence is given approximately by $R = R_0e^{B/T}$, where R is in ohms (Ω), T is in kelvins, and R_0 and B are constants that can be determined by measuring R at calibration points such as the ice point and the steam point. (*a*) If $R = 7360\ \Omega$ at the ice point and $153\ \Omega$ at the steam point, find R_0 and B. (*b*) What is the resistance of the thermistor at $t = 98.6$°F? (*c*) What is the rate of change of the resistance with temperature (dR/dT) at the ice point and the steam point? (*d*) At which temperature is the thermistor most sensitive?

Picture the Problem We can use the temperature dependence of the resistance of the thermistor and the given data to determine R_0 and B. Once we know these quantities, we can use the temperature-dependence equation to find the resistance at any temperature in the calibration range. Differentiation of R with respect to T will allow us to express the rate of change of resistance with temperature at both the ice point and the steam point temperatures.

(*a*) Express the resistance at the ice point as a function of temperature of the ice point:

$$7360\,\Omega = R_0e^{B/273\,\text{K}} \quad (1)$$

Express the resistance at the steam point as a function of temperature of the steam point:

$$153\,\Omega = R_0e^{B/373\,\text{K}} \quad (2)$$

Divide equation (1) by equation (2) to obtain:

$$\frac{7360\,\Omega}{153\,\Omega} = 48.10 = e^{B/273\,\text{K} - B/373\,\text{K}}$$

Solve for B by taking the logarithm of both sides of the equation:

$$\ln 48.1 = B\left(\frac{1}{273} - \frac{1}{373}\right)\text{K}^{-1}$$

and

$$\boldsymbol{B} = \frac{\ln 48.1}{\left(\frac{1}{273} - \frac{1}{373}\right)\text{K}^{-1}} = 3.944\times10^3\ \text{K}$$

$$= \boxed{3.94\times10^3\ \text{K}}$$

Solve equation (1) for R_0 and substitute for B:

$$R_0 = \frac{7360\,\Omega}{e^{B/273\,\text{K}}} = (7360\,\Omega)e^{-B/273\,\text{K}}$$
$$= (7360\,\Omega)e^{-3.944\times10^3\,\text{K}/273\,\text{K}}$$
$$= 3.913\times10^{-3}\,\Omega$$
$$= \boxed{3.91\times10^{-3}\,\Omega}$$

(*b*) From (*a*) we have:

$$R = (3.913\times10^{-3}\,\Omega)e^{3.944\times10^3\,\text{K}/T}$$

Convert 98.6°F to kelvins to obtain:

$$T = 310\,\text{K}$$

Substitute for T to obtain:

$$R(310\,\text{K}) = (3.913\times10^{-3}\,\Omega)e^{3.944\times10^3\,\text{K}/310\,\text{K}}$$
$$= \boxed{1.31\,\text{k}\Omega}$$

(*c*) Differentiate R with respect to T to obtain:

$$\frac{dR}{dT} = \frac{d}{dT}\left(R_0 e^{B/T}\right) = R_0 e^{B/T}\frac{d}{dT}\left(\frac{B}{T}\right)$$
$$= \frac{-B}{T^2}R_0 e^{B/T} = -\frac{RB}{T^2}$$

Evaluate dR/dT at the ice point:

$$\left(\frac{dR}{dT}\right)_{\text{ice point}} = -\frac{(7360\,\Omega)(3.943\times10^3\,\text{K})}{(273.16\,\text{K})^2}$$
$$= \boxed{-389\,\Omega/\text{K}}$$

Evaluate dR/dT at the steam point:

$$\left(\frac{dR}{dT}\right)_{\text{steam point}} = -\frac{(153\,\Omega)(3.943\times10^3\,\text{K})}{(373.16\,\text{K})^2}$$
$$= \boxed{-4.33\,\Omega/\text{K}}$$

(*d*) The thermistor is more sensitive (has greater sensitivity) at lower temperatures.

The Ideal-Gas Law

47 • **[SSM]** A 10.0-L vessel contains gas at a temperature of 0.00°C and a pressure of 4.00 atm. How many moles of gas are in the vessel? How many molecules?

Picture the Problem We can use the ideal-gas law to find the number of moles of gas in the vessel and the definition of Avogadro's number to find the number of molecules.

Apply the ideal-gas law to the gas:

$$PV = nRT \Rightarrow n = \frac{PV}{RT}$$

Substitute numerical values and evaluate n:

$$n = \frac{(4.00\,\text{atm})(10.0\,\text{L})}{(8.206\times10^{-2}\,\text{L}\cdot\text{atm/mol}\cdot\text{K})(273\,\text{K})}$$

$$= 1.786\ \text{mol} = \boxed{1.79\,\text{mol}}$$

Relate the number of molecules N in the gas in terms of the number of moles n:

$$N = nN_\text{A}$$

Substitute numerical values and evaluate N:

$$N = (1.786\,\text{mol})(6.022\times10^{23}\ \text{molecules/mol}) = \boxed{1.08\times10^{24}\ \text{molecules}}$$

55 •• **[SSM]** After nitrogen (N_2) and oxygen (O_2), the most abundant molecule in Earth's atmosphere is water, H_2O. However, the fraction of H_2O molecules in a given volume of air varies dramatically, from practically zero percent under the driest conditions to as high as 4 percent where it is very humid. (*a*) At a given temperature and pressure, would air be denser when its water vapor content is large or small? (*b*) What is the difference in mass, at room temperature and atmospheric pressure, between a cubic meter of air with no water vapor molecules, and a cubic meter of air in which 4 percent of the molecules are water vapor molecules?

Picture the Problem (*a*) At a given temperature and pressure, the ideal-gas law tells us that the total number of molecules per unit volume, N/V, is constant. The denser gas will, therefore, be the one in which the average mass *per molecule* is greater. (*b*) We can apply the ideal-gas law and use the relationship between the masses of the dry and humid air and their molar masses to find the difference in mass, at room temperature and atmospheric pressure, between a cubic meter of air containing no water vapor, and a cubic meter of air containing 4% water vapor. See Appendix C for the molar masses of N_2, O_2, and H_2O.

(*a*) The molecular mass of N_2 is 28.014 amu (see Appendix C), that of O_2 is 31.999 amu, and that of H_2O is 18.0152 amu. Because H_2O molecules are lighter than the predominant molecules in air, a given volume of air will be less when its water vapor content is lower.

(*b*) Express the difference in mass between a cubic meter of air containing no water vapor, and a cubic meter of air containing 4% water vapor:

$$\Delta m = m_{\text{dry}} - m_{\text{humid}}$$

The masses of the dry air and humid air are related to their molar masses according to:

$$m_{\text{dry}} = 0.04nM_{\text{dry}}$$

and

$$m_{\text{humid}} = 0.04nM_{\text{humid}}$$

Substitute for m_{dry} and m_{humid} and simplify to obtain:

$$\Delta m = 0.04nM_{\text{dry}} - 0.04nM_{\text{humid}} = 0.04n(M_{\text{dry}} - M_{\text{humid}})$$

From the ideal-gas law we have:

$$n = \frac{P_{\text{atm}}V}{RT}$$

and so Δm can be written as

$$\Delta m = \frac{0.04P_{\text{atm}}V}{RT}(M_{\text{dry}} - M_{\text{humid}})$$

Substitute numerical values (28.811 g/mol is an 80% nitrogen and 20% oxygen weighted average) and evaluate Δm:

$$\Delta m = \frac{(0.04)(101.325\,\text{kPa})(1.00\,\text{m}^3)}{(8.314\,\text{J/mol}\cdot\text{K})(300\,\text{K})}(28.811\,\text{g/mol} - 18.0152\,\text{g/mol}) \approx \boxed{18\,\text{g}}$$

57 •• [SSM] A hot-air balloon is open at the bottom. The balloon has a volume of 446 m^3 is filled with air with an average temperature of 100°C. The air outside the balloon has a temperature of 20.0°C and a pressure of 1.00 atm. How large a payload (including the envelope of the balloon itself) can the balloon lift? Use 29.0 g/mol for the molar mass of air. (Neglect the volume of both the payload and the envelope of the balloon.)

Picture the Problem Assume that the volume of the balloon is not changing. Then the air inside and outside the balloon must be at the same pressure of about 1.00 atm. The contents of the balloon are the air molecules inside it. We can use Archimedes principle to express the buoyant force on the balloon and we can find the weight of the air molecules inside the balloon. You'll need to determine the molar mass of air. See Appendix C for the molar masses of oxygen and nitrogen.

Express the net force on the balloon and its contents:

$$F_{\text{net}} = B - w_{\text{air inside the balloon}} \qquad (1)$$

Using Archimedes principle, express the buoyant force on the balloon:

$$B = w_{\text{displaced fluid}} = m_{\text{displaced fluid}}g$$

or

$$B = \rho_o V_{\text{balloon}} g$$

where ρ_o is the density of the air outside the balloon.

Express the weight of the air inside the balloon:

$$w_{\text{air inside the balloon}} = \rho_i V_{\text{balloon}} g$$

where ρ_i is the density of the air inside the balloon.

Substitute in equation (1) for B and $w_{\text{air inside the balloon}}$ to obtain:

$$\begin{aligned} F_{\text{net}} &= \rho_o V_{\text{balloon}} g - \rho_i V_{\text{balloon}} g \\ &= (\rho_o - \rho_i) V_{\text{balloon}} g \end{aligned} \qquad (2)$$

Express the densities of the air molecules in terms of their number densities, molecular mass, and Avogadro's number:

$$\rho = \frac{M}{N_A}\left(\frac{N}{V}\right)$$

Using the ideal-gas law, relate the number density of air N/V to its temperature and pressure:

$$PV = NkT \text{ and } \frac{N}{V} = \frac{P}{kT}$$

Substitute for $\frac{N}{V}$ to obtain:

$$\rho = \frac{M}{N_A}\left(\frac{P}{kT}\right)$$

Substitute in equation (2) and simplify to obtain:

$$\begin{aligned} F_{\text{net}} &= \frac{MP}{N_A k}\left(\frac{1}{T_o} - \frac{1}{T_i}\right) V_{\text{balloon}} g \\ &= \frac{MP}{N_A k}\left(\frac{1}{T_o} - \frac{1}{T_i}\right)\left(\tfrac{1}{6}\pi d^3\right) g \end{aligned}$$

Assuming that the average molar mass of air is 28.81 g/mol, substitute numerical values and evaluate F_{net}:

$$F_{\text{net}} = \frac{(28.81\,\text{g/mol})(101.325\,\text{kPa})\left(\dfrac{1}{297\,\text{K}} - \dfrac{1}{348\,\text{K}}\right)\left[\tfrac{1}{6}\pi(15.0\,\text{m})^3\right](9.81\,\text{m/s}^2)}{(6.022\times10^{23}\,\text{particles/mol})(1.381\times10^{-23}\,\text{J/K})}$$

$$= \boxed{3.0\,\text{kN}}$$

Kinetic Theory of Gases

59 • **[SSM]** (*a*) One mole of argon gas is confined to a 1.0-liter container at a pressure of 10 atm. What is the rms speed of the argon atoms? (*b*) Compare your answer to the rms speed for helium atoms under the same conditions.

Picture the Problem We can express the rms speeds of argon and helium atoms by combining $PV = nRT$ and $v_{rms} = \sqrt{3RT/M}$ to obtain an expression for v_{rms} in terms of *P, V,* and *M*. See Appendix C for the molar masses of argon and helium.

Express the rms speed of an atom as a function of the temperature and its molar mass:

$$v_{rms} = \sqrt{\frac{3RT}{M}}$$

From the ideal-gas law we have:

$$RT = \frac{PV}{n}$$

Substitute for RT to obtain:

$$v_{rms} = \sqrt{\frac{3PV}{nM}}$$

(*a*) Substitute numerical values and evaluate v_{rms} for argon atoms:

$$v_{rms,\,Ar} = \sqrt{\frac{3(10\,\text{atm})(101.325\,\text{kPa/atm})(1.0\times10^{-3}\,\text{m}^3)}{(1\,\text{mol})(39.948\times10^{-3}\,\text{kg/mol})}} = \boxed{0.28\,\text{km/s}}$$

(*b*) Substitute numerical values and evaluate v_{rms} for helium atoms:

$$v_{rms,\,He} = \sqrt{\frac{3(10\,\text{atm})(101.325\,\text{kPa/atm})(1.0\times10^{-3}\,\text{m}^3)}{(1\,\text{mol})(4.003\times10^{-3}\,\text{kg/mol})}} = \boxed{0.87\,\text{km/s}}$$

The rms speed of argon atoms is slightly less than one third the rms speed of helium atoms.

65 •• **[SSM]** Oxygen (O_2) is confined to a cube-shaped container 15 cm on an edge at a temperature of 300 K. Compare the average kinetic energy of a molecule of the gas to the change in its gravitational potential energy if it falls 15 cm (the height of the container).

Picture the Problem We can use $K = \frac{3}{2}kT$ and $\Delta U = mgh = Mgh/N_A$ to express the ratio of the average kinetic energy of a molecule of the gas to the change in its gravitational potential energy if it falls from the top of the container to the bottom. See Appendix C for the molar mass of oxygen.

Express the average kinetic energy of a molecule of the gas as a function of its temperature:

$$K_{av} = \tfrac{3}{2}kT$$

Letting h represent the height of the container, express the change in the potential energy of a molecule as it falls from the top of the container to the bottom:

$$\Delta U = mgh = \frac{M_{O_2}gh}{N_A}$$

Express the ratio of K_{av} to ΔU and simplify to obtain:

$$\frac{K_{av}}{\Delta U} = \frac{\tfrac{3}{2}kT}{\frac{M_{O_2}gh}{N_A}} = \frac{3N_A kT}{2M_{O_2}gh}$$

Substitute numerical values and evaluate $K_{av}/\Delta U$:

$$\frac{K_{av}}{\Delta U} = \frac{3(6.022\times10^{23}\ \text{particles/mol})(1.381\times10^{-23}\ \text{J/K})(300\,\text{K})}{2(32.0\times10^{-3}\ \text{kg/mol})(9.81\,\text{m/s}^2)(0.15\,\text{m})} = \boxed{7.9\times10^{4}}$$

*The Distribution of Molecular Speeds

67 •• **[SSM]** The fractional distribution function $f(v)$ is defined in Equation 17-36. Because $f(v)\,dv$ gives the fraction of molecules that have speeds in the range between v and $v + dv$, the integral of $f(v)\,dv$ over all the possible ranges of speeds must equal 1. Given that the integral $\int_0^\infty v^2 e^{-av^2}dv = \frac{\sqrt{\pi}}{4}a^{-3/2}$, show that $\int_0^\infty f(v)dv = 1$, where $f(v)$ is given by Equation 17-36.

Picture the Problem We can show that $f(v)$ is normalized by using the given integral to integrate it over all possible speeds.

Express the integral of Equation 17-36:

$$\int_0^\infty f(v)dv = \frac{4}{\sqrt{\pi}}\left(\frac{m}{2kT}\right)^{3/2}\int_0^\infty v^2 e^{-mv^2/2kT}dv$$

Let $a = m/2kT$ to obtain:

$$\int_0^\infty f(v)dv = \frac{4}{\sqrt{\pi}}a^{3/2}\int_0^\infty v^2 e^{-av^2}dv$$

Use the given integral to obtain:

$$\int_0^\infty f(v)dv = \frac{4}{\sqrt{\pi}}a^{3/2}\left(\frac{\sqrt{\pi}}{4}a^{-3/2}\right) = \boxed{1}$$

That is, $f(v)$ is normalized.

General Problems

73 •• **[SSM]** The Maxwell-Boltzmann distribution applies not just to gases, but also to the molecular motions within liquids. The fact that not all molecules have the same speed helps us understand the process of evaporation. (*a*) Explain in terms of molecular motion why a drop of water becomes cooler as molecules evaporate from the drop's surface. (Evaporative cooling is an important mechanism for regulating our body temperatures, and is also used to cool buildings in hot, dry locations.) (*b*) Use the Maxwell-Boltzmann distribution to explain why even a slight increase in temperature can greatly increase the rate at which a drop of water evaporates.

Determine the Concept
(*a*) To escape from the surface of a droplet of water, molecules must have enough kinetic energy to overcome the attractive forces from their neighbors. Therefore the molecules that escape will be those that are moving faster, leaving the slower molecules behind. The slower molecules have less kinetic energy, so the temperature of the droplet, which is proportional to the average kinetic energy per molecule, decreases.

(*b*) As long as the temperature isn't too high, the molecules that evaporate from a surface will be only those with the most extreme speeds, at the high-energy "tail" of the Maxwell-Boltzmann distribution. Within this part of the distribution, increasing the temperature only slightly can greatly increase the percentage of molecules with speeds above a certain threshold. For example, suppose that we set an initial threshold at $E = 5kT_1$, then imagine increasing the temperature by 10% so $T_2 = 1.1T_1$. At the threshold, the ratio of the new energy distribution to the old one is

$$\frac{F(T_2)}{F(T_1)} = \left(\frac{T_1}{T_2}\right)e^{\frac{-E}{kT_2}}e^{\frac{+E}{kT_1}} = (1.1)^{\frac{-3}{2}}e^{\frac{-5}{1.1}}e^5 = 1.366$$

an increase of almost 37%.

75 •• **[SSM]** In attempting to create liquid hydrogen for fuel, one of the proposals is to convert plain old water (H_2O) into H_2 and O_2 gases by *electrolysis*. How many moles of each of these gases result from the electrolysis of 2.0 L of water?

Picture the Problem We can use the molar mass of water to find the number of moles in 2.0 L of water. Because there are two hydrogen atoms in each molecule of water, there must be as many hydrogen molecules in the gas formed by electrolysis as there were molecules of water and, because there is one oxygen atom in each molecule of water, there must be half as many oxygen molecules in the gas formed by electrolysis as there were molecules of water. See Appendix C for the molar masses of hydrogen and oxygen.

Express the electrolysis of water into H_2 and O_2:

$$n_{\mathrm{H_2O}} \rightarrow n_{\mathrm{H_2}} + \tfrac{1}{2} n_{\mathrm{O_2}}$$

Express the number of moles in 2.0 L of water:

$$n_{\mathrm{H_2O}} = \frac{2.0\,\mathrm{L} \times 1000\,\dfrac{\mathrm{g}}{\mathrm{L}}}{18.02\,\mathrm{g/mol}} = 110\,\mathrm{mol}$$

Because there are two hydrogen atoms for each water molecule:

$$n_{\mathrm{H_2}} = \boxed{110\,\mathrm{mol}}$$

Because there is one oxygen atom for each water molecule:

$$n_{\mathrm{O_2}} = \tfrac{1}{2} n_{\mathrm{H_2O}} = \tfrac{1}{2}(110\,\mathrm{mol}) = \boxed{55\,\mathrm{mol}}$$

79 •• **[SSM]** Current experiments in atomic trapping and cooling can create low-density gases of rubidium and other atoms with temperatures in the nanokelvin (10^{-9} K) range. These atoms are trapped and cooled using magnetic fields and lasers in ultrahigh vacuum chambers. One method that is used to measure the temperature of a trapped gas is to turn the trap off and measure the time it takes for molecules of the gas to fall a given distance. Consider a gas of rubidium atoms at a temperature of 120 nK. Calculate how long it would take an atom traveling at the rms speed of the gas to fall a distance of 10.0 cm if (*a*) it were initially moving directly downward and (*b*) if it were initially moving directly upward. Assume that the atom doesn't collide with any others along its trajectory.

Picture the Problem Choose a coordinate system in which downward is the positive direction. We can use a constant-acceleration equation to relate the fall distance to the initial velocity of the molecule, the acceleration due to gravity, the fall time, and $v_{\mathrm{rms}} = \sqrt{3kT/m}$ to find the initial velocity of the rubidium molecules.

(*a*) Using a constant-acceleration equation, relate the fall distance to the initial velocity of a molecule, the acceleration due to gravity, and the fall time:

$$y = v_0 t + \tfrac{1}{2} g t^2 \qquad (1)$$

Relate the rms speed of rubidium atoms to their temperature and mass:

$$v_{\text{rms}} = \sqrt{\frac{3kT}{m_{\text{Rb}}}}$$

Substitute numerical values and evaluate v_{rms}:

$$v_{\text{rms}} = \sqrt{\frac{3(1.381\times10^{-23}\ \text{J/K})(120\,\text{nK})}{(85.47\,\text{u})(1.6606\times10^{-27}\ \text{kg/u})}} = 5.918\times10^{-3}\ \text{m/s}$$

Letting $v_{\text{rms}} = v_0$, substitute in equation (1) to obtain:

$$0.100\,\text{m} = (5.918\times10^{-3}\ \text{m/s})\,t + \tfrac{1}{2}(9.81\,\text{m/s}^2)\,t^2$$

Use the quadratic formula or your graphing calculator to solve this equation for its positive root:

$$t = 0.14218\,\text{s} = \boxed{142\ \text{ms}}$$

(*b*) If the atom is initially moving upward:

$$v_{\text{rms}} = v_0 = -5.918\times10^{-3}\ \text{m/s}$$

Substitute in equation (1) to obtain:

$$0.100\,\text{m} = (-5.918\times10^{-3}\ \text{m/s})\,t + \tfrac{1}{2}(9.81\,\text{m/s}^2)\,t^2$$

Use the quadratic formula or your graphing calculator to solve this equation for its positive root:

$$t = 0.1464\ \text{s} = \boxed{146\ \text{ms}}$$

Chapter 18
Heat and the First Law of Thermodynamics

Conceptual Problems

3 • **[SSM]** The specific heat of aluminum is more than twice the specific heat of copper. A block of copper and a block of aluminum have the same mass and temperature (20°C). The blocks are simultaneously dropped into a single calorimeter containing water at 40°C. Which statement is true when thermal equilibrium is reached? (*a*) The aluminum block is at a higher temperature than the copper block. (*b*) The aluminum block has absorbed less energy than the copper block. (*c*) The aluminum block has absorbed more energy than the copper block. (*d*) Both (*a*) and (*c*) are correct statements.

Picture the Problem We can use the relationship $Q = mc\Delta T$ to relate the amount of energy absorbed by the aluminum and copper blocks to their masses, specific heats, and temperature changes.

Express the energy absorbed by the aluminum block:

$$Q_{\text{Al}} = m_{\text{Al}}c_{\text{Al}}\Delta T$$

Express the energy absorbed by the copper block:

$$Q_{\text{Cu}} = m_{\text{Cu}}c_{\text{Cu}}\Delta T$$

Divide the second of these equations by the first to obtain:

$$\frac{Q_{\text{Cu}}}{Q_{\text{Al}}} = \frac{m_{\text{Cu}}c_{\text{Cu}}\Delta T}{m_{\text{Al}}c_{\text{Al}}\Delta T}$$

Because the block's masses are the same and they experience the same change in temperature:

$$\frac{Q_{\text{Cu}}}{Q_{\text{Al}}} = \frac{c_{\text{Cu}}}{c_{\text{Al}}} < 1$$

or

$Q_{\text{Cu}} < Q_{\text{Al}}$ and $\boxed{(c)}$ is correct.

11 • **[SSM]** A real gas cools during a free expansion, while an ideal gas does not cool during a free expansion. Explain the reason for this difference.

Determine the Concept $\Delta E_{\text{int}} = Q_{\text{in}} + W_{\text{on}}$. For an ideal gas, ΔE_{int} is a function of T only. Because $W_{\text{on}} = 0$ and $Q_{\text{in}} = 0$ in a free expansion, $\Delta E_{\text{int}} = 0$ and T is constant. For a real gas, E_{int} depends on the density of the gas because the molecules exert weak attractive forces on each other. In a free expansion, the repulsive forces between the ions increase the potential energy as the volume of the gas increases and the average kinetic energy decreases. Consequently, the temperature decreases and the real gas cools.

21 •• **[SSM]** An ideal gas in a cylinder is at pressure P and volume V. During a quasi-static adiabatic process, the gas is compressed until its volume has decreased to $V/2$. Then, in a quasi-static isothermal process, the gas is allowed to expand until its volume again has a value of V. What kind of process will return the system to its original state? Sketch the cycle on a graph.

Determine the Concept Adiabatic processes are steeper than isothermal processes. During an adiabatic process the pressure varies as $V^{-\gamma}$, whereas during an isothermal processes the pressure varies as V^{-1}. Thus when a gas expands isothermally, the pressure doesn't quite go down as far as the starting point. Thus the final process necessary is a constant-volume process dropping back to the initial pressure. Heat must be removed from the system in order for this to occur.

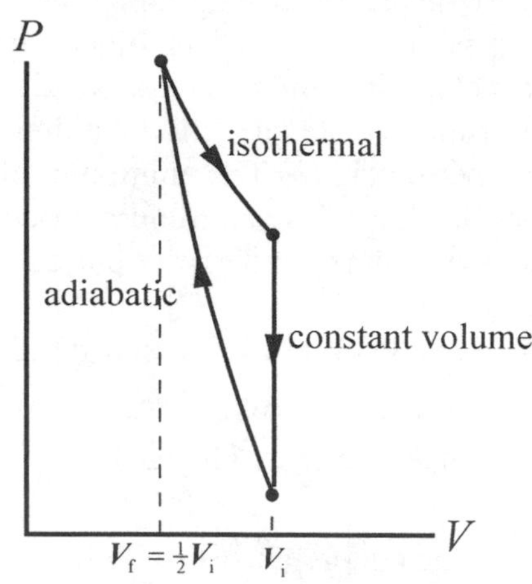

Estimation and Approximation

25 •• **[SSM]** A "typical" microwave oven has a power consumption of about 1200 W. Estimate how long it should take to boil a cup of water in the microwave assuming that 50% of the electrical power consumption goes into heating the water. How does this estimate correspond to everyday experience?

Picture the Problem Assume that the water is initially at 30°C and that the cup contains 200 g of water. We can use the definition of power to express the required time to bring the water to a boil in terms of its mass, heat capacity, change in temperature, and the rate at which energy is supplied to the water by the microwave oven.

Use the definition of power to relate the energy needed to warm the water to the elapsed time:

$$P = \frac{\Delta W}{\Delta t} = \frac{mc\Delta T}{\Delta t} \Rightarrow \Delta t = \frac{mc\Delta T}{P}$$

Substitute numerical values and evaluate Δt:

$$\Delta t = \frac{(0.200\,\text{kg})\left(4.184\frac{\text{kJ}}{\text{kg}\cdot\text{K}}\right)(100°\text{C} - 30°\text{C})}{600\,\text{W}} = 97.63\,\text{s} \approx \boxed{1.6\,\text{min}},$$

an elapsed time that seems to be consistent with experience.

Heat Capacity, Specific Heat, Latent Heat

29 • **[SSM]** How much heat must be absorbed by 60.0 g of ice at −10.0°C to transform it into 60.0 g of water at 40.0°C?

Picture the Problem We can find the amount of heat that must be absorbed by adding the heat required to warm the ice from −10.0°C to 0°C, the heat required to melt the ice, and the heat required to warm the water formed from the ice to 40.0°C.

Express the total heat required:

$$Q = Q_{\text{warm ice}} + Q_{\text{melt ice}} + Q_{\text{warm water}}$$

Substitute for each term to obtain:

$$\begin{aligned} Q &= mc_{\text{ice}}\Delta T_{\text{ice}} + mL_{\text{f}} + mc_{\text{water}}\Delta T_{\text{water}} \\ &= m\left(c_{\text{ice}}\Delta T_{\text{ice}} + L_{\text{f}} + c_{\text{water}}\Delta T_{\text{water}}\right) \end{aligned}$$

Substitute numerical values (See Tables 18-1 and 18-2) and evaluate Q:

$$\begin{aligned} Q &= (0.0600\,\text{kg})\left[\left(2.05\frac{\text{kJ}}{\text{kg}\cdot\text{K}}\right)(0°\text{C} - (-10.0°\text{C})) + 333.5\frac{\text{kJ}}{\text{kg}}\right. \\ &\qquad \left. + \left(4.184\frac{\text{kJ}}{\text{kg}\cdot\text{K}}\right)(40.0°\text{C} - 0°\text{C})\right] \\ &= \boxed{31.3\,\text{kJ}} \end{aligned}$$

Calorimetry

33 • **[SSM]** While spending the summer on your uncle's horse farm, you spend a week apprenticing with his farrier (a person who makes and fits horseshoes). You observe the way he cools a shoe after pounding the hot, pliable shoe into the correct size and shape. Suppose a 750-g iron horseshoe is taken from the farrier's fire, shaped, and at a temperature of 650°C, dropped into a 25.0-L bucket of water at 10.0°C. What is the final temperature of the water after the horseshoe and water arrive at equilibrium? Neglect any heating of the bucket and assume the specific heat of iron is $460\ \text{J/(kg}\cdot\text{K)}$.

Picture the Problem During this process the water will gain energy at the expense of the horseshoe. We can use conservation of energy to find the equilibrium temperature. See Table 18-1 for the specific heat of water.

Apply conservation of energy to obtain:

$$\sum_i Q_i = Q_{\text{warm the water}} + Q_{\text{cool the horseshoe}} = 0$$

or

$$m_{\text{water}} c_{\text{water}} (t_f - 10.0°\text{C}) + m_{\text{Fe}} c_{\text{Fe}} (t_f - 650°\text{C}) = 0$$

Solve for t_f to obtain:

$$t_f = \frac{m_{\text{water}} c_{\text{water}} (10.0°\text{C}) + m_{\text{Fe}} c_{\text{Fe}} (650°\text{C})}{m_{\text{water}} c_{\text{water}} + m_{\text{Fe}} c_{\text{Fe}}}$$

Substitte numerical values and evaluate t_f:

$$t_f = \frac{(25.0\,\text{kg})\left(4.184\,\frac{\text{kJ}}{\text{kg}\cdot\text{K}}\right)(10.0°\text{C}) + (0.750\,\text{kg})\left(0.460\,\frac{\text{kJ}}{\text{kg}\cdot\text{K}}\right)(650°\text{C})}{(25.0\,\text{kg})\left(4.184\,\frac{\text{kJ}}{\text{kg}\cdot\text{K}}\right) + (0.750\,\text{kg})\left(0.460\,\frac{\text{kJ}}{\text{kg}\cdot\text{K}}\right)}$$

$$= \boxed{12.1°\text{C}}$$

37 •• **[SSM]** A 200-g piece of ice at 0°C is placed in 500 g of water at 20°C. This system is in a container of negligible heat capacity and is insulated from its surroundings. (*a*) What is the final equilibrium temperature of the system? (*b*) How much of the ice melts?

Picture the Problem Because we can not tell, without performing a couple of calculations, whether there is enough heat available in the 500 g of water to melt all of the ice, we'll need to resolve this question first. See Tables 18-1 and 18-2 for specific heats and the latent heat of fusion of water.

(*a*) Determine the energy required to melt 200 g of ice:

$$Q_{\text{melt ice}} = m_{\text{ice}} L_f = (0.200\,\text{kg})\left(333.5\,\frac{\text{kJ}}{\text{kg}}\right) = 66.70\,\text{kJ}$$

The energy available from 500 g of water at 20°C is:

$$Q_{\text{available, max}} = m_{\text{water}} c_{\text{water}} \Delta T_{\text{water}} = (0.500\,\text{kg})\left(4.184\,\frac{\text{kJ}}{\text{kg}\cdot\text{K}}\right)(0°\text{C} - 20°\text{C}) = -41.84\,\text{kJ}$$

Because $|Q_{\text{available, max}}| < Q_{\text{melt ice}}$: $\boxed{\text{The final temperature is 0°C.}}$

(*b*) Equate the energy available from the water $|Q_{\text{available, max}}|$ to $m_{\text{ice}}L_f$ and solve for m_{ice} to obtain:

$$m_{\text{ice}} = \frac{|Q_{\text{available, max}}|}{L_f}$$

Substitute numerical values and evaluate m_{ice}:

$$m_{\text{ice}} = \frac{41.84\,\text{kJ}}{333.5\,\dfrac{\text{kJ}}{\text{kg}}} = \boxed{125\,\text{g}}$$

43 •• **[SSM]** A 100-g piece of copper is heated in a furnace to a temperature t_C. The copper is then inserted into a 150-g copper calorimeter containing 200 g of water. The initial temperature of the water and calorimeter is 16.0°C, and the temperature after equilibrium is established is 38.0°C. When the calorimeter and its contents are weighed, 1.20 g of water are found to have evaporated. What was the temperature t_C?

Picture the Problem We can find the temperature *t* by applying conservation of energy to this calorimetry problem. See Tables 18-1 and 18-2 for specific heats and the heat of vaporization of water.

Use conservation of energy to obtain:

$$\sum_i Q_i = Q_{\substack{\text{vaporize}\\ \text{water}}} + Q_{\substack{\text{warm}\\ \text{the water}}} + Q_{\substack{\text{warm the}\\ \text{calorimeter}}} + Q_{\substack{\text{cool the}\\ \text{Cu sample}}} = 0$$

or

$$m_{\text{H}_2\text{O, vaporized}}L_{\text{f, w}} + m_{\text{H}_2\text{O}}c_{\text{H}_2\text{O}}\Delta T_{\text{H}_2\text{O}} + m_{\text{cal}}c_{\text{cal}}\Delta T_{\text{w}} + m_{\text{Cu}}c_{\text{Cu}}\Delta T_{\text{Cu}} = 0$$

Substituting numerical values yields:

$$(1.20\,\text{g})\left(2257\frac{\text{kJ}}{\text{kg}\cdot\text{K}}\right) + (200\,\text{g})\left(4.184\frac{\text{kJ}}{\text{kg}\cdot\text{K}}\right)(38.0°\text{C} - 16.0°\text{C})$$
$$+ (150\,\text{g})\left(0.386\frac{\text{kJ}}{\text{kg}\cdot\text{K}}\right)(38.0°\text{C} - 16.0°\text{C}) + (100\,\text{g})\left(0.386\frac{\text{kJ}}{\text{kg}\cdot\text{K}}\right)(38.0°\text{C} - t_C) = 0$$

Solving for t_C yields:

$$t_C = \boxed{618°\text{C}}$$

Work and the *PV* Diagram for a Gas

51 • **[SSM]** The gas is first cooled at constant volume until it reaches its final pressure. It is then allowed to expand at constant pressure until it reaches its final volume. (*a*) Illustrate this process on a *PV* diagram and calculate the work done by the gas. (*b*) Find the heat absorbed by the gas during this process.

Picture the Problem We can find the work done by the gas during this process from the area under the curve. Because no work is done along the constant volume (vertical) part of the path, the work done by the gas is done during its isobaric expansion. We can then use the first law of thermodynamics to find the heat absorbed by the gas during this process

(*a*) The path from the initial state (1) to the final state (2) is shown on the *PV* diagram.

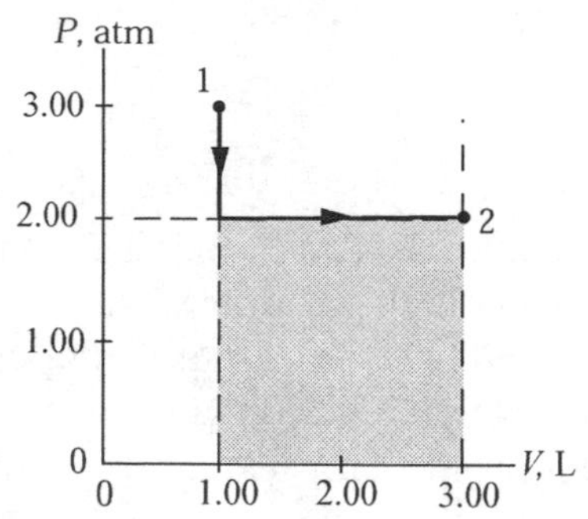

The work done by the gas equals the area under the curve:

$$W_{\text{by gas}} = P\Delta V = (2.00\,\text{atm})(2.00\,\text{L}) = \left(2.00\,\text{atm} \times \frac{101.325\,\text{kPa}}{\text{atm}}\right)\left(2.00\,\text{L} \times \frac{10^{-3}\,\text{m}^3}{\text{L}}\right)$$

$$= \boxed{405\,\text{J}}$$

(*b*) The work done by the gas is the negative of the work done on the gas. Apply the first law of thermodynamics to the system to obtain:

$$\begin{aligned} Q_{\text{in}} &= \Delta E_{\text{int}} - W_{\text{on}} \\ &= \left(E_{\text{int},2} - E_{\text{int},1}\right) - \left(-W_{\text{by gas}}\right) \\ &= \left(E_{\text{int},2} - E_{\text{int},1}\right) + W_{\text{by gas}} \end{aligned}$$

Substitute numerical values and evaluate Q_{in}:

$$Q_{\text{in}} = (912\,\text{J} - 456\,\text{J}) + 405\,\text{J} = \boxed{861\,\text{J}}$$

57 •• **[SSM]** An ideal gas initially at 20°C and 200 kPa has a volume of 4.00 L. It undergoes a quasi-static, isothermal expansion until its pressure is reduced to 100 kPa. Find (*a*) the work done by the gas, and (*b*) the heat absorbed by the gas during the expansion.

Picture the Problem The *PV* diagram shows the isothermal expansion of the ideal gas from its initial state 1 to its final state 2. We can use the ideal-gas law for a fixed amount of gas to find V_2 and then evaluate $\int PdV$ for an isothermal process to find the work done by the gas. In Part (*b*) of the problem we can apply the first law of thermodynamics to find the heat added to the gas during the expansion.

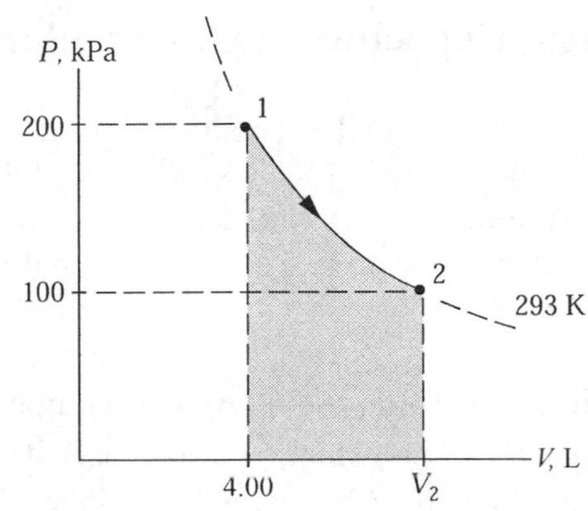

(*a*) Express the work done by a gas during an isothermal process:

$$W_{\text{by gas}} = \int_{V_1}^{V_2} PdV = nRT\int_{V_1}^{V_2}\frac{dV}{V} = P_1V_1\int_{V_1}^{V_2}\frac{dV}{V}$$

Apply the ideal-gas law for a fixed amount of gas undergoing an isothermal process:

$$P_1V_1 = P_2V_2 \Rightarrow V_2 = \frac{P_1}{P_2}V_1$$

Substitute numerical values and evaluate V_2:

$$V_2 = \frac{200\,\text{kPa}}{100\,\text{kPa}}(4.00\,\text{L}) = 8.00\,\text{L}$$

Substitute numerical values and evaluate W:

$$\begin{aligned}W_{\text{by gas}} &= (200\text{ kPa})(4.00\text{ L})\int_{4.00\text{ L}}^{8.00\text{ L}}\frac{dV}{V} = (800\text{ kPa}\cdot\text{L})[\ln V]_{4.00\text{ L}}^{8.00\text{ L}} \\ &= (800\text{ kPa}\cdot\text{L})\ln\left(\frac{8.00\text{ L}}{4.00\text{ L}}\right) = 554.5\text{ kPa}\cdot\text{L}\times\frac{10^{-3}\text{ m}^3}{\text{L}} \\ &= \boxed{555\text{ J}}\end{aligned}$$

(*b*) Apply the first law of thermodynamics to the system to obtain:

$Q_{\text{in}} = \Delta E_{\text{int}} - W_{\text{on}}$

or, because $\Delta E_{\text{int}} = 0$ for an isothermal process,

$Q_{\text{in}} = -W_{\text{on}}$

Because the work done by the gas is the negative of the work done on the gas:

$$Q_{\text{in}} = -(-W_{\text{by gas}}) = W_{\text{by gas}} = \boxed{555\text{ J}}$$

Remarks: in an isothermal expansion the heat added to the gas is always equal to the work done by the gas ($\Delta E_{\text{int}} = 0$).

Heat Capacities of Gases and the Equipartition Theorem

59 •• **[SSM]** The heat capacity at constant pressure of a certain amount of a diatomic gas is 14.4 J/K. (*a*) Find the number of moles of the gas. (*b*) What is the internal energy of the gas at $T = 300$ K? (*c*) What is the molar heat capacity of this gas at constant volume? (*d*) What is the heat capacity of this gas at constant volume?

Picture the Problem (*a*) The number of moles of the gas is related to its heat capacity at constant pressure and its molar heat capacity at constant pressure according to $C_P = nc'_P$. For a diatomic gas, the molar heat capacity at constant pressure is $c'_P = \frac{7}{2}R$. (*b*) The internal energy of a gas depends on its number of degrees of freedom and, for a diatomic gas, is given by $E_{int} = \frac{5}{2}nRT$. (*c*) The molar heat capacity of this gas at constant volume is related to its molar heat capacity at constant pressure according to $\boldsymbol{c'_V = c'_P - R}$. (*d*) The heat capacity of this gas at constant volume is the product of the number of moles in the gas and its molar heat capacity at constant volume.

(*a*) The number of moles of the gas is the ratio of its heat capacity at constant pressure to its molar heat capacity at constant pressure:

$$n = \frac{C_P}{c'_P}$$

For a diatomic gas, the molar heat capacity is given by:

$$c'_P = \frac{7}{2}R = 29.1\,\frac{\text{J}}{\text{mol}\cdot\text{K}}$$

Substitute numerical values and evaluate n:

$$n = \frac{14.4\,\frac{\text{J}}{\text{K}}}{29.1\,\frac{\text{J}}{\text{mol}\cdot\text{K}}} = 0.4948\text{ mol}$$

$$= \boxed{0.495\text{ mol}}$$

(*b*) With 5 degrees of freedom at this temperature:

$$E_{int} = \frac{5}{2}nRT$$

Substitute numerical values and evaluate E_{int}:

$$\boldsymbol{E_{int}} = \frac{5}{2}(0.4948\text{ mol})\left(8.314\,\frac{\text{J}}{\text{mol}\cdot\text{K}}\right)(300\text{ K}) = \boxed{3.09\text{ kJ}}$$

(*c*) The molar heat capacity of this gas at constant volume is the difference between the molar heat capacity at constant pressure and the gas constant *R*:

$$c'_V = c'_P - R$$

Because $c'_P = \frac{7}{2}R$ for a diatomic gas:

$$c'_V = \frac{7}{2}R - R = \frac{5}{2}R$$

Substitute the numerical value of *R* to obtain:

$$c'_V = \frac{5}{2}\left(8.314\frac{\text{J}}{\text{mol}\cdot\text{K}}\right) = 20.79\frac{\text{J}}{\text{mol}\cdot\text{K}}$$

$$= \boxed{20.8\frac{\text{J}}{\text{mol}\cdot\text{K}}}$$

(*d*) The heat capacity of this gas at constant volume is given by:

$$C'_V = nc'_V$$

Substitute numerical values and evaluate C'_V:

$$C'_V = (0.4948\text{ mol})\left(20.79\frac{\text{J}}{\text{mol}\cdot\text{K}}\right)$$

$$= \boxed{10.3\frac{\text{J}}{\text{K}}}$$

65 •• **[SSM]** Carbon dioxide (CO_2) at a pressure of 1.00 atm and a temperature of −78.5°C sublimates directly from a solid to a gaseous state without going through a liquid phase. What is the change in the heat capacity at constant pressure per mole of CO_2 when it undergoes sublimation? (Assume that the gas molecules can rotate but do not vibrate.) Is the change in the heat capacity positive or negative during sublimation? The CO_2 molecule is pictured in Figure 18-22.

Picture the Problem We can find the change in the heat capacity at constant pressure as CO_2 undergoes sublimation from the energy per molecule of CO_2 in the solid and gaseous states.

Express the change in the heat capacity (at constant pressure) per mole as the CO_2 undergoes sublimation:

$$\Delta C_P = C_{P,\text{gas}} - C_{P,\text{solid}} \qquad (1)$$

Express $C_{P,\text{gas}}$ in terms of the number of degrees of freedom per molecule:

$$C_{P,\text{gas}} = f\left(\tfrac{1}{2}Nk\right) = \tfrac{5}{2}Nk$$

because each molecule has three translational and two rotational degrees of freedom in the gaseous state.

We know, from the Dulong-Petit Law, that the molar specific heat of most solids is $3R = 3Nk$. This result is essentially a per-atom result as it was obtained for a monatomic solid with six degrees of freedom. Use this result and the fact CO_2 is triatomic to express $C_{P,solid}$:

$$C_{P,solid} = \frac{3Nk}{atom} \times 3\,atoms = 9Nk$$

Substitute in equation (1) to obtain:

$$\Delta C_P = \tfrac{5}{2}Nk - \tfrac{18}{2}Nk = \boxed{-\tfrac{13}{2}Nk}$$

Quasi-Static Adiabatic Expansion of a Gas

69 •• **[SSM]** A 0.500-mol sample of an ideal monatomic gas at 400 kPa and 300 K, expands quasi-statically until the pressure decreases to 160 kPa. Find the final temperature and volume of the gas, the work done by the gas, and the heat absorbed by the gas if the expansion is (*a*) isothermal and (*b*) adiabatic.

Picture the Problem We can use the ideal-gas law to find the initial volume of the gas. In Part (*a*) we can apply the ideal-gas law for a fixed amount of gas to find the final volume and the expression for the work done in an isothermal process. Application of the first law of thermodynamics will allow us to find the heat absorbed by the gas during this process. In Part (*b*) we can use the relationship between the pressures and volumes for a quasi-static adiabatic process to find the final volume of the gas. We can apply the ideal-gas law to find the final temperature and, as in (*a*), apply the first law of thermodynamics, this time to find the work done by the gas.

Use the ideal-gas law to express the initial volume of the gas:

$$V_i = \frac{nRT_i}{P_i}$$

Substitute numerical values and evaluate V_i:

$$V_i = \frac{(0.500\,mol)\left(8.314\frac{J}{mol\cdot K}\right)(300\,K)}{400\,kPa} = 3.118\times10^{-3}\,m^3$$

(*a*) Because the process is isothermal:

$$T_f = T_i = \boxed{300\,K}$$

Use the ideal-gas law for a fixed amount of gas to express V_f:

$$\frac{P_iV_i}{T_i} = \frac{P_fV_f}{T_f}$$

or, because T = constant,

$$V_f = V_i\frac{P_i}{P_f}$$

Substitute numerical values and evaluate V_f:

$$V_f = (3.118\,\text{L})\left(\frac{400\,\text{kPa}}{160\,\text{kPa}}\right) = 7.795\,\text{L}$$

$$= \boxed{7.80\,\text{L}}$$

Express the work done by the gas during the isothermal expansion:

$$W_{\text{by gas}} = nRT\ln\frac{V_f}{V_i}$$

Substitute numerical values and evaluate $W_{\text{by gas}}$:

$$W_{\text{by gas}} = (0.500\,\text{mol})\left(8.314\frac{\text{J}}{\text{mol}\cdot\text{K}}\right) \times (300\,\text{K})\ln\left(\frac{7.795\,\text{L}}{3.118\,\text{L}}\right)$$

$$= \boxed{1.14\,\text{kJ}}$$

Noting that the work done by the gas during the process equals the negative of the work done on the gas, apply the first law of thermodynamics to find the heat absorbed by the gas:

$$Q_{\text{in}} = \Delta E_{\text{int}} - W_{\text{on}} = 0 - (-1.14\,\text{kJ})$$

$$= \boxed{1.14\,\text{kJ}}$$

(*b*) Using $\gamma = 5/3$ and the relationship between the pressures and volumes for a quasi-static adiabatic process, express V_f:

$$P_iV_i^\gamma = P_fV_f^\gamma \Rightarrow V_f = V_i\left(\frac{P_i}{P_f}\right)^{1/\gamma}$$

Substitute numerical values and evaluate V_f:

$$V_f = (3.118\,\text{L})\left(\frac{400\,\text{kPa}}{160\,\text{kPa}}\right)^{3/5} = 5.403\,\text{L}$$

$$= \boxed{5.40\,\text{L}}$$

Apply the ideal-gas law to find the final temperature of the gas:

$$T_f = \frac{P_f V_f}{nR}$$

Substitute numerical values and evaluate T_f:

$$T_f = \frac{(160\,\text{kPa})(5.403\times10^{-3}\,\text{m}^3)}{(0.500\,\text{mol})\left(8.314\frac{\text{J}}{\text{mol}\cdot\text{K}}\right)} = \boxed{208\,\text{K}}$$

For an adiabatic process:

$$Q_{in} = \boxed{0}$$

Apply the first law of thermodynamics to express the work done on the gas during the adiabatic process:

$$W_{on} = \Delta E_{int} - Q_{in} = C_V \Delta T - 0 = \tfrac{3}{2} nR\Delta T$$

Substitute numerical values and evaluate W_{on}:

$$W_{on} = \tfrac{3}{2}(0.500\,\text{mol})(8.314\,\text{J/mol}\cdot\text{K}) \times (208\,\text{K} - 300\,\text{K}) = -574\,\text{J}$$

Because the work done by the gas equals the negative of the work done on the gas:

$$W_{by\ gas} = -(-574\,\text{J}) = \boxed{574\,\text{J}}$$

Cyclic Processes

73 •• **[SSM]** A 1.00-mol sample of an ideal diatomic gas is allowed to expand. This expansion is represented by the straight line from 1 to 2 in the *PV* diagram (Figure 18-23). The gas is then compressed isothermally. This compression is represented by the straight line from 2 to 1 in the *PV* diagram. Calculate the work per cycle done by the gas.

Picture the Problem The total work done as the gas is taken through this cycle is the area bounded by the two processes. Because the process from 1→2 is linear, we can use the formula for the area of a trapezoid to find the work done during this expansion. We can use $W_{\text{isothermal process}} = nRT\ln(V_f/V_i)$ to find the work done on the gas during the process 2→1. The net work done during this cycle is then the sum of these two terms.

Express the net work done per cycle:

$$W_{net} = W_{\text{by the gas}} + W_{\text{on the gas}} = W_{1\to2} + W_{2\to1} \quad (1)$$

Work is done by the gas during its expansion from 1 to 2 and hence is equal to the negative of the area of the trapezoid defined by this path and the vertical lines at $V_1 = 11.5$ L and $V_2 = 23$ L. Use the formula for the area of a trapezoid to express $W_{1\to2}$:

$$\begin{aligned}W_{1\to2} &= -A_{\text{trap}}\\ &= -\tfrac{1}{2}(23\,\text{L} - 11.5\,\text{L})\\ &\quad\times(2.0\,\text{atm} + 1.0\,\text{atm})\\ &= -17.3\,\text{L}\cdot\text{atm}\end{aligned}$$

Work is done on the gas during the isothermal compression from V_2 to V_1 and hence is equal to the area under the curve representing this process. Use the expression for the work done during an isothermal process to express $W_{2\to1}$:

$$W_{2\to1} = nRT\ln\left(\frac{V_\text{f}}{V_\text{i}}\right)$$

Apply the ideal-gas law at point 1 to find the temperature along the isotherm 2→1:

$$T = \frac{PV}{nR} = \frac{(2.0\,\text{atm})(11.5\,\text{L})}{(1.00\,\text{mol})(8.206\times10^{-2}\,\text{L}\cdot\text{atm/mol}\cdot\text{K})} = 280\,\text{K}$$

Substitute numerical values and evaluate $W_{2\to1}$:

$$W_{2\to1} = \left|(1.00\,\text{mol})(8.206\times10^{-2}\,\text{L}\cdot\text{atm/mol}\cdot\text{K})(280\,\text{K})\ln\left(\frac{11.5\,\text{L}}{23\,\text{L}}\right)\right| = 15.9\,\text{L}\cdot\text{atm}$$

Substitute in equation (1) and evaluate W_net:

$$\begin{aligned}W_\text{net} &= -17.3\,\text{L}\cdot\text{atm} + 15.9\,\text{L}\cdot\text{atm}\\ &= -1.40\,\text{L}\cdot\text{atm}\times\frac{101.325\,\text{J}}{\text{L}\cdot\text{atm}}\\ &= \boxed{-0.14\,\text{kJ}}\end{aligned}$$

75 ••• **[SSM]** At point D in Figure 18-24 the pressure and temperature of 2.00 mol of an ideal monatomic gas are 2.00 atm and 360 K, respectively. The volume of the gas at point B on the *PV* diagram is three times that at point D and its pressure is twice that at point C. Paths AB and CD represent isothermal processes. The gas is carried through a complete cycle along the path DABCD. Determine the total amount of work done by the gas and the heat absorbed by the gas along each portion of the cycle.

Picture the Problem We can find the temperatures, pressures, and volumes at all points for this ideal monatomic gas (3 degrees of freedom) using the ideal-gas law and the work for each process by finding the areas under each curve. We can find the heat exchanged for each process from the heat capacities and the initial and final temperatures for each process.

Express the total work done by the gas per cycle:

$$W_{\text{by gas,tot}} = W_{\text{D}\to\text{A}} + W_{\text{A}\to\text{B}} + W_{\text{B}\to\text{C}} + W_{\text{C}\to\text{D}}$$

1. Use the ideal-gas law to find the volume of the gas at point D:

$$V_{\text{D}} = \frac{nRT_{\text{D}}}{P_{\text{D}}} = \frac{(2.00\,\text{mol})\left(8.314\frac{\text{J}}{\text{mol}\cdot\text{K}}\right)(360\,\text{K})}{(2.00\,\text{atm})\left(101.325\frac{\text{kPa}}{\text{atm}}\right)} = 29.54\,\text{L}$$

2. We're given that the volume of the gas at point B is three times that at point D:

$$V_{\text{B}} = V_{\text{C}} = 3V_{\text{D}} = 3(29.54\,\text{L}) = 88.62\,\text{L}$$

Use the ideal-gas law to find the pressure of the gas at point C:

$$P_{\text{C}} = \frac{nRT_{\text{C}}}{V_{\text{C}}} = \frac{(2.00\,\text{mol})\left(8.206\times10^{-2}\,\frac{\text{L}\cdot\text{atm}}{\text{mol}\cdot\text{K}}\right)(360\,\text{K})}{88.62\,\text{L}} = 0.6667\,\text{atm}$$

We're given that the pressure at point B is twice that at point C:

$$P_{\text{B}} = 2P_{\text{C}} = 2(0.6667\,\text{atm}) = 1.333\,\text{atm}$$

3. Because path DC represents an isothermal process:

$$T_{\text{D}} = T_{\text{C}} = 360\,\text{K}$$

Use the ideal-gas law to find the temperatures at points B and A:

$$T_{\text{A}} = T_{\text{B}} = \frac{P_{\text{B}}V_{\text{B}}}{nR} = \frac{(1.333\,\text{atm})(88.62\,\text{L})}{(2.00\,\text{mol})\left(8.206\times10^{-2}\,\frac{\text{L}\cdot\text{atm}}{\text{mol}\cdot\text{K}}\right)} = 719.8\,\text{K}$$

Because the temperature at point A is twice that at D and the volumes are the same, we can conclude that:

$$P_A = 2P_D = 4.00\,\text{atm}$$

The pressure, volume, and temperature at points A, B, C, and D are summarized in the table to the right.

Point	P	V	T
	(atm)	(L)	(K)
A	4.00	29.5	720
B	1.33	88.6	720
C	0.667	88.6	360
D	2.00	29.5	360

4. For the path D→A, $W_{D\to A} = 0$ and:

$$Q_{D\to A} = \Delta E_{\text{int, D}\to\text{A}} = \tfrac{3}{2}nR\Delta T_{D\to A} = \tfrac{3}{2}nR(T_A - T_D)$$

Substitute numerical values and evaluate $Q_{D\to A}$:

$$Q_{D\to A} = \tfrac{3}{2}(2.00\,\text{mol})\left(8.314\frac{\text{J}}{\text{mol}\cdot\text{K}}\right)(720\,\text{K} - 360\,\text{K}) = 8.979\,\text{kJ}$$

For the path A→B:

$$W_{A\to B} = Q_{A\to B} = nRT_{A,B}\ln\left(\frac{V_B}{V_A}\right)$$

Substitute numerical values and evaluate $W_{A\to B}$:

$$W_{A\to B} = (2.00\,\text{mol})\left(8.314\frac{\text{J}}{\text{mol}\cdot\text{K}}\right)(720\,\text{K})\ln\left(\frac{88.62\,\text{L}}{29.54\,\text{L}}\right) = 13.15\,\text{kJ}$$

and, because process A→B is isothermal, $\Delta E_{\text{int, A}\to\text{B}} = 0$

For the path B→C, $W_{B\to C} = 0$, and:

$$Q_{B\to C} = \Delta U_{B\to C} = C_V\Delta T = \tfrac{3}{2}nR(T_C - T_B)$$

Substitute numerical values and evaluate $Q_{B\to C}$:

$$Q_{B\to C} = \tfrac{3}{2}(2.00\,\text{mol})(8.314\,\text{J/mol}\cdot\text{K})(360\,\text{K} - 720\,\text{K}) = -8.979\,\text{kJ}$$

For the path C→D:

$$W_{C\to D} = nRT_{C,D}\ln\left(\frac{V_D}{V_C}\right)$$

Substitute numerical values and evaluate $W_{C\to D}$:

$$W_{C\to D} = (2.00\,\text{mol})\left(8.314\frac{\text{J}}{\text{mol}\cdot\text{K}}\right)(360\,\text{K})\ln\left(\frac{29.54\,\text{L}}{88.62\,\text{L}}\right) = -6.576\,\text{kJ}$$

Also, because process A→B is isothermal, $\Delta E_{\text{int,A}\to\text{B}} = 0$, and

$Q_{C\to D} = W_{C\to D} = -6.58\,\text{kJ}$

Q_{in}, W_{on}, and ΔE_{int} are summarized for each of the processes in the table to the right.

Process	Q_{in}	W_{on}	ΔE_{int}
	(kJ)	(kJ)	(kJ)
D→A	$\boxed{8.98}$	0	8.98
A→B	$\boxed{13.2}$	−13.2	0
B→C	$\boxed{-8.98}$	0	−8.98
C→D	$\boxed{-6.58}$	6.58	0

Referring to the table, find the total work done by the gas per cycle:

$$\begin{aligned}W_{\text{by gas, tot}} &= W_{D\to A} + W_{A\to B} + W_{B\to C} + W_{C\to D} \\ &= 0 + 13.2\,\text{kJ} + 0 - 6.58\,\text{kJ} \\ &= \boxed{6.6\,\text{kJ}}\end{aligned}$$

Remarks: Note that, as it should be, ΔE_{int} is zero for the complete cycle.

General Problems

79 • **[SSM]** The *PV* diagram in Figure 18-25 represents 3.00 mol of an ideal monatomic gas. The gas is initially at point A. The paths AD and BC represent isothermal changes. If the system is brought to point C along the path AEC, find (*a*) the initial and final temperatures of the gas, (*b*) the work done by the gas, and (*c*) the heat absorbed by the gas.

Picture the Problem We can use the ideal-gas law to find the temperatures T_A and T_C. Because the process EDC is isobaric, we can find the area under this line geometrically and then use the 1st law of thermodynamics to find Q_{AEC}.

(*a*) Using the ideal-gas law, find the temperature at point A:

$$T_A = \frac{P_A V_A}{nR} = \frac{(4.0\,\text{atm})(4.01\,\text{L})}{(3.00\,\text{mol})\left(8.206\times10^{-2}\,\frac{\text{L}\cdot\text{atm}}{\text{mol}\cdot\text{K}}\right)} = 65.2\,\text{K} = \boxed{65\,\text{K}}$$

Using the ideal-gas law, find the temperature at point C:

$$T_C = \frac{P_C V_C}{nR} = \frac{(1.0\,\text{atm})(20.0\,\text{L})}{(3.00\,\text{mol})\left(8.206\times10^{-2}\,\frac{\text{L}\cdot\text{atm}}{\text{mol}\cdot\text{K}}\right)} = 81.2\,\text{K} = \boxed{81\,\text{K}}$$

(*b*) Express the work done by the gas along the path AEC:

$$\begin{aligned} W_{AEC} &= W_{AE} + W_{EC} = 0 + P_{EC}\Delta V_{EC} \\ &= (1.0\,\text{atm})(20.0\,\text{L} - 4.01\,\text{L}) \\ &= 15.99\,\text{L}\cdot\text{atm}\times\frac{101.325\,\text{J}}{\text{L}\cdot\text{atm}} \\ &= 1.62\,\text{kJ} = \boxed{1.6\,\text{kJ}} \end{aligned}$$

(*c*) Apply the first law of thermodynamics to express Q_{AEC}:

$$\begin{aligned} Q_{AEC} &= W_{AEC} + \Delta E_{int} = W_{AEC} + C_V\Delta T \\ &= W_{AEC} + \tfrac{3}{2}nR\Delta T \\ &= W_{AEC} + \tfrac{3}{2}nR(T_C - T_A) \end{aligned}$$

Substitute numerical values and evaluate Q_{AEC}:

$$Q_{AEC} = 1.62\,\text{kJ} + \tfrac{3}{2}(3.00\,\text{mol})(8.314\,\text{J/mol}\cdot\text{K})(81.2\,\text{K} - 65.2\,\text{K}) = \boxed{2.2\,\text{kJ}}$$

Remarks The difference between W_{AEC} and Q_{AEC} is the change in the internal energy $\Delta E_{int,AEC}$ during this process.

83 •• **[SSM]** As part of a laboratory experiment, you test the calorie content of various foods. Assume that when you eat these foods, 100% of the energy released by the foods is absorbed by your body. Suppose you burn a 2.50-g potato chip, and the resulting flame warms a small aluminum can of water. After burning the potato chip, you measure its mass to be 2.20 g. The mass of the can is 25.0 g, and the volume of water contained in the can is 15.0 ml. If the temperature increase in the water is 12.5°C, how many kilocalories (1 kcal = 1 dietary calorie) per 150-g serving of these potato chips would you estimate there are? Assume the can of water captures 50.0 percent of the heat released during the burning of the potato chip. *Note: Although the joule is the SI unit of choice in most thermodynamic situations, the food industry in the United States currently expresses the energy released during metabolism in terms of the "dietary calorie," which is our kilocalorie.*

Picture the Problem The ratio of the energy in a 150-g serving to the energy in 0.30 g of potato chip is the same as the ratio of the masses of the serving and the amount of the chip burned while heating the aluminum can and the water in it.

The ratio of the energy in a 150-g serving to the energy in 0.30 g of potato chip is the same as the ratio of the masses of the serving and the amount of the chip burned while heating the aluminum can and the water in it:

$$\frac{Q_{\text{150-g serving}}}{Q_{0.30\,\text{g}}} = \frac{150\ \text{g}}{0.30\ \text{g}} = 500$$

or

$$Q_{\text{150 g serving}} = 500 Q_{0.30\,\text{g}}$$

Letting f represent the fraction of the heat captured by the can of water, express the energy transferred to the aluminum can and the water in it during the burning of the potato chip:

$$\begin{aligned} fQ_{0.30\,\text{g}} &= Q_{\text{Al}} + Q_{\text{H}_2\text{O}} \\ &= m_{\text{Al}}c_{\text{Al}}\Delta T + m_{\text{H}_2\text{O}}c_{\text{H}_2\text{O}}\Delta T \\ &= \left(m_{\text{Al}}c_{\text{Al}} + m_{\text{H}_2\text{O}}c_{\text{H}_2\text{O}}\right)\Delta T \end{aligned}$$

where ΔT is the common temperature change of the aluminum cup and the water it contains.

Substituting for $Q_{0.30\,\text{g}}$ yields and solving for $Q_{\text{150-g serving}}$ yields:

$$Q_{\text{150-g serving}} = \frac{500\left(m_{\text{Al}}c_{\text{Al}} + m_{\text{H}_2\text{O}}c_{\text{H}_2\text{O}}\right)\Delta T}{f}$$

Substitute numerical values and evaluate $Q_{\text{150-g serving}}$:

$$Q_{\text{150-g serving}} = \frac{500\left[(0.0250\ \text{kg})\left(0.900\frac{\text{kJ}}{\text{kg}\cdot\text{K}}\right) + (0.0150\ \text{kg})\left(4.184\frac{\text{kJ}}{\text{kg}\cdot\text{K}}\right)\right](12.5\ \text{C}^\circ)}{0.500}$$

$$= 1.07\times10^{6}\ \text{J}\times\frac{1\ \text{cal}}{4.184\ \text{J}} = 256\times10^{3}\ \text{cal} \approx \boxed{256\ \text{kcal}}$$

89 •• **[SSM]** A thermally insulated system consists of 1.00 mol of a diatomic gas at 100 K and 2.00 mol of a solid at 200 K that are separated by a rigid insulating wall. Find the equilibrium temperature of the system after the insulating wall is removed, assuming that the gas obeys the ideal-gas law and that the solid obeys the Dulong–Petit law.

Picture the Problem We can use conservation of energy to relate the equilibrium temperature to the heat capacities of the gas and the solid. We can apply the Dulong-Petit law to find the heat capacity of the solid at constant volume and use the fact that the gas is diatomic to find its heat capacity at constant volume.

Apply conservation of energy to this process:

$$\Delta Q = Q_{\text{gas}} + Q_{\text{solid}} = 0$$

Use $Q = C_V \Delta T$ to substitute for Q_{gas} and Q_{solid}:

$$C_{\text{V,gas}}\left(T_{\text{equil}} - 100\,\text{K}\right) + C_{\text{V,solid}}\left(T_{\text{equil}} - 200\,\text{K}\right) = 0$$

Solving for T_{equil} yields:

$$T_{\text{equil}} = \frac{(100\,\text{K})\left(C_{\text{V,gas}}\right) + (200\,\text{K})\left(C_{\text{V,solid}}\right)}{C_{\text{V,gas}} + C_{\text{V,solid}}}$$

Using the Dulong-Petit law, determine the heat capacity of the solid at constant volume:

$$C_{\text{V,solid}} = 3n_{\text{solid}}R$$

The heat capacity of the gas at constant volume is given by:

$$C_{\text{V,gas}} = \tfrac{5}{2}n_{\text{gas}}R$$

Substitute for $C_{\text{V,solid}}$ and $C_{\text{V,gas}}$ and simplify to obtain:

$$T_{\text{equil}} = \frac{(100\,\text{K})\left(\frac{5}{2}n_{\text{gas}}R\right) + (200\,\text{K})\left(3n_{\text{solid}}R\right)}{\frac{5}{2}n_{\text{gas}}R + 3n_{\text{solid}}R} = \frac{(100\,\text{K})\left(\frac{5}{2}n_{\text{gas}}\right) + (200\,\text{K})\left(3n_{\text{solid}}\right)}{\frac{5}{2}n_{\text{gas}} + 3n_{\text{solid}}}$$

Substitute numerical values for n_{gas} and n_{solid} and evaluate T_{equil}:

$$T_{\text{equil}} = \frac{(100\,\text{K})\left(\frac{5}{2}\right)(1.00\,\text{mol}) + (200\,\text{K})(3)(2.00\,\text{mol})}{\frac{5}{2}(1.00\,\text{mol}) + 3(2.00\,\text{mol})} = \boxed{171\,\text{K}}$$

95 ••• **[SSM]** (*a*) Use the results of Problem 94 to show that in the limit that $T >> T_{\text{E}}$, the Einstein model gives the same expression for specific heat that the Dulong–Petit law does. (*b*) For diamond, T_{E} is approximately 1060 K. Integrate numerically $dE_{\text{int}} = c'_{\text{V}}\,dT$ to find the increase in the internal energy if 1.00 mol of diamond is heated from 300 to 600 K.

Picture the Problem (*a*) We can rewrite our expression for c'_{V} by dividing its numerator and denominator by $e^{T_{\text{E}}/T}$ and then using the power series for e^x to show that, for $T > T_E$, $c'_V \approx 3R$. In part (*b*), we can use the result of Problem 108 to obtain values for c'_V every 100 K between 300 K and 600 K and use this data to find ΔU numerically.

(*a*) From Problem 94 we have:

$$c'_V = 3R\left(\frac{T_E}{T}\right)^2 \frac{e^{T_E/T}}{\left(e^{T_E/T}-1\right)^2}$$

Divide the numerator and denominator by $e^{T_E/T}$ to obtain:

$$c'_V = 3R\left(\frac{T_E}{T}\right)^2 \frac{1}{\dfrac{e^{2T_E/T}-2e^{T_E/T}+1}{e^{T_E/T}}}$$

$$= 3R\left(\frac{T_E}{T}\right)^2 \frac{1}{e^{T_E/T}-2+e^{-T_E/T}}$$

Express the exponential terms in their power series to obtain:

$$e^{T_E/T}-2+e^{-T_E/T} = 1+\frac{T_E}{T}+\frac{1}{2}\left(\frac{T_E}{T}\right)^2+\ldots-2+1-\frac{T_E}{T}+\frac{1}{2}\left(\frac{T_E}{T}\right)^2+\ldots$$

$$\approx \left(\frac{T_E}{T}\right)^2 \text{ for } T >> T_E$$

Substitute for $e^{T_E/T}-2+e^{-T_E/T}$ to obtain:

$$c'_V \approx 3R\left(\frac{T_E}{T}\right)^2 \frac{1}{\left(\dfrac{T_E}{T}\right)^2} = \boxed{3R}$$

(*b*) Use the result of Problem 94 to verify the following table:

T	(K)	300	400	500	600
c_V	(J/mol·K)	9.65	14.33	17.38	19.35

The following graph of specific heat as a function of temperature was plotted using a spreadsheet program:

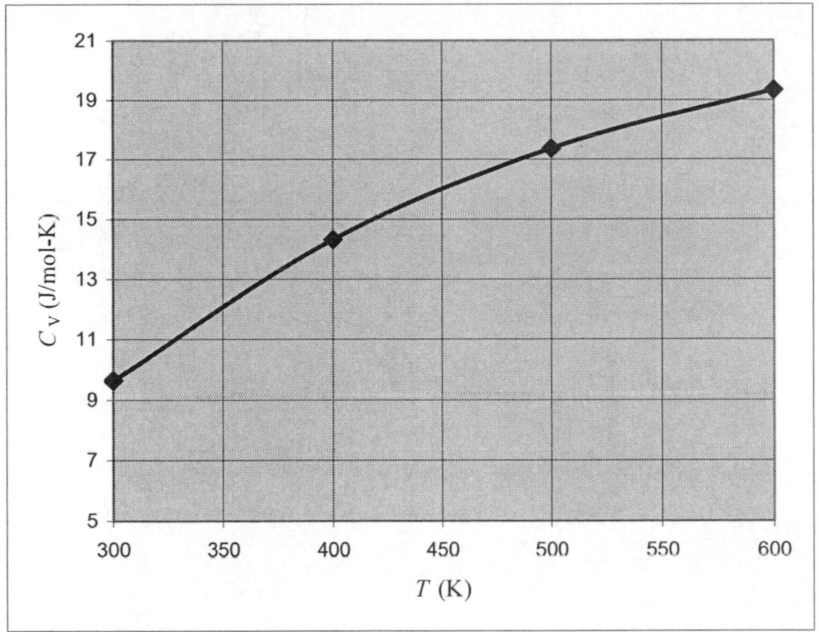

Integrate numerically, using the formula for the area of a trapezoid, to obtain:

$$\Delta U = \tfrac{1}{2}(1.00\text{ mol})(100\text{ K})(9.65+14.33)\frac{\text{J}}{\text{mol}\cdot\text{K}}$$
$$+\tfrac{1}{2}(1.00\text{ mol})(100\text{ K})(14.33+17.38)\frac{\text{J}}{\text{mol}\cdot\text{K}}$$
$$+\tfrac{1}{2}(1.00\text{ mol})(100\text{ K})(17.38+19.35)\frac{\text{J}}{\text{mol}\cdot\text{K}}$$
$$= \boxed{4.62\text{ kJ}}$$

Chapter 19
The Second Law of Thermodynamics

Conceptual Problems

5 • **[SSM]** An air conditioner's COP is mathematically identical to that of a refrigerator, that is, $\text{COP}_{\text{AC}} = \text{COP}_{\text{ref}} = \frac{Q_\text{c}}{W}$. However a heat pump's COP is defined differently, as $\text{COP}_{\text{hp}} = \frac{Q_\text{h}}{W}$. Explain clearly *why* the two COPs are defined differently. *Hint*: *Think of the end use of the three different devices.*

Determine the Concept The COP is defined so as to be a measure of the effectiveness of the device. For a refrigerator or air conditioner, the important quantity is the heat drawn from the already colder interior, Q_c. For a heat pump, the ideas is to focus on the heat drawn into the warm interior of the house, Q_h.

9 •• **[SSM]** Explain why the following statement is true: To increase the efficiency of a Carnot engine, you should make the difference between the two operating temperatures as large as possible; but to increase the efficiency of a Carnot cycle *refrigerator*, you should make the difference between the two operating temperatures as small as possible.

Determine the Concept A refrigerator can be more efficient when the temperatures are close together because it is easier to extract heat from an already cold interior if the temperature of the exterior is close to the temperature of the interior of the refrigerator.

17 •• **[SSM]** Sketch an *SV* diagram of the Carnot cycle for an ideal gas.

Determine the Concept Referring to Figure 19-8, process 1→2 is an isothermal expansion. In this process heat is added to the system and the entropy and volume increase. Process 2→3 is adiabatic, so S is constant as V increases. Process 3→4 is an isothermal compression in which S decreases and V also decreases. Finally, process 4→1 is adiabatic, that is, isentropic, and S is constant while V decreases.

During the isothermal expansion (from point 1 to point 2) the work done by the gas equals the heat added to the gas. The change in entropy of the gas from point 1 (where the temperature is T_1) to an arbitrary point on the curve is given by:

$$\Delta S = \frac{Q}{T_1}$$

For an isothermal expansion, the work done by the gas, and thus the heat added to the gas, are given by:

$$Q = W = nRT_1 \ln\left(\frac{V}{V_1}\right)$$

Substituting for Q yields:

$$\Delta S = nR \ln\left(\frac{V}{V_1}\right)$$

Since $S = S_1 + \Delta S$, we have:

$$S = S_1 + nR \ln\left(\frac{V}{V_1}\right)$$

The graph of S as a function of V for an isothermal expansion shown to the right was plotted using a spreadsheet program. This graph establishes the curvature of the 1→2 and 3→4 paths for the SV graph.

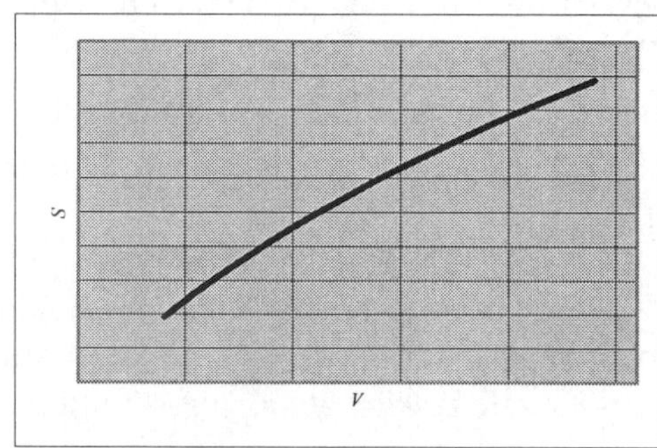

An SV graph for the Carnot cycle (see Figure 19-8) is shown to the right.

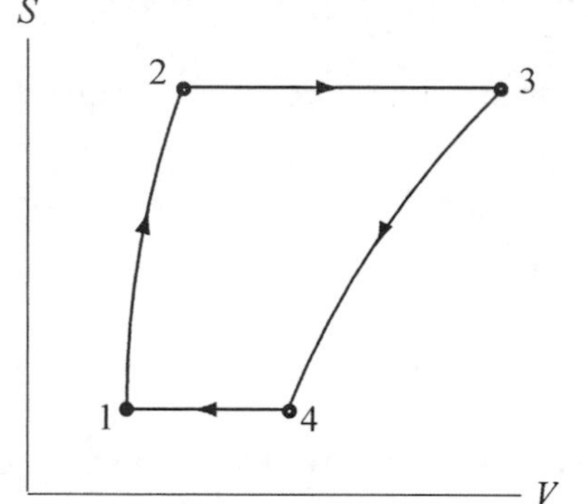

Estimation and Approximation

23 •• **[SSM]** Estimate the maximum efficiency of an automobile engine that has a compression ratio of 8.0:1.0. Assume the engine operates according to the Otto cycle and assume $\gamma = 1.4$. (The Otto cycle is discussed in Section 19-1.)

Picture the Problem The maximum efficiency of an automobile engine is given by the efficiency of a Carnot engine operating between the same two temperatures. We can use the expression for the Carnot efficiency and the equation relating V and T for a quasi-static adiabatic expansion to express the Carnot efficiency of the engine in terms of its compression ratio.

Express the Carnot efficiency of an engine operating between the temperatures T_c and T_h:

$$\varepsilon_C = 1 - \frac{T_c}{T_h}$$

Relate the temperatures T_c and T_h to the volumes V_c and V_h for a quasi-static adiabatic compression from V_c to V_h:

$$T_c V_c^{\gamma-1} = T_h V_h^{\gamma-1} \Rightarrow \frac{T_c}{T_h} = \frac{V_h^{\gamma-1}}{V_c^{\gamma-1}} = \left(\frac{V_h}{V_c}\right)^{\gamma-1}$$

Substitute for $\frac{T_c}{T_h}$ to obtain:

$$\varepsilon_C = 1 - \left(\frac{V_h}{V_c}\right)^{\gamma-1}$$

Express the compression ratio r:

$$r = \frac{V_c}{V_h}$$

Substituting for r yields:

$$\varepsilon_C = 1 - \frac{1}{r^{\gamma-1}}$$

Substitute numerical values for r and γ(1.4 for diatomic gases) and evaluate ε_C:

$$\varepsilon_C = 1 - \frac{1}{(8.0)^{1.4-1}} \approx \boxed{56\%}$$

25 •• **[SSM]** The average temperature of the surface of the Sun is about 5400 K, the average temperature of the surface of Earth is about 290 K. The solar constant (the intensity of sunlight reaching Earth's atmosphere) is about $1.37\ \text{kW/m}^2$. (*a*) Estimate the total power of the sunlight hitting Earth. (*b*) Estimate the net rate at which Earth's entropy is increasing due to this solar radiation.

Picture the Problem We can use the definition of intensity to find the total power of sunlight hitting the earth and the definition of the change in entropy to find the changes in the entropy of Earth and the Sun resulting from the radiation from the Sun.

(*a*) Using its definition, express the intensity of the Sun's radiation on Earth in terms of the power P delivered to Earth and Earth's cross sectional area A:

$$I = \frac{P}{A}$$

Solve for P and substitute for A to obtain:

$$P = IA = I\pi R^2$$

where R is the radius of Earth.

Substitute numerical values and evaluate P:

$$P = \pi\left(1.37\,\text{kW/m}^2\right)\left(6.37\times10^6\ \text{m}\right)^2 = 1.746\times10^{17}\ \text{W} = \boxed{1.75\times10^{17}\ \text{W}}$$

(*b*) Express the rate at which Earth's entropy S_{Earth} changes due to the flow of solar radiation:

$$\frac{dS_{\text{Earth}}}{dt} = \frac{P}{T_{\text{Earth}}}$$

Substitute numerical values and evaluate $\frac{dS_{\text{Earth}}}{dt}$:

$$\frac{dS_{\text{Earth}}}{dt} = \frac{1.746\times10^{17}\ \text{W}}{290\,\text{K}} = \boxed{6.02\times10^{14}\ \text{J/K}\cdot\text{s}}$$

Heat Engines and Refrigerators

27 • **[SSM]** A heat engine with 20.0% efficiency does 0.100 kJ of work during each cycle. (*a*) How much heat is absorbed from the hot reservoir during each cycle? (*b*) How much heat is released to the cold reservoir during each cycle?

Picture the Problem (*a*) The efficiency of the engine is defined to be $\varepsilon = W/Q_{\text{h}}$ where W is the work done per cycle and Q_{h} is the heat absorbed from the hot reservoir during each cycle. (*b*) Because, from conservation of energy, $Q_{\text{h}} = W + Q_{\text{c}}$, we can express the efficiency of the engine in terms of the heat Q_{c} released to the cold reservoir during each cycle.

(*a*) Q_{h} absorbed from the hot reservoir during each cycle is given by:

$$Q_{\text{h}} = \frac{W}{\varepsilon} = \frac{100\,\text{J}}{0.200} = \boxed{500\,\text{J}}$$

(*b*) Use $Q_{\text{h}} = W + Q_{\text{c}}$ to obtain:

$$Q_{\text{c}} = Q_{\text{h}} - W = 500\ \text{J} - 100\ \text{J} = \boxed{400\ \text{J}}$$

31 •• **[SSM]** The working substance of an engine is 1.00 mol of a monatomic ideal gas. The cycle begins at P_1 = 1.00 atm and V_1 = 24.6 L. The gas is heated at constant volume to P_2 = 2.00 atm. It then expands at constant pressure until its volume is 49.2 L. The gas is then cooled at constant volume until its pressure is again 1.00 atm. It is then compressed at constant pressure to its original state. All the steps are quasi-static and reversible. (*a*) Show this cycle on a *PV* diagram. For each step of the cycle, find the work done by the gas, the heat absorbed by the gas, and the change in the internal energy of the gas. (*b*) Find the efficiency of the cycle.

Picture the Problem To find the heat added during each step we need to find the temperatures in states 1, 2, 3, and 4. We can then find the work done on the gas during each process from the area under each straight-line segment and the heat that enters the system from $Q = C_V\Delta T$ and $Q = C_P\Delta T$. We can use the 1st law of thermodynamics to find the change in internal energy for each step of the cycle. Finally, we can find the efficiency of the cycle from the work done each cycle and the heat that *enters* the system each cycle.

(*a*) The cycle is shown to the right:

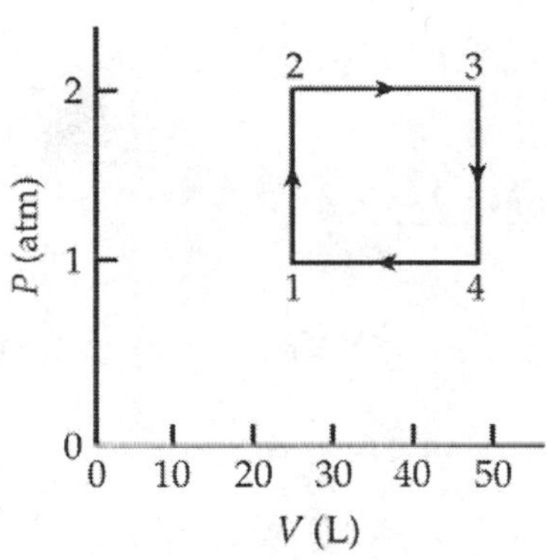

Apply the ideal-gas law to state 1 to find T_1:

$$T_1 = \frac{P_1V_1}{nR} = \frac{(1.00\,\text{atm})(24.6\,\text{L})}{(1.00\,\text{mol})\left(8.206\times10^{-2}\,\frac{\text{L}\cdot\text{atm}}{\text{mol}\cdot\text{K}}\right)} = 300\,\text{K}$$

The pressure doubles while the volume remains constant between states 1 and 2. Hence:

$$T_2 = 2T_1 = 600\,K$$

The volume doubles while the pressure remains constant between states 2 and 3. Hence:

$$T_3 = 2T_2 = 1200\,K$$

The pressure is halved while the volume remains constant between states 3 and 4. Hence:

$$T_4 = \tfrac{1}{2}T_3 = 600\,K$$

For path 1→2:

$$W_{12} = P\Delta V_{12} = \boxed{0}$$

and

$$Q_{12} = C_V\Delta T_{12} = \tfrac{3}{2}R\Delta T_{12} = \tfrac{3}{2}\left(8.314\frac{\text{J}}{\text{mol}\cdot\text{K}}\right)(600\,\text{K} - 300\,\text{K}) = \boxed{3.74\,\text{kJ}}$$

The change in the internal energy of the system as it goes from state 1 to state 2 is given by the 1st law of thermodynamics:

$$\Delta E_{int} = Q_{in} + W_{on}$$

Because $W_{12} = 0$:

$$\Delta E_{int,12} = Q_{12} = \boxed{3.74\text{ kJ}}$$

For path 2→3:

$$W_{on} = -W_{23} = -P\Delta V_{23} = -(2.00\text{ atm})(49.2\text{ L} - 24.6\text{ L})\left(\frac{101.325\text{ J}}{\text{L}\cdot\text{atm}}\right) = \boxed{-4.99\text{ kJ}}$$

$$Q_{23} = C_P\Delta T_{23} = \tfrac{5}{2}R\Delta T_{23} = \tfrac{5}{2}\left(8.314\frac{\text{J}}{\text{mol}\cdot\text{K}}\right)(1200\text{ K} - 600\text{ K}) = \boxed{12.5\text{ kJ}}$$

Apply $\Delta \boldsymbol{E}_{int} = \boldsymbol{Q}_{in} + \boldsymbol{W}_{on}$ to obtain:

$$\Delta E_{int,\,23} = 12.5\text{ kJ} - 4.99\text{ kJ} = \boxed{7.5\text{ kJ}}$$

For path 3→4:

$$W_{34} = P\Delta V_{34} = \boxed{0}$$

and

$$Q_{34} = \Delta E_{int,34} = C_V\Delta T_{34} = \tfrac{3}{2}R\Delta T_{34} = \tfrac{3}{2}\left(8.314\frac{\text{J}}{\text{mol}\cdot\text{K}}\right)(600\text{ K} - 1200\text{ K}) = \boxed{-7.48\text{ kJ}}$$

Apply $\Delta \boldsymbol{E}_{int} = \boldsymbol{Q}_{in} + \boldsymbol{W}_{on}$ to obtain:

$$\Delta E_{int,\,34} = -7.48\text{ kJ} + 0 = \boxed{-7.48\text{ kJ}}$$

For path 4→1:

$$W_{on} = -W_{41} = -P\Delta V_{41} = -(1.00\text{ atm})(24.6\text{ L} - 49.2\text{ L})\left(\frac{101.325\text{ J}}{\text{L}\cdot\text{atm}}\right) = \boxed{2.49\text{ kJ}}$$

and

$$Q_{41} = C_P\Delta T_{41} = \tfrac{5}{2}R\Delta T_{41} = \tfrac{5}{2}\left(8.314\frac{\text{J}}{\text{mol}\cdot\text{K}}\right)(300\text{ K} - 600\text{ K}) = \boxed{-6.24\text{ kJ}}$$

Apply $\Delta E_{int} = Q_{in} + W_{on}$ to obtain:

$$\Delta E_{int,\,41} = -6.24\text{ kJ} + 2.49\text{ kJ} = \boxed{-3.75\text{ kJ}}$$

For easy reference, the results of the preceding calculations are summarized in the following table:

Process	W_{on}, kJ	Q_{in}, kJ	$\Delta E_{\text{int}}\left(=Q_{\text{in}}+W_{\text{on}}\right)$, kJ
1→2	0	3.74	3.74
2→3	−4.99	12.5	7.5
3→4	0	−7.48	−7.48
4→1	2.49	−6.24	−3.75

(*b*) The efficiency of the cycle is given by:

$$\varepsilon=\frac{W_{\text{by}}}{Q_{\text{in}}}=\frac{-W_{23}+\left(-W_{41}\right)}{Q_{12}+Q_{23}}$$

Substitute numerical values and evaluate ε:

$$\varepsilon=\frac{4.99\,\text{kJ}-2.49\,\text{kJ}}{3.74\,\text{kJ}+12.5\,\text{kJ}}\approx\boxed{15\%}$$

Remarks: Note that the work done per cycle is the area bounded by the rectangular path. Note also that, as expected because the system returns to its initial state, the sum of the changes in the internal energy for the cycle is zero.

Second Law of Thermodynamics

39 •• **[SSM]** A refrigerator absorbs 500 J of heat from a cold reservoir and releases 800 J to a hot reservoir. Assume that the heat-engine statement of the second law of thermodynamics is false, and show how a perfect engine working with this refrigerator can violate the refrigerator statement of the second law of thermodynamics.

Determine the Concept The following diagram shows an ordinary refrigerator that uses 300 J of work to remove 500 J of heat from a cold reservoir and releases 800 J of heat to a hot reservoir (see (*a*) in the diagram). Suppose the heat-engine statement of the second law is false. Then a "perfect" heat engine could remove energy from the hot reservoir and convert it completely into work with 100 percent efficiency. We could use this perfect heat engine to remove 300 J of energy from the hot reservoir and do 300 J of work on the ordinary refrigerator (see (*b*) in the diagram). Then, the combination of the perfect heat engine and the ordinary refrigerator would be a perfect refrigerator; transferring 500 J of heat from the cold reservoir to the hot reservoir without requiring any work (see (*c*) in the diagram).This violates the refrigerator statement of the second law.

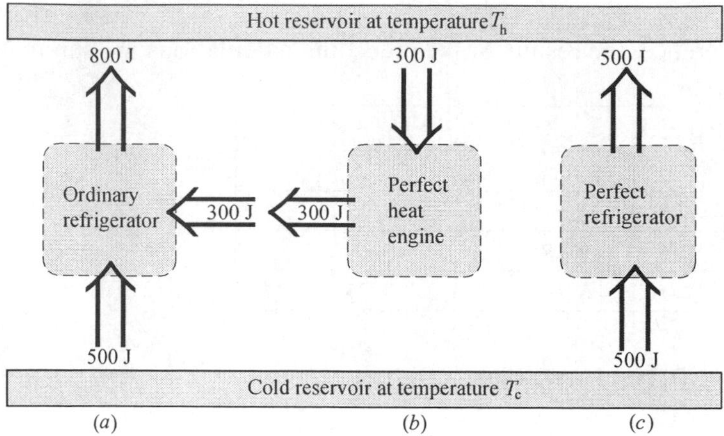

Carnot Cycles

41 • **[SSM]** A Carnot engine works between two heat reservoirs at temperatures $T_h = 300$ K and $T_c = 200$ K. (*a*) What is its efficiency? (*b*) If it absorbs 100 J of heat from the hot reservoir during each cycle, how much work does it do each cycle? (*c*) How much heat does it release during each cycle? (*d*) What is the COP of this engine when it works as a refrigerator between the same two reservoirs?

Picture the Problem We can find the efficiency of the Carnot engine using $\varepsilon = 1 - T_c / T_h$ and the work done per cycle from $\varepsilon = W / Q_h$. We can apply conservation of energy to find the heat rejected each cycle from the heat absorbed and the work done each cycle. We can find the COP of the engine working as a refrigerator from its definition.

(*a*) The efficiency of the Carnot engine depends on the temperatures of the hot and cold reservoirs:

$$\varepsilon_C = 1 - \frac{T_c}{T_h} = 1 - \frac{200\,\text{K}}{300\,\text{K}} = \boxed{33.3\%}$$

(*b*) Using the definition of efficiency, relate the work done each cycle to the heat absorbed from the hot reservoir:

$$W = \varepsilon_C Q_h = (0.333)(100\,\text{J}) = \boxed{33.3\,\text{J}}$$

(*c*) Apply conservation of energy to relate the heat given off each cycle to the heat absorbed and the work done:

$$Q_c = Q_h - W = 100\,\text{J} - 33.3\,\text{J} = 66.7\,\text{J}$$

$$= \boxed{67\,\text{J}}$$

(*d*) Using its definition, express and evaluate the refrigerator's coefficient of performance:

$$\text{COP} = \frac{Q_c}{W} = \frac{66.7\,\text{J}}{33.3\,\text{J}} = \boxed{2.0}$$

47 •• **[SSM]** In the cycle shown 1 in Figure 19-19, 1.00 mol of an ideal diatomic gas is initially at a pressure of 1.00 atm and a temperature of 0.0°C. The gas is heated at constant volume to $T_2 = 150°\text{C}$ and is then expanded adiabatically until its pressure is again 1.00 atm. It is then compressed at constant pressure back to its original state. Find (*a*) the temperature after the adiabatic expansion, (*b*) the heat absorbed or released by the system during each step, (*c*) the efficiency of this cycle, and (*d*) the efficiency of a Carnot cycle operating between the temperature extremes of this cycle.

Picture the Problem We can use the ideal-gas law for a fixed amount of gas and the equations of state for an adiabatic process to find the temperatures, volumes, and pressures at the end points of each process in the given cycle. We can use $Q = C_V\Delta T$ and $Q = C_P\Delta T$ to find the heat entering and leaving during the constant-volume and isobaric processes and the first law of thermodynamics to find the work done each cycle. Once we've calculated these quantities, we can use its definition to find the efficiency of the cycle and the definition of the Carnot efficiency to find the efficiency of a Carnot engine operating between the extreme temperatures.

(*a*) Apply the ideal-gas law for a fixed amount of gas to relate the temperature at point 3 to the temperature at point 1:

$$\frac{P_1V_1}{T_1} = \frac{P_3V_3}{T_3}$$

or, because $P_1 = P_3$,

$$T_3 = T_1\frac{V_3}{V_1} \qquad (1)$$

Apply the ideal-gas law for a fixed amount of gas to relate the pressure at point 2 to the temperatures at points 1 and 2 and the pressure at 1:

$$\frac{P_1V_1}{T_1} = \frac{P_2V_2}{T_2} \Rightarrow P_2 = \frac{P_1V_1T_2}{V_2T_1}$$

Because $V_1 = V_2$:

$$P_2 = P_1\frac{T_2}{T_1} = (1.00\,\text{atm})\frac{423\,\text{K}}{273\,\text{K}} = 1.55\,\text{atm}$$

Apply an equation for an adiabatic process to relate the pressures and volumes at points 2 and 3:

$$P_1V_1^{\gamma} = P_3V_3^{\gamma} \Rightarrow V_3 = V_1\left(\frac{P_1}{P_3}\right)^{\frac{1}{\gamma}}$$

Noting that $V_1 = 22.4$ L, evaluate V_3:

$$V_3 = (22.4\,\text{L})\left(\frac{1.55\,\text{atm}}{1\,\text{atm}}\right)^{\frac{1}{1.4}} = 30.6\,\text{L}$$

Substitute numerical values in equation (1) and evaluate T_3 and t_3:

$$T_3 = (273\,\text{K})\frac{30.6\,\text{L}}{22.4\,\text{L}} = 373\,\text{K}$$

and

$$t_3 = T_3 - 273 = \boxed{100°\text{C}}$$

(*b*) Process 1→2 takes place at constant volume (note that $\gamma = 1.4$ corresponds to a diatomic gas and that $C_P - C_V = R$):

$$Q_{12} = C_V \Delta T_{12} = \tfrac{5}{2} R \Delta T_{12}$$
$$= \tfrac{5}{2}\left(8.314 \frac{\text{J}}{\text{mol}\cdot\text{K}}\right)(423\,\text{K} - 273\,\text{K})$$
$$= \boxed{3.12\,\text{kJ}}$$

Process 2→3 takes place adiabatically:

$$Q_{23} = \boxed{0}$$

Process 3→1 is isobaric (note that $C_P = C_V + R$):

$$Q_{31} = C_P \Delta T_{31} = \tfrac{7}{2} R \Delta T_{12}$$
$$= \tfrac{7}{2}\left(8.314 \frac{\text{J}}{\text{mol}\cdot\text{K}}\right)(273\,\text{K} - 373\,\text{K})$$
$$= \boxed{-2.91\,\text{kJ}}$$

(*c*) The efficiency of the cycle is given by:

$$\varepsilon = \frac{W}{Q_{\text{in}}} \qquad (2)$$

Apply the first law of thermodynamics to the cycle:

$$\Delta E_{\text{int}} = Q_{\text{in}} + W_{\text{on}}$$

or, because $\Delta E_{\text{int, cycle}} = 0$ (the system begins and ends in the same state) and $W_{\text{on}} = -W_{\text{by the gas}} = Q_{\text{in}}$.

Evaluating W yields:

$$W = \sum Q = Q_{12} + Q_{23} + Q_{31}$$
$$= 3.12\,\text{kJ} + 0 - 2.91\,\text{kJ} = 0.21\,\text{kJ}$$

Substitute numerical values in equation (2) and evaluate ε:

$$\varepsilon = \frac{0.21\,\text{kJ}}{3.12\,\text{kJ}} = \boxed{6.7\%}$$

(*d*) Express and evaluate the efficiency of a Carnot cycle operating between 423 K and 273 K:

$$\varepsilon_C = 1 - \frac{T_c}{T_h} = 1 - \frac{273\,\text{K}}{423\,\text{K}} = \boxed{35.5\%}$$

*Heat Pumps

49 • **[SSM]** As an engineer, you are designing a heat pump that is capable of delivering heat at the rate of 20 kW to a house. The house is located where, in January, the average outside temperature is –10°C. The temperature of the air in the air handler inside the house is to be 40°C. (*a*) What is maximum possible COP for a heat pump operating between these temperatures? (*b*) What must be the minimum power of the electric motor driving the heat pump? (*c*) In reality, the COP of the heat pump will be only 60 percent of the ideal value. What is the minimum power of the engine when the COP is 60 percent of the ideal value?

Picture the Problem We can use the definition of the COP_{HP} and the Carnot efficiency of an engine to express the maximum efficiency of the refrigerator in terms of the reservoir temperatures. We can apply the definition of power to find the minimum power needed to run the heat pump.

(*a*) Express the COP_{HP} in terms of T_h and T_c:

$$\text{COP}_{\text{HP}} = \frac{Q_h}{W} = \frac{Q_h}{Q_h - Q_c} = \frac{1}{1 - \dfrac{Q_c}{Q_h}} = \frac{1}{1 - \dfrac{T_c}{T_h}} = \frac{T_h}{T_h - T_c}$$

Substitute numerical values and evaluate COP_{HP}:

$$\text{COP}_{\text{HP}} = \frac{313\,\text{K}}{313\,\text{K} - 263\,\text{K}} = 6.26 = \boxed{6.3}$$

(*b*) The COP_{HP} is also given by:

$$\text{COP}_{\text{HP}} = \frac{P_{\text{out}}}{P_{\text{motor}}} \Rightarrow P_{\text{motor}} = \frac{P_{\text{out}}}{\text{COP}_{\text{HP}}}$$

Substitute numerical values and evaluate P_{motor}:

$$\boldsymbol{P}_{\text{motor}} = \frac{20\,\text{kW}}{6.26} = \boxed{3.2\,\text{kW}}$$

(*c*) The minimum power of the engine is given by:

$$P_{\min} = \frac{\dfrac{dQ_c}{dt}}{\varepsilon_{\text{HP}}} = \frac{\dfrac{dQ_c}{dt}}{\varepsilon\left(\text{COP}_{\text{HP,max}}\right)}$$

where ε_{HP} is the efficiency of the heat pump.

Substitute numerical values and evaluate P_{min}:

$$P_{\min} = \frac{20\,\text{kW}}{(0.60)(6.26)} = \boxed{5.3\,\text{kW}}$$

Entropy Changes

53 • **[SSM]** You inadvertently leave a pan of water boiling away on the hot stove. You return just in time to see the last drop converted into steam. The pan originally held 1.00 L of boiling water. What is the change in entropy of the water associated with its change of state from liquid to gas?

Picture the Problem Because the water absorbed heat in the vaporization process its change in entropy is positive and given by $\Delta S_{H_2O} = \dfrac{Q_{\substack{\text{absorbed}\\ \text{by } H_2O}}}{T}$. See Table 18-2 for the latent heat of vaporization of water.

The change in entropy of the water is given by:

$$\Delta S_{H_2O} = \frac{Q_{\substack{\text{absorbed}\\ \text{by } H_2O}}}{T}$$

The heat absorbed by the water as it vaporizes is the product of its mass and latent heat of vaporization:

$$Q_{\substack{\text{absorbed}\\ \text{by } H_2O}} = mL_v = \rho V L_v$$

Substituting for $Q_{\substack{\text{absorbed}\\ \text{by } H_2O}}$ yields:

$$\Delta S_{H_2O} = \frac{\rho V L_v}{T}$$

Substitute numerical values and evaluate ΔS_{H_2O}:

$$\Delta S_{H_2O} = \frac{\left(1.00\,\frac{\text{kg}}{\text{L}}\right)(1.00\,\text{L})\left(2257\,\frac{\text{kJ}}{\text{kg}}\right)}{373\,\text{K}} = \boxed{6.05\,\frac{\text{kJ}}{\text{K}}}$$

57 •• **[SSM]** A system completes a cycle consisting of six quasi-static steps, during which the total work done by the system is 100 J. During step 1 the system absorbs 300 J of heat from a reservoir at 300 K, during step 3 the system absorbs 200 J of heat from a reservoir at 400 K, and during step 5 it absorbs heat from a reservoir at temperature T_3. (During steps 2, 4 and 6 the system undergoes adiabatic processes in which the temperature of the system changes from one reservoir's temperature to that of the next.) (*a*) What is the entropy change of the system for the complete cycle? (*b*) If the cycle is reversible, what is the temperature T_3?

Picture the Problem We can use the fact that the system *returns to its original state* to find the entropy change for the complete cycle. Because the entropy change for the complete cycle is the sum of the entropy changes for each process, we can find the temperature T_3 from the entropy changes during the 1st two processes and the heat released during the third.

(*a*) Because S is a state function of the system, and because the system's final state is identical to its initial state:

$$\Delta S_{\substack{\text{system}\\ \text{1 complete cycle}}} = \boxed{0}$$

(*b*) Relate the entropy changes for each of the three heat reservoirs and the system for one complete cycle of the system:

$$\Delta S_1 + \Delta S_2 + \Delta S_3 + \Delta S_{\text{system}} = 0$$

or

$$\frac{Q_1}{T_1} + \frac{Q_2}{T_2} + \frac{Q_3}{T_3} + 0 = 0$$

Substitute numerical values. Heat is rejected by the two high-temperature reservoirs and absorbed by the cold reservoir:

$$\frac{-300\,\text{J}}{300\,\text{K}} + \frac{-200\,\text{J}}{400\,\text{K}} + \frac{400\,\text{J}}{T_3} = 0$$

Solving for T_3 yields:

$$T_3 = \boxed{267\,\text{K}}$$

61 •• **[SSM]** A 1.00-kg block of copper at 100°C is placed in an insulated calorimeter of negligible heat capacity containing 4.00 L of liquid water at 0.0°C. Find the entropy change of (*a*) the copper block, (*b*) the water, and (*c*) the universe.

Picture the Problem We can use conservation of energy to find the equilibrium temperature of the water and apply the equations for the entropy change during a constant pressure process to find the entropy changes of the copper block, the water, and the universe.

(*a*) Use the equation for the entropy change during a constant-pressure process to express the entropy change of the copper block:

$$\Delta S_{\text{Cu}} = m_{\text{Cu}} c_{\text{Cu}} \ln\left(\frac{T_{\text{f}}}{T_{\text{i}}}\right) \qquad (1)$$

Apply conservation of energy to obtain:

$$\sum_{\text{i}} Q_{\text{i}} = 0$$

or

$$Q_{\text{copper block}} + Q_{\text{warming water}} = 0$$

Substitute to relate the masses of the block and water to their temperatures, specific heats, and the final temperature T_f of the water:

$$(1.00\,\text{kg})\left(0.386\frac{\text{kJ}}{\text{kg}\cdot\text{K}}\right)(T_f - 373\,\text{K}) + (4.00\,\text{L})\left(1.00\frac{\text{kg}}{\text{L}}\right)\left(4.18\frac{\text{kJ}}{\text{kg}\cdot\text{K}}\right)(T_f - 273\,\text{K}) = 0$$

Solve for T_f to obtain: $T_f = 275.26\,\text{K}$

Substitute numerical values in equation (1) and evaluate ΔS_{Cu}:

$$\Delta S_{\text{Cu}} = (1.00\,\text{kg})\left(0.386\frac{\text{kJ}}{\text{kg}\cdot\text{K}}\right)\ln\left(\frac{275.26\,\text{K}}{373\,\text{K}}\right) = \boxed{-117\frac{\text{J}}{\text{K}}}$$

(*b*) The entropy change of the water is given by:

$$\Delta S_{\text{water}} = m_{\text{water}}c_{\text{water}}\ln\left(\frac{T_f}{T_i}\right)$$

Substitute numerical values and evaluate ΔS_{water}:

$$\Delta S_{\text{water}} = (4.00\,\text{kg})\left(4.18\frac{\text{kJ}}{\text{kg}\cdot\text{K}}\right)\ln\left(\frac{275.26\,\text{K}}{273\,\text{K}}\right) = \boxed{138\frac{\text{J}}{\text{K}}}$$

(*c*) Substitute for ΔS_{Cu} and ΔS_{water} and evaluate the entropy change of the universe:

$$\Delta S_u = \Delta S_{\text{Cu}} + \Delta S_{\text{water}} = -117\frac{\text{J}}{\text{K}} + 138\frac{\text{J}}{\text{K}} = \boxed{20\frac{\text{J}}{\text{K}}}$$

Remarks: The result that $\Delta S_u > 0$ tells us that this process is irreversible.

Entropy and "Lost" Work

63 •• **[SSM]** A a reservoir at 300 K absorbs 500 J of heat from a second reservoir at 400 K. (*a*) What is the change in entropy of the universe, and (*b*) how much work is lost during the process?

Picture the Problem We can find the entropy change of the universe from the entropy changes of the high- and low-temperature reservoirs. The maximum amount of the 500 J of heat that could be converted into work can be found from the maximum efficiency of an engine operating between the two reservoirs.

(*a*) The entropy change of the universe is the sum of the entropy changes of the two reservoirs:

$$\Delta S_{\mathrm{u}} = \Delta S_{\mathrm{h}} + \Delta S_{\mathrm{c}} = -\frac{Q}{T_{\mathrm{h}}} + \frac{Q}{T_{\mathrm{c}}}$$

$$= -Q\left(\frac{1}{T_{\mathrm{h}}} - \frac{1}{T_{\mathrm{c}}}\right)$$

Substitute numerical values and evaluate ΔS_{u}:

$$\Delta S_{\mathrm{u}} = (-500\,\mathrm{J})\left(\frac{1}{400\,\mathrm{K}} - \frac{1}{300\,\mathrm{K}}\right)$$

$$= \boxed{0.42\,\mathrm{J/K}}$$

(*b*) Relate the heat that could have been converted into work to the maximum efficiency of an engine operating between the two reservoirs:

$$W = \varepsilon_{\mathrm{max}} Q_{\mathrm{h}}$$

The maximum efficiency of an engine operating between the two reservoir temperatures is the efficiency of a Carnot device operating between the reservoir temperatures:

$$\varepsilon_{\mathrm{max}} = \varepsilon_{\mathrm{C}} = 1 - \frac{T_{\mathrm{c}}}{T_{\mathrm{h}}}$$

Substitute for $\varepsilon_{\mathrm{max}}$ to obtain:

$$W = \left(1 - \frac{T_{\mathrm{c}}}{T_{\mathrm{h}}}\right) Q_{\mathrm{h}}$$

Substitute numerical values and evaluate W:

$$W = \left(1 - \frac{300\,\mathrm{K}}{400\,\mathrm{K}}\right)(500\,\mathrm{J}) = \boxed{125\,\mathrm{J}}$$

General Problems

67 • **[SSM]** An engine absorbs 200 kJ of heat per cycle from a reservoir at 500 K and releases heat to a reservoir at 200 K. Its efficiency is 85 percent of that of a Carnot engine working between the same reservoirs. (*a*) What is the efficiency of this engine? (*b*) How much work is done in each cycle? (*c*) How much heat is released to the low-temperature reservoir during each cycle?

Picture the Problem We can use the definition of efficiency to find the work done by the engine during each cycle and the first law of thermodynamics to find the heat released to the low-temperature reservoir during each cycle.

(*a*) Express the efficiency of the engine in terms of the efficiency of a Carnot engine working between the same reservoirs:

$$\varepsilon = 0.85\varepsilon_C = 0.85\left(1 - \frac{T_c}{T_h}\right)$$

Substitute numerical values and evaluate ε:

$$\varepsilon = 0.85\left(1 - \frac{200\,\text{K}}{500\,\text{K}}\right) = 0.510 = \boxed{51\%}$$

(*b*) Use the definition of efficiency to find the work done in each cycle:

$$W = \varepsilon Q_h = (0.510)(200\,\text{kJ}) = 102\,\text{kJ} = \boxed{0.10\,\text{MJ}}$$

(*c*) Apply the first law of thermodynamics to the cycle to obtain:

$$Q_{c,\text{cycle}} = Q_{h,\text{cycle}} - W = 200\,\text{kJ} - 102\,\text{kJ} = \boxed{98\,\text{kJ}}$$

73 •• **[SSM]** (*a*) Which of these two processes is more wasteful? (1) A block moving with 500 J of kinetic energy being slowed to rest by sliding (kinetic) friction when the temperature of the environment is 300 K, or (2) A reservoir at 400 K releasing 1.00 kJ of heat to a reservoir at 300 K? Explain your choice. *Hint: How much of the 1.00 kJ of heat could be converted into work by an ideal cyclic process?* (*b*) What is the change in entropy of the universe for each process?

Picture the Problem All 500 J of mechanical energy are lost, i.e., transformed into heat in process (1). For process (2), we can find the heat that would be converted to work by a Carnot engine operating between the given temperatures and subtract that amount of work from 1.00 kJ to find the energy that is lost. In Part (*b*) we can use its definition to find the change in entropy for each process.

(*a*) For process (2):

$$W_{2,\text{max}} = W_{\text{recovered}} = \varepsilon_C Q_{\text{in}}$$

The efficiency of a Carnot engine operating between temperatures T_h and T_c is given by:

$$\varepsilon_C = 1 - \frac{T_c}{T_h}$$

and hence

$$W_{\text{recovered}} = \left(1 - \frac{T_c}{T_h}\right)Q_{\text{in}}$$

Substitute for ε_C to obtain:

$$W_{\text{recovered}} = \left(1 - \frac{300\text{ K}}{400\text{ K}}\right)(1.00\text{ kJ}) = 250\text{ J}$$

or 750 J are lost.

Process (1) is more wasteful of *mechanical* energy. Process (2) is more wasteful of *total* energy.

(*b*) Find the change in entropy of the universe for process (1):

$$\Delta S_1 = \frac{\Delta Q}{T} = \frac{500\text{ J}}{300\text{ K}} = \boxed{1.67\text{ J/K}}$$

Express the change in entropy of the universe for process (2):

$$\Delta S_2 = \Delta S_h + \Delta S_c = -\frac{\Delta Q}{T_h} + \frac{\Delta Q}{T_c}$$

$$= \Delta Q\left(\frac{1}{T_c} - \frac{1}{T_h}\right)$$

Substitute numerical values and evaluate ΔS_2:

$$\Delta S_2 = (1.00\text{ kJ})\left(\frac{1}{300\text{ K}} - \frac{1}{400\text{ K}}\right)$$

$$= \boxed{0.833\text{ J/K}}$$

75 •• **[SSM]** A heat engine that does the work of blowing up a balloon at a pressure of 1.00 atm absorbs 4.00 kJ from a reservoir at 120°C. The volume of the balloon increases by 4.00 L, and heat is released to a reservoir at a temperature T_c, where $T_c < 120$°C. If the efficiency of the heat engine is 50% of the efficiency of a Carnot engine working between the same two reservoirs, find the temperature T_c.

Picture the Problem We can express the temperature of the cold reservoir as a function of the Carnot efficiency of an ideal engine and, given that the efficiency of the heat engine is half that of a Carnot engine, relate T_c to the work done by and the heat input to the real heat engine.

Using its definition, relate the efficiency of a Carnot engine working between the same reservoirs to the temperature of the cold reservoir:

$$\varepsilon_C = 1 - \frac{T_c}{T_h} \Rightarrow T_c = T_h(1 - \varepsilon_C)$$

Relate the efficiency of the heat engine to that of a Carnot engine working between the same temperatures:

$$\varepsilon = \frac{W}{Q_{in}} = \tfrac{1}{2}\varepsilon_C \Rightarrow \varepsilon_C = \frac{2W}{Q_{in}}$$

Substitute for ε_C to obtain:

$$T_c = T_h\left(1 - \frac{2W}{Q_{in}}\right)$$

The work done by the gas in expanding the balloon is:

$$W = P\Delta V = (1.00\,\text{atm})(4.00\,\text{L}) = 4.00\,\text{atm}\cdot\text{L}$$

Substitute numerical values and evaluate T_c:

$$T_c = (393\,\text{K})\left(1 - \frac{2\left(4.00\,\text{atm}\cdot\text{L}\times\frac{101.325\,\text{J}}{\text{atm}\cdot\text{L}}\right)}{4.00\,\text{kJ}}\right) = \boxed{313\,\text{K}}$$

79 •• **[SSM]** In a heat engine, 2.00 mol of a diatomic gas are carried through the cycle ABCDA shown in Figure 19-21. (The *PV* diagram is not drawn to scale.) The segment AB represents an isothermal expansion, the segment BC an adiabatic expansion. The pressure and temperature at A are 5.00 atm and 600 K. The volume at B is twice the volume at A. The pressure at D is 1.00 atm. (*a*) What is the pressure at B? (*b*) What is the temperature at C? (*c*) Find the total work done by the gas in one cycle.

Picture the Problem We can use the ideal-gas law to find the unknown temperatures, pressures, and volumes at points B, C, and D. We can then find the work done by the gas and the efficiency of the cycle by using the expressions for the work done on or by the gas and the heat that enters the system for the various thermodynamic processes of the cycle.

(*a*) Apply the ideal-gas law for a fixed amount of gas to the isothermal process AB to find the pressure at B:

$$P_B = P_A\frac{V_A}{V_B} = (5.00\,\text{atm})\frac{V_A}{2V_A} = 2.50\,\text{atm}\times\frac{101.325\,\text{kPa}}{1\,\text{atm}} = 253.3\,\text{kPa} = \boxed{253\,\text{kPa}}$$

(*b*) Apply the ideal-gas law for a fixed amount of gas to the adiabatic process BC to express the temperature at C:

$$T_C = T_B\frac{P_C V_C}{P_B V_B} \qquad (1)$$

Use the ideal-gas law to find the volume of the gas at B:

$$V_B = \frac{nRT_B}{P_B} = \frac{(2.00\,\text{mol})\left(8.314\frac{\text{J}}{\text{mol}\cdot\text{K}}\right)(600\,\text{K})}{253.3\,\text{kPa}} = 39.39\,\text{L}$$

Use the equation of state for an adiabatic process and $\gamma = 1.4$ to find the volume occupied by the gas at C:

$$V_C = V_B\left(\frac{P_B}{P_C}\right)^{1/\gamma} = (39.39\,\text{L})\left(\frac{2.50\,\text{atm}}{1.00\,\text{atm}}\right)^{1/1.4} = 75.78\,\text{L}$$

Substitute numerical values in equation (1) and evaluate T_C:

$$T_C = (600\,\text{K})\frac{(1.00\,\text{atm})(75.78\,\text{L})}{(2.50\,\text{atm})(39.39\,\text{L})} = \boxed{462\,\text{K}}$$

(*c*) The work done by the gas in one cycle is given by:

$$W = W_{AB} + W_{BC} + W_{CD} + W_{DA}$$

The work done during the isothermal expansion AB is:

$$W_{AB} = nRT_A \ln\left(\frac{V_B}{V_A}\right) = (2.00\,\text{mol})\left(8.314\frac{\text{J}}{\text{mol}\cdot\text{K}}\right)(600\,\text{K})\ln\left(\frac{2\text{V}_A}{\text{V}_A}\right) = 6.915\,\text{kJ}$$

The work done during the adiabatic expansion BC is:

$$W_{BC} = -C_V\Delta T_{BC} = -\tfrac{5}{2}nR\Delta T_{BC} = -\tfrac{5}{2}(2.00\,\text{mol})\left(8.314\frac{\text{J}}{\text{mol}\cdot\text{K}}\right)(462\,\text{K} - 600\,\text{K}) = 5.737\,\text{kJ}$$

The work done during the isobaric compression CD is:

$$W_{CD} = P_C(V_D - V_C) = (1.00\,\text{atm})(19.7\,\text{L} - 75.78\,\text{L}) = -56.09\,\text{atm}\cdot\text{L}\times\frac{101.325\,\text{J}}{\text{atm}\cdot\text{L}} = -5.680\,\text{kJ}$$

Express and evaluate the work done during the constant-volume process DA:

$$W_{DA} = 0$$

Substitute numerical values and evaluate W:

$$W = 6.915\,\text{kJ} + 5.737\,\text{kJ} - 5.680\,\text{kJ} + 0$$
$$= 6.972\,\text{kJ} = \boxed{6.97\,\text{kJ}}$$

83 ••• **[SSM]** A common practical cycle, often used in refrigeration, is the *Brayton cycle*, which involves (1) an adiabatic compression, (2) an isobaric (constant pressure) expansion,(3) an adiabatic expansion, and (4) an isobaric compression back to the original state. Assume the system begins the adiabatic compression at temperature T_1, and transitions to temperatures T_2, T_3 and T_4 after each leg of the cycle. (*a*) Sketch this cycle on a *PV* diagram. (*b*) Show that the efficiency of the overall cycle is given by $\varepsilon = 1 - \frac{(T_4 - T_1)}{(T_3 - T_2)}$. (*c*) Show that this efficiency, can be written as $\varepsilon = 1 - r^{(1-\gamma)/\gamma}$, where r is the pressure ratio P_{high}/P_{low} of the maximum and minimum pressures in the cycle.

Picture the Problem The efficiency of the cycle is the ratio of the work done to the heat that flows into the engine. Because the adiabatic transitions in the cycle do not have heat flow associated with them, all we must do is consider the heat flow in and out of the engine during the isobaric transitions.

(*a*) The *Brayton heat engine cycle* is shown to the right. The paths 1→2 and 3→4 are adiabatic. Heat Q_h enters the gas during the isobaric transition from state 2 to state 3 and heat Q_c leaves the gas during the isobaric transition from state 4 to state 1.

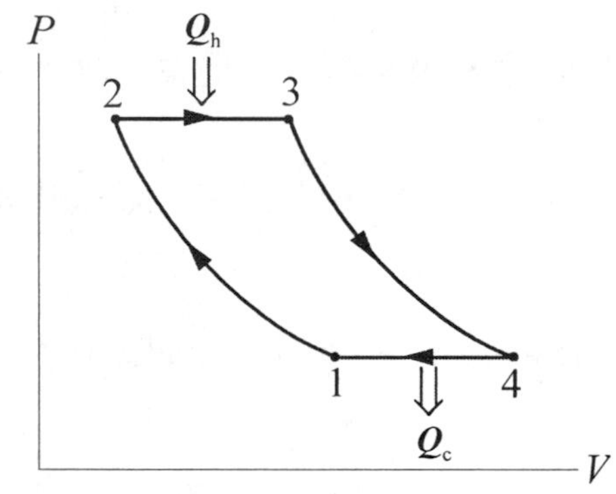

(*b*) The efficiency of a heat engine is given by:

$$\varepsilon = \frac{W}{Q_{in}} = \frac{Q_h - Q_c}{Q_{in}} \quad (1)$$

During the constant-pressure expansion from state 1 to state 2 heat enters the system:

$$Q_{23} = Q_h = nC_P\Delta T = nC_P(T_3 - T_2)$$

During the constant-pressure compression from state 3 to state 4 heat enters the system:

$$Q_{41} = -Q_c = -nC_P\Delta T = -nC_P(T_1 - T_4)$$

Substituting in equation (1) and simplifying yields:

$$\varepsilon=\frac{nC_P(T_3-T_2)-(-nC_P(T_1-T_4))}{nC_P(T_3-T_2)}$$

$$=\frac{(T_3-T_2)+(T_1-T_4)}{(T_3-T_2)}$$

$$=\boxed{1-\frac{(T_4-T_1)}{(T_3-T_2)}}$$

(*c*) Given that, for an adiabatic transition, $TV^{\gamma-1}=\text{constant}$, use the ideal-gas law to eliminate V and obtain:

$$\frac{T^{\gamma}}{P^{\gamma-1}}=\text{constant}$$

Let the pressure for the transition from state 1 to state 2 be P_{low} and the pressure for the transition from state 3 to state 4 be P_{high}. Then for the adiabatic transition from state 1 to state 2:

$$\frac{T_1^{\gamma}}{P_{\text{low}}^{\gamma-1}}=\frac{T_2^{\gamma}}{P_{\text{high}}^{\gamma-1}}\Rightarrow T_1=\left(\frac{P_{\text{low}}}{P_{\text{high}}}\right)^{\frac{\gamma-1}{\gamma}}T_2$$

Similarly, for the adiabatic transition from state 3 to state 4:

$$T_4=\left(\frac{P_{\text{low}}}{P_{\text{high}}}\right)^{\frac{\gamma-1}{\gamma}}T_3$$

Subtract T_1 from T_4 and simplify to obtain:

$$T_4-T_1=\left(\frac{P_{\text{low}}}{P_{\text{high}}}\right)^{\frac{\gamma-1}{\gamma}}T_3-\left(\frac{P_{\text{low}}}{P_{\text{high}}}\right)^{\frac{\gamma-1}{\gamma}}T_2$$

$$=\left(\frac{P_{\text{low}}}{P_{\text{high}}}\right)^{\frac{\gamma-1}{\gamma}}(T_3-T_2)$$

Dividing both sides of the equation by T_3-T_2 yields:

$$\frac{T_4-T_1}{T_3-T_2}=\left(\frac{P_{\text{low}}}{P_{\text{high}}}\right)^{\frac{\gamma-1}{\gamma}}$$

Substitute in the result of Part (*b*) and simplify to obtain:

$$\varepsilon = 1-\left(\frac{P_{\text{low}}}{P_{\text{high}}}\right)^{\frac{\gamma-1}{\gamma}} = 1-\left(\frac{P_{\text{high}}}{P_{\text{low}}}\right)^{\frac{1-\gamma}{\gamma}}$$

$$= \boxed{1-(r)^{\frac{1-\gamma}{\gamma}}}$$

where $r = \dfrac{P_{\text{high}}}{P_{\text{low}}}$

89 ••• **[SSM]** The English mathematician and philosopher Bertrand Russell (1872-1970) once said that if a million monkeys were given a million typewriters and typed away at random for a million years, they would produce all of Shakespeare's works. Let us limit ourselves to the following fragment of Shakespeare (*Julius Caesar* III:ii):

Friends, Romans, countrymen! Lend me your ears.
I come to bury Caesar, not to praise him.
The evil that men do lives on after them,
The good is oft interred with the bones.
So let it be with Caesar.
The noble Brutus hath told you that Caesar was ambitious,
And, if so, it were a grievous fault,
And grievously hath Caesar answered it . . .

Even with this small fragment, it will take a lot longer than a million years! By what factor (roughly speaking) was Russell in error? Make any reasonable assumptions you want. (You can even assume that the monkeys are immortal.)

Picture the Problem There are 26 letters and four punctuation marks (space, comma, period, and exclamation point) used in the English language, disregarding capitalization, so we have a grand total of 30 characters to choose from. This fragment is 330 characters (including spaces) long; there are then 30^{330} different possible arrangements of the character set to form a fragment this long. We can use this number of possible arrangements to express the probability that one monkey will write out this passage and then an estimate of a monkey's typing speed to approximate the time required for one million monkeys to type the passage from Shakespeare.

Assuming the monkeys type at random, express the probability P that one monkey will write out this passage:

$$P = \frac{1}{30^{330}}$$

Use the approximation $30 \approx \sqrt{1000} = 10^{1.5}$ to obtain:

$$P = \frac{1}{10^{(1.5)(330)}} = \frac{1}{10^{495}} = 10^{-495}$$

Assuming the monkeys can type at a rate of 1 character per second, it would take about 330 s to write a passage of length equal to the quotation from Shakespeare. Find the time T required for a million monkeys to type this particular passage by accident:

$$T = \frac{(330\,\text{s})(10^{495})}{10^6} = (3.30\times10^{491}\,\text{s})\left(\frac{1\,\text{y}}{3.16\times10^{7}\,\text{s}}\right) \approx \boxed{10^{484}\,\text{y}}$$

Express the ratio of T to Russell's estimate:

$$\frac{T}{T_{\text{Russell}}} = \frac{10^{484}\,\text{y}}{10^{6}\,\text{y}} = 10^{478}$$

or

$$T \approx \boxed{10^{478}\,T_{\text{Russell}}}$$

Chapter 20
Thermal Properties and Processes

Conceptual Problems

3 • **[SSM]** Why is it a bad idea to place a tightly sealed glass bottle that is completely full of water, into your kitchen freezer in order to make ice?

Determine the Concept Water expands greatly as it freezes. If a sealed glass bottle full of water is placed in a freezer, as the water freezes there will be no room for the expansion to take place. Because the glass contracts as it gets cold, making a slightly smaller volume available for the freezing water, ultimately the bottle will shatter.

9 •• **[SSM]** The phase diagram in Figure 20-15 can be interpreted to yield information on how the boiling and melting points of water change with altitude. (*a*) Explain how this information can be obtained. (*b*) How might this information affect cooking procedures in the mountains?

Determine the Concept
(*a*) With increasing altitude, P decreases; from curve OF, T of the liquid-gas interface diminishes, so the boiling temperature decreases. Likewise, from curve OH, the melting temperature increases with increasing altitude.

(*b*) Boiling at a lower temperature means that the cooking time will have to be increased.

13 •• **[SSM]** Two solid cylinders made of materials A and B have the same lengths; their diameters are related by $d_A = 2d_B$. When the same temperature difference is maintained between the ends of the cylinders, they conduct heat at the same rate. Their thermal conductivities are therefore related by which of the following equations? (*a*) $k_A = k_B/4$, (*b*) $k_A = k_B/2$, (*c*) $k_A = k_B$, (*d*) $k_A = 2k_B$, (*e*) $k_A = 4k_B$

Picture the Problem The rate at which heat is conducted through a cylinder is given by $I = dQ/dt = kA\Delta T/\Delta x$ where A is the cross-sectional area of the cylinder.

Express the rate at which heat is conducted through cylinder A:

$$I_A = kA_A\frac{\Delta T}{\Delta x} = \tfrac{1}{4}k_A\pi d_A^2\frac{\Delta T}{\Delta x}$$

Express the rate at which heat is conducted through cylinder B:

$$I_B = kA_B\frac{\Delta T}{\Delta x} = \tfrac{1}{4}k_B\pi d_B^2\frac{\Delta T}{\Delta x}$$

Equate these expressions and simplify to obtain:

$$\tfrac{1}{4}k_{A}\pi d_{A}^{2}\frac{\Delta T}{\Delta x}=\tfrac{1}{4}k_{B}\pi d_{B}^{2}\frac{\Delta T}{\Delta x}$$

or

$$k_{A}d_{A}^{2}=k_{B}d_{B}^{2}$$

Because $d_A = 2d_B$:

$$k_{A}(2d_{B})^{2}=k_{B}d_{B}^{2}$$

and

$$k_{A}=k_{B}/4\Rightarrow\boxed{(a)}\text{ is correct.}$$

Estimation and Approximation

17 •• **[SSM]** You are using a 4.0-L cooking pot to boil water for a pasta dish. The recipe calls for at least 4.0 L of water to be used. You fill the pot with 4.0 L of room temperature water and note that this amount of water filled the pot to the brim. Knowing some physics, you are counting on the volume expansion of the steel pot to keep all of the water in the pot while the water is heated to a boil. Is your assumption correct? Explain. If your assumption is not correct, how much water runs over the sides of the pot due to the thermal expansion of the water?

Determine the Concept The volume of water overflowing is the difference between the change in volume of the water and the change in volume of the pot. See Table 20-1 for the coefficient of volume expansion of water and the coefficient of linear expansion of steel.

Express the volume of water that overflows when the pot and the water are heated:

$$\begin{aligned}V_{\text{overflow}}&=\Delta V_{H_2O}-\Delta V_{\text{pot}}\\&=\beta_{H_2O}V_0\Delta T-\beta_{\text{steel}}V_0\Delta T\\&=(\beta_{H_2O}-\beta_{\text{steel}})V_0\Delta T\end{aligned}$$

Because the coefficient of volume expansion of steel is three times its coefficient of linear expansion:

$$\beta_{\text{steel}}=3\alpha_{\text{steel}}$$

Substituting for β_{H_2O} and β_{steel} yields:

$$V_{\text{overflow}}=(\alpha_{H_2O}-3\alpha_{\text{steel}})V_0\Delta T$$

Substitute numerical values and evaluate V_{overflow}:

$$V_{\text{overflow}}=\left(0.207\times10^{-3}\text{ K}^{-1}-3\left(11\times10^{-6}\text{ K}^{-1}\right)\right)(4.0\text{ L})(100^{\circ}\text{C}-20^{\circ}\text{C})=\boxed{56\text{ mL}}$$

Your assumption was not correct and 56 mL of water overflowed.

19 •• **[SSM]** Estimate the thermal conductivity of human skin.

Picture the Problem We can use the thermal current equation for the thermal conductivity of the skin. If we model a human body as a rectangular parallelepiped that is 1.5 m high × 7 cm thick × 50 cm wide, then its surface area is about 1.8 m^2. We'll also assume that a typical human, while resting, produces energy at the rate of 120 W, that normal internal and external temperatures are 33°C and 37°C, respectively, and that an average skin thickness is 1.0 mm.

Use the thermal current equation to express the rate of conduction of thermal energy:

$$I = kA\frac{\Delta T}{\Delta x} \Rightarrow k = \frac{I}{A\dfrac{\Delta T}{\Delta x}}$$

Substitute numerical values and evaluate k:

$$k = \frac{120\text{ W}}{\left(1.8\text{ m}^2\right)\dfrac{37°\text{C} - 33°\text{C}}{1.0\times10^{-3}\text{ m}}} = \boxed{17\frac{\text{mW}}{\text{m}\cdot\text{K}}}$$

25 ••• **[SSM]** You are in charge of transporting a liver from New York, New York to Los Angeles, California for a transplant surgery. The liver is kept cold in a Styrofoam ice chest initially filled with 1.0 kg of ice. It is crucial that the liver temperature is never warmer than 5.0°C. Assuming the trip from the hospital in New York to the hospital in Los Angeles takes 7.0 h, estimate the R-value the Styrofoam walls of the ice chest must have.

Picture the Problem The R factor is the thermal resistance per unit area of a slab of material. We can use the thermal current equation to express the thermal resistance of the styrofoam in terms of the maximum amount of heat that can enter the chest in 7.0 h without raising the temperature above 5.0°C. We'll assume that the surface area of the ice chest is 1.0 m^2 and that the ambient temperature is 25°C

The R-factor needed for the styrofoam walls of the ice chest is the product of their thermal resistance and area:

$$R_\text{f} = RA \qquad (1)$$

Use the thermal current equation to express R:

$$R = \frac{\Delta T}{I} = \frac{\Delta T}{\dfrac{Q_\text{tot}}{\Delta t}} = \frac{\Delta T\Delta t}{Q_\text{tot}}$$

Substitute for R in equation (1) to obtain:

$$R_\text{f} = \frac{A\Delta T\Delta t}{Q_\text{tot}}$$

The total heat entering the chest in 7 h is given by:

$$Q_{tot} = Q_{\substack{melt\\ice}} + Q_{\substack{warm\\ice\ water}}$$
$$= m_{ice}L_f + m_{ice}c_{H_2O}\Delta T_{H_2O}$$

Substitute for Q_{tot} and simplify to obtain:

$$R_f = \frac{A\Delta T\Delta t}{m_{ice}\left(L_f + c_{H_2O}\Delta T_{H_2O}\right)}$$

Substitute numerical values and evaluate R_f:

$$R_f = \frac{\left(1.0\,\text{m}^2 \times \dfrac{1\,\text{ft}^2}{9.29\times10^{-2}\,\text{m}^2}\right)\left(20\,\text{C}°\times\dfrac{9\,\text{F}°}{5\,\text{C}°}\right)(7\,\text{h})\left(\dfrac{1054.35\,\text{J}}{\text{Btu}}\right)}{(1\,\text{kg})\left(333.5\dfrac{\text{kJ}}{\text{kg}} + \left(4.18\dfrac{\text{kJ}}{\text{kg}\cdot\text{K}}\right)(5\,\text{C}°)\right)} \approx \boxed{8\frac{\text{F}°\cdot\text{h}\cdot\text{ft}^2}{\text{Btu}}}$$

Thermal Expansion

27 •• **[SSM]** You need to fit a copper collar tightly around a steel shaft that has a diameter of 6.0000 cm at 20°C. The inside diameter of the collar at that temperature is 5.9800 cm. What temperature must the copper collar have so that it will just slip on the shaft, assuming the shaft itself remains at 20°C?

Picture the Problem Because the temperature of the steel shaft does not change, we need consider just the expansion of the copper collar. We can express the required temperature in terms of the initial temperature and the change in temperature that will produce the necessary increase in the diameter D of the copper collar. This increase in the diameter is related to the diameter at 20°C and the increase in temperature through the definition of the coefficient of linear expansion.

Express the temperature to which the copper collar must be raised in terms of its initial temperature and the increase in its temperature:

$$T = T_i + \Delta T$$

Apply the definition of the coefficient of linear expansion to express the change in temperature required for the collar to fit on the shaft:

$$\Delta T = \frac{\left(\dfrac{\Delta D}{D}\right)}{\alpha} = \frac{\Delta D}{\alpha D}$$

Substitute for ΔT to obtain:

$$T = T_i + \frac{\Delta D}{\alpha D}$$

Substitute numerical values and evaluate T:

$$T = 20°\text{C} + \frac{6.0000\,\text{cm} - 5.9800\,\text{cm}}{\left(17\times10^{-6}\,\text{K}^{-1}\right)\left(5.9800\,\text{cm}\right)} = \boxed{220°\text{C}}$$

The van der Waals Equation, Liquid-Vapor Isotherms, and Phase Diagrams

33 •• [SSM] Using Figure 20-16, find the following quantities. (*a*) The temperature at which water boils on a mountain where the atmospheric pressure is 70.0 kPa, (*b*) the temperature at which water boils in a container where the pressure inside the container is 0.500 atm, and (*c*) the pressure at which water boils at 115°C.

Picture the Problem Consulting Figure 20-16, we see that:

(*a*) At 70.0 kPa, water boils at approximately $\boxed{90°\text{C}}$.

(*b*) At 0.500 atm (about 51 kPa), water boils at approximately $\boxed{82°\text{C}}$.

(*c*) The pressure at which water boils at 115°C is approximately $\boxed{170\,\text{kPa}}$.

Conduction

35 • [SSM] A 20-ft × 30-ft slab of insulation has an R factor of 11. At what rate is heat conducted through the slab if the temperature on one side is a constant 68°F and the temperature of the other side is a constant 30°F?

Picture the Problem We can use its definition to express the thermal current in the slab in terms of the temperature differential across it and its thermal resistance and use the definition of the R factor to express I as a function of ΔT, the cross-sectional area of the slab, and R_f.

Express the thermal current through the slab in terms of the temperature difference across it and its thermal resistance:

$$I = \frac{\Delta T}{R}$$

Substitute to express R in terms of the insulation's R factor:

$$I = \frac{\Delta T}{R_f / A} = \frac{A\Delta T}{R_f}$$

Substitute numerical values and evaluate I:

$$I=\frac{(20\,\text{ft})(30\,\text{ft})(68°\text{F}-30°\text{F})}{11\dfrac{\text{h}\cdot\text{ft}^2\cdot\text{F°}}{\text{Btu}}}$$

$$=\boxed{2.1\frac{\text{kBtu}}{\text{h}}}$$

Radiation

41 • **[SSM]** The universe is filled with radiation that is believed to be remaining from the Big Bang. If the entire universe is considered to be a blackbody with a temperature equal to 2.3 K, what is the λ_{max} (the wavelength at which the power of the radiation is maximum) of this radiation?

Picture the Problem We can use Wein's law to find the peak wavelength of this radiation.

Wein's law relates the maximum wavelength of the background radiation to the temperature of the universe:

$$\lambda_{max}=\frac{2.898\,\text{mm}\cdot\text{K}}{T}$$

Substitute the universe's temperature to obtain:

$$\lambda_{max}=\frac{2.898\,\text{mm}\cdot\text{K}}{2.3\,\text{K}}=\boxed{1.3\,\text{mm}}$$

General Problems

49 •• **[SSM]** The solar constant is the power received from the Sun per unit area perpendicular to the Sun's rays at the mean distance of Earth from the Sun. Its value at the upper atmosphere of Earth is about 1.37 kW/m^2. Calculate the effective temperature of the Sun if it radiates like a blackbody. (The radius of the Sun is 6.96×10^8 m.).

Picture the Problem We can apply the Stefan-Boltzmann law to express the effective temperature of the Sun in terms of the total power it radiates. We can, in turn, use the intensity of the Sun's radiation in the upper atmosphere of Earth to approximate the total power it radiates.

Apply the Stefan-Boltzmann law to relate the energy radiated by the Sun to its temperature:

$$P_r=e\sigma AT^4\Rightarrow T=\sqrt[4]{\frac{P_r}{e\sigma A}}$$

Express the surface area of the sun:

$$A=4\pi R_S^2$$

Relate the intensity of the sun's radiation in the upper atmosphere to the total power radiated by the sun:

$$I = \frac{P_r}{4\pi R^2} \Rightarrow P_r = 4\pi R^2 I$$

where R is the earth-sun distance.

Substitute for P_r and A in the expression for T and simplify to obtain:

$$T = \sqrt[4]{\frac{4\pi R^2 I}{e\sigma 4\pi R_S^2}} = \sqrt[4]{\frac{R^2 I}{e\sigma R_S^2}}$$

Substitute numerical values and evaluate T:

$$T = \sqrt[4]{\frac{(1.5\times10^{11}\text{ m})^2\left(1.35\dfrac{\text{kW}}{\text{m}^2}\right)}{(1)\left(5.67\times10^{-8}\dfrac{\text{W}}{\text{m}^2\cdot\text{K}}\right)(6.96\times10^{8}\text{ m})^2}} = \boxed{5800\text{ K}}$$

53 •• **[SSM]** On the average, the temperature of Earth's crust increases 1.0°C for every increase in depth of 30 m. The average thermal conductivity of Earth's crust material is 0.74 J/m·s·K. What is the heat loss of Earth per second due to conduction from the core? How does this heat loss compare with the average power received from the Sun (which is about 1.37 kW/m^2)?

Picture the Problem We can apply the thermal-current equation to calculate the heat loss of Earth per second due to conduction from its core. We can also use the thermal-current equation to find the power per unit area radiated from Earth and compare this quantity to the solar constant.

Express the heat loss of Earth per unit time as a function of the thermal conductivity of Earth and its temperature gradient:

$$I = \frac{dQ}{dt} = kA\frac{\Delta T}{\Delta x} \qquad (1)$$

or

$$\frac{dQ}{dt} = 4\pi R_E^2 k\frac{\Delta T}{\Delta x}$$

Substitute numerical values and evaluate dQ/dt:

$$\frac{dQ}{dt} = 4\pi(6.37\times10^6\text{ m})^2\left(0.74\frac{\text{J}}{\text{m}\cdot\text{s}\cdot\text{K}}\right)\left(\frac{1.0\text{C}^\circ}{30\text{m}}\right) = \boxed{1.3\times10^{10}\text{ kW}}$$

Rewrite equation (1) to express the thermal current per unit area:

$$\frac{I}{A} = k\frac{\Delta T}{\Delta x}$$

Substitute numerical values and evaluate I/A:

$$\frac{I}{A} = (0.74\,\text{J/m}\cdot\text{s}\cdot\text{K})\left(\frac{1.0\,\text{C}^\circ}{30\,\text{m}}\right) = 0.0247\,\text{W/m}^2$$

Express the ratio of I/A to the solar constant:

$$\frac{I/A}{\text{solar constant}} = \frac{0.0247\,\text{W/m}^2}{1.37\,\text{kW/m}^2} < \boxed{0.002\%}$$

59 ••• **[SSM]** A small pond has a layer of ice 1.00 cm thick floating on its surface. (*a*) If the air temperature is –10°C on a day when there is a breeze, find the rate in centimeters per hour at which ice is added to the bottom of the layer. The density of ice is 0.917 g/cm^3. (*b*) How long do you and your friends have to wait for a 20.0-cm layer to be built up so you can play hockey?

Picture the Problem (*a*) We can differentiate the expression for the heat that must be removed from water in order to form ice to relate dQ/dt to the rate of ice build-up dm/dt. We can apply the thermal-current equation to express the rate at which heat is removed from the water to the temperature gradient and solve this equation for dm/dt. In Part (*b*) we can separate the variables in the differential equation relating dm/dt and ΔT and integrate to find out how long it takes for a 20.0-cm layer of ice to be built up.

(*a*) Relate the heat that must be removed from the water to freeze it to its mass and heat of fusion:

$$Q = mL_\text{f} \Rightarrow \frac{dQ}{dt} = L_\text{f}\frac{dm}{dt}$$

Using the definition of density, relate the mass of the ice added to the bottom of the layer to its density and volume:

$$m = \rho V = \rho A x$$

Differentiate with respect to time to express the rate of build-up of the ice:

$$\frac{dm}{dt} = \rho A \frac{dx}{dt}$$

Substitute for $\frac{dm}{dt}$ to obtain:

$$\frac{dQ}{dt} = L_\text{f}\rho A \frac{dx}{dt}$$

The thermal-current equation is:

$$\frac{dQ}{dt} = kA\frac{\Delta T}{x}$$

Equate these expressions and solve for $\frac{dx}{dt}$:

$$L_f \rho A \frac{dx}{dt} = kA\frac{\Delta T}{x} \Rightarrow \frac{dx}{dt} = \frac{k}{L_f \rho}\frac{\Delta T}{x} \quad (1)$$

Substitute numerical values and evaluate $\frac{dx}{dt}$:

$$\frac{dx}{dt} = \frac{\left(0.592\frac{\text{W}}{\text{m}\cdot\text{K}}\right)(0°\text{C} - (-10°\text{C}))}{\left(333.5\frac{\text{kJ}}{\text{kg}}\right)(917\,\text{kg/m}^3)(0.0100\,\text{m})}$$

$$= 1.94\frac{\mu\text{m}}{\text{s}} \times \frac{3600\,\text{s}}{\text{h}} = 0.697\,\text{cm/h}$$

$$= \boxed{0.70\,\text{cm/h}}$$

(*b*) Separating the variables in equation (1) gives:

$$xdx = \frac{k\Delta T}{L_f \rho}dt$$

Integrate x from x_i to x_f and t' from 0 to t:

$$\int_{x_i}^{x_f} xdx = \frac{k\Delta T}{L_f \rho}\int_0^t dt'$$

and

$$\tfrac{1}{2}\left(x_f^2 - x_i^2\right) = \frac{k\Delta T}{\rho L_f}t \Rightarrow t = \frac{\rho L_f\left(x_f^2 - x_i^2\right)}{2k\Delta T}$$

Substitute numerical values and evaluate t:

$$t = \frac{\left(917\frac{\text{kg}}{\text{m}^3}\right)\left(333.5\frac{\text{kJ}}{\text{kg}}\right)}{2\left(0.592\frac{\text{W}}{\text{m}\cdot\text{K}}\right)(0°\text{C} - (-10°\text{C}))}\left[(0.200\,\text{m})^2 - (0.010\,\text{m})^2\right]$$

$$= 1.03\times10^6\,\text{s} \times \frac{1\,\text{h}}{3600\,\text{s}} \times \frac{1\,\text{d}}{24\,\text{h}} = \boxed{12\,\text{d}}$$